CASS LIBRARY OF SCIENCE CLASSICS

No. 12

General Editor: Dr. L. L. LAUDAN, University College London

T0187671

# CORRESPONDENCE

OF

# SIR ISAAC NEWTON

AND

# PROFESSOR COTES

# CORRESPONDENCE

OF

# SIR ISAAC NEWTON

AND

## PROFESSOR COTES

## J. EDLESTON

LONDON AND NEW YORK

First published 1850 by
**FRANK CASS AND COMPANY LIMITED**

Published 2013 by Routledge
2 Park Square, Milton Park, Abingdon, Oxfordshire OX14 4RN
711 Third Avenue, New York, NY 10017

First issued in paperback 2014

*Routledge is an imprint of the Taylor & Francis Group,
an informa business*

ISBN 13: 978-0-714-61597-4 (hbk)
ISBN 13: 978-0-415-76061-4 (pbk)

# Publisher's Note to the 1969 Edition

This is an exact facsimile reproduction of the 1850 edition of Edleston's *Correspondence of Sir Isaac Newton and Professor Cotes*. An index, prepared under the General Editor's supervision, has been added at the end of the volume.

# CONTENTS.

*₊* Words, or parts of words, enclosed within { } have, with one exception in p. 218, been added by the Editor for the purpose of illustration or to supply omissions in the original MS.

# CONTENTS. V

---

* Since the sheet containing note * p. 171 was printed off, I have seen a paper by Brinkley on the origin of the error in Newton's 1st solution of the resistance problem (*Royal Irish Transactions*, 1810, p. 45) in which the mistake is traced to its true source.

# vi CONTENTS.

# APPENDIX.

# PREFACE.

NEWTON'S *Philosophiæ Naturalis Principia Mathematica*, the most remarkable production of the human intellect that has yet been seen on the earth, whose mysterious path through space was first explained in its pages, was published about the middle of the year 1687, a few weeks after his appearance before James's Ecclesiastical Commission, as the upholder of the rights of his University and the laws of the realm, against the aggressions of arbitrary power. We are not informed how many copies of the work were printed, but the number probably was not large. If the extent of the impression had been rigorously limited to the number of persons likely to comprehend its contents, the volume would now have been one of excessive rarity. The work, however, seems to have found a readier sale than the abstruse nature of the subject and the engrossing interest of politics at that crisis of our history might have prepared us to expect; and the sensation which it produced was long remembered, even by those who saw but darkly that the veil was now raised from the face of nature, which successive generations of philosophers, from the first dawn of science, had vainly endeavoured to draw aside. It is true that, in a legal argument by Lord Mansfield, when Solicitor-General, the names of Locke and Newton are coupled with that of the author of *Paradise Lost*, as affording instances of the neglect shewn to works of

genius for a considerable time after their being given to
the world. Dugald Stewart has assigned good reasons
for doubting the correctness of the statement with respect to the *Essay on the Human Understanding*, and
I believe the assertion to be equally unfounded as predi-
cated of the *Principia*, except so far as the slow reception of the Newtonian doctrines, in some parts of the
continent, may be considered as supplying ground for
affirming the fact. Doubtless there were others besides
Locke who tried to master the first principles, read the
enunciations of the propositions, and accepted them
either on the faith of the author's own word, or in re-
liance upon the judgment of some known mathematician;
nor was Bentley, we may rest assured, the only person
in that inquisitive age who was struck with the wonderful truths developed by the new philosophy, and strove
to attain to an intellectual appreciation of them. Locke's
more popular book appeared in 1690, and a second edition was published in 1694. The *Principia* seems to
have been sold off with almost equal rapidity. In 1691
we hear of an improved edition of it as being in contem-
plation. In 1694 Newton renewed his attack on the
lunar and planetary theories with a view to a new edition of his book. And if Flamsteed, the Astronomer-
Royal, had cordially co-operated with him in the humble
capacity of an observer in the way that Newton pointed
out and requested of him, (and for his almost unpardon-
able omission to do so I know of no better apology that
can be offered than that he did not understand the real
nature and, consequently, the importance of the researches in which Newton was engaged, his purely empi-
rical and tabular views never having been replaced in his

mind by a clear conception of the Principle of Universal Gravitation,) the lunar theory would, if its creator did not overrate his own powers, have been completely investigated, so far as he could do it, in the first few months of 1695, and a second edition of the *Principia* would probably have followed the execution of the task at no long interval. But science and the world were not destined to such good fortune. Flamsteed's infirmities of temper and bodily health conspired to thwart Newton's plans for the first half of the year just mentioned; and the imperfect manner in which the Astronomer-Royal then met his wishes, leaves it uncertain whether we are to attribute the entire blame of the non-completion of the lunar theory in the latter half of the year to the circumstance of steps being at last taken by Newton's friends to provide for his material interests. His appointment to the Wardenship of the Mint in March, 1696, was a bar to the further prosecution of his researches in physical astronomy. Henceforward his official duties made it impossible for him to work continuously at his former pursuits: his studies in mathematics and natural philosophy were by snatches and in the intervals of business. We shall accordingly find, when at length his consent to a new edition of the *Principia* was wrung from him, that his necessary avocations seriously interfered with the progress of the work through the press. But his removal to a new sphere of labour did not abate his zeal for the promotion of science : the starving mathematician found in him a kind and liberal patron, and he was always ready with his purse and counsel to encourage any rational attempt to extract from nature more of her secrets.

Probably as good an idea may be formed of the actual feeling which prevailed with reference to the demand for a republication of the *Principia*, until the time when a new edition was finally determined upon, as would be conveyed by any description that I could give, if I cite a few notices referring to the subject, extracted from various contemporary letters and journals.

1691 Dec. 18. Fatio writing to Huygens from London says: "Mr. Il est assez inutile de prier M$^r$ Newton de faire une nouvelle édition de son livre. Je l'ai importuné plusieurs fois sur ce sujet, sans l'avoir jamais pu flechir. Mais il n'est pas impossible que j'entreprenne cette édition; à quoi je me sens d'autant plus porté, que je ne crois pas qu'il y ait personne qui entende à fond une si grande partie de ce livre que moi, graces aux peines que j'ai prises et au temps que j'ay employé pour en surmonter l'obscurité. D'ailleurs je pourrois facilement aller faire un tour à Cambridge, et recevoir de Mr Newton même l'explication de ce que je n'ai point entendu....."

Again, on Feb. 5, 1692 he writes : " Je n'ai encore ni abandonné, ni embrassé absolutement la pensée de faire une seconde édition du livre de Mr Newton."

1692 "Mr Newton is preparing *a new System of Philosophy*, which will be much larger and plainer than his Principia Philosophiæ Naturalis Physico-Mathematica." (De la Croze's *Works of the Learned* for Jan. 169$\frac{1}{2}$. p. 269, under the head of " Cambridge.")

" According to the best of our advices nothing considerable is doing new at Cambridge, but

Mr Newton's new System of Philosophy, and Mr Barnes's edition of Euripides." (Id. for March and April 1692, p. 398.)

1694 May 29. Huygens, in a letter to Leibniz, speaks of "la nouvelle édition" of the *Principia*, "que doit procurer D. Gregorius."

Nov. 1. "I desire only such observations as tend to perfecting the theory of the planets, in order to a second edition of my book." Newton to Flamsteed (Baily, p. 138.)

1697 Dublin, Nov. 4. "I hear Mr Newton's *Phil. Nat. Prin. Math.* is out of press, and that he designs a 2nd Edition. Pray advise him to make it a little more plain to Readers not so well versed in Abstruse Mathematicks, a few Marginal Notes and references and Quotations would doe the business." (P. S. to a letter from W. Molyneux to Sloane. *Orig. Lett. Bk.* Roy. Soc. M. I. 99.)

1699 July 15. J. Monroe, writing from Paris, says that Malebranche "mightily commends Mr Newton, adding at the same time that there were many things in his book that passed the bounds of his penetration, and that he would be very glad to see Dr Gregory's critick upon it." *Orig. Lett. Bk.* Roy. Soc. M. II. 10. (Comp. Addison's account of his visit to Malebranche at Paris, in the latter half of the year 1700. "His book is now reprinted with many additions, among which he shewed me a very pretty hypothesis of colours, which is different from that of Cartesius or Mr Newton, tho they may all three be true. He very much praised Mr Newton's mathematics, shook his head at the name of Hobbes and told me he thought him a *pauvre esprit*." Letter to Bp. Hough from Lyons, Aikin's Life, I. 91.)

1700   Febr. $\frac{2}{13}$. "J'ai appris aussi (je ne sçai où) qu'il
       donnera encore quelque chose sur le mouvement
       de la Lune; et on m'a dit aussi qu'il y aura
       une nouvelle édition de ses principes de la
       nature." (Leibniz to T. Burnet, *Opp.* Tom. vi.
       pars i. p. 266.)

       July 4. "The Royal Society have laboured to get
       his Theory of the Moon, Book of Colours &c.
       printed, but his excessive modesty has hitherto
       hindered him, but the Society will do what
       further they can with him." (Sloane to Leibniz,
       *Orig. Lett. Bk.* Roy. Soc. S. ii. 14.)

1701   In some MS. memoranda by David Gregory, dated
       Oxon. 21 May, of a variety of points upon which
       he wished to consult Newton we find the follow-
       ing : "To see if he has any design of reprinting
       his *Principia Mathematica* or any other thing."
       (Rigaud, Appendix to Essay, p. 80.)

{1702, Monday} Nov. 30. "He owns there are a great
       many faults in his book, and has crossed it,
       and interleaved it, and writ in the margin of
       it, in a great many places. It is talked he
       designs to reprint it, though he would not
       own it. I asked him about his proof of a
       vacuum, and said that if there is such a matter
       as escapes through the pores of all sensible
       bodies, this could not be weighed.....I find he
       designs to alter that part, for he has writ in
       the margin, Materia sensibilis ; perceiving his
       reasons do not conclude in all matter what-
       soever." Bd. Greves to Lord Aston (Tixall
       Letters, ii. 152), giving an account of a visit
       which he had paid to Newton the preceding
       Thursday in company with Sir E. Southcote

at the request of Lord Aston, "a great lover
of the mathematics, who would gladly be satisfied in a difficulty or two of that science."
1704  Nov. 15.  "The book {Newton's Optics} makes
no noise in town, as the Principia did, which
I hear he is preparing again for the press with
necessary corrections." (Flamsteed to Pound,
Greenwich MSS. xxxiii. 81.)

The book had now become extremely scarce, and
proportionately dear. Sir William Browne, who took
his B.A. degree in 1711, states that when he was at
Cambridge, he gave two guineas for a copy, "which was
then esteemed a very cheap purchase." (Speech at
Royal Society, Nov. 19, 1772, when he was eighty years
of age, in Nichols's *Literary Anecdotes*, iii. 322.) Its
original price seems to have been 10s. At last, in the
beginning of 1709, Bentley's importunity prevailed over
the scruples of the author, and induced him to entrust
the superintendence of a new edition to the care of a
promising young mathematician, Roger Cotes, Fellow
of Trinity College, and recently appointed Professor of
Astronomy and Experimental Philosophy. "Itaque cum
Exemplaria prioris Editionis rarissima admodum et im-
mani pretio coemenda superessent; suasit Ille crebris
efflagitationibus et tantum non objurgando perpulit denique Virum Præstantissimum, nec modestia minus quam
eruditione summa Insignem, ut novam hanc Operis Edi-
tionem, per omnia elimatam denuo et egregiis insuper
accessionibus ditatam, suis sumptibus et auspiciis prodire
pateretur: Mihi vero, pro jure suo, pensum non ingratum demandavit, ut quam posset emendate id fieri cura-
rem." (Cotes, Pref. to 2nd ed.) In a letter to Professor

Sike, dated March 31, 1706 (the true date of which, I apprehend, from internal and external evidence, which it is not necessary to adduce here, to be 1709), Bentley says: "Pray tell Professor Cotes, that the book in your parcel, directed to him, is presented by Sir Isaac Newton; let him read it over with care, and I will tell him further of it in a particular letter. The bundle of wood cuts were found by Sir Isaac in his study, some of which he thinks may belong to the future sheets of his book. In the printed book are folded the MS. sheets that Sir Isaac has now finished." (Bentley's *Correspondence*, p. 231. Lond. 1842.) The book here alluded to was probably a copy of the *Principia*, containing Newton's MS. corrections and additions. This does not seem to have been the copy from which the second edition was printed, unless it was sent back to Newton for further modification. In May following, Cotes received intimation from Bentley that Newton would be glad to see him in town, and to put into his hands part of his revised copy of the *Principia*. The reader is now at the point where the Correspondence now offered to the public commences.

This Correspondence, consisting of the letters which passed between Newton and Cotes relative to questions that arose connected with the new edition of the *Principia*, in the course of its passage through the press, is preserved, with some of the MS. sheets of Newton's interleaved copy of the first edition, and various mathematical papers in Cotes's handwriting, in the library of Trinity College. It was "collected from amongst the loose papers bequeathed" by Dr Robert Smith to the Rev. Edward Howkins, Fellow of Trinity College, who in 1779 demised the Collection, with a profile of Newton, a lock of his hair, and other objects of interest, to the Society. The papers had come into Smith's possession on the death of Cotes, who was his cousin. In their original state they contained among other things, which were afterwards lost, about twenty or thirty letters, written by Newton to Cotes "during the printing of the 2nd edition of the *Principia*," which were borrowed from Smith by Conduitt, who was collecting materials for a *Life of Newton*, and were never returned. They will, I suppose, be found among the papers which have descended with other property of Newton's from his niece, Catharine Barton (who married Conduitt), to the Earl of Portsmouth. Smith, in 1757, endeavoured, with the assistance of a friend, to obtain a clue to these letters which had belonged to him, and instituted inquiries, which were equally unsuccessful, respecting a commonplace book of Newton's, "bound in green parchment," which he had formerly seen in the hands of William Jones, the father of the celebrated orientalist. Some correspondence which took place with reference to this subject is bound up with the Newtonian Letters and Papers.

The late Mr Kidd, in 1796, saw in the possession of the Rev. Thomas Jones, Fellow of Trinity College, a copy of the *Principia*, "with an astonishing quantity of additions and corrections" in Newton's hand. "Numerous loose papers of 4to form covered with diagrams and writing were placed between the leaves in different parts of the volume," which contained also "a loose copy of Halley's laudatory verses on the *Principia*, corrected throughout by the hand of D$^r$ Bentley." Jones stated that this interesting volume was given to him by Mr

Davies, Senior Fellow of Trinity College, who received it from Smith, and he from Newton. All attempts that have been recently made to discover its existence have hitherto failed. I am inclined to think that it may be the identical volume alluded to in Bentley's letter to Sike, quoted above, in which case a link must be inserted in the chain of its transmission between Newton and Smith.

Of the other letters in the Trinity College Newtonian Collection which have been admitted into this publication, those which were not written by or to Newton will be found, with few exceptions, to refer to him in some way or other, and to throw light upon the scientific history of the time.

The Appendix contains various letters and papers, of more or less interest, from Newton's pen, collected principally from original sources. For details of these, and of other matter which is placed before the Correspondence, the reader is referred to the Table of Contents.

The Portrait which accompanies this Work is taken, by the obliging permission of the Master and Fellows of Magdalen College, from an original drawing in Indian ink, which is preserved in the Pepysian Collection. It is uncertain when Pepys first became acquainted with Newton, but there is reason to think that their acquaintance began a short time previous to the Revolution, and they are known to have been on intimate terms in 1691 and 1693. The absence of Newton's name from the long list of persons who received at Pepys's funeral, in 1703, some token in memory of the deceased, may create a suspicion that their intimacy did not ripen into a friendship that continued unbroken to the last; a circumstance which need not excite much surprise when we reflect that neither the politics nor the morality of the Secretary of the Admiralty, under the two last Stuart kings, were at all congenial to Newton's taste. In assigning, therefore, the date of the portrait to the period of a few years on either side of 1691, we shall not perhaps be very wide of the truth.  If this supposition be well-founded, this portrait may be considered as the most interesting of all the known portraits of our philosopher, as representing him at a time of his life the least remote from those memorable eighteen months which it cost him to produce the great work that has immortalized his name.

The public is indebted to the liberality of the Master and Seniors of Trinity College for the appearance of the present volume.

TRINITY COLLEGE, CAMBRIDGE,
   *October* 1850.

# SYNOPTICAL VIEW

OF

# NEWTON'S LIFE.

Quo fit ut omnis
Votivâ pateat veluti descripta tabellâ
Vita senis.

1642    Dec. 25.   Isaac Newton born at Woolsthorpe, near Grantham, Lincolnshire([1]).

1655    Sent to Grantham School.

1656    Taken away from school and put to agricultural employment. Reads mathematics while watching the sheep, and in consequence

1660    Sent back to school with the view of his going to College.

1661    Jun. 5.   Admitted Subsizar at Trin. Coll.

        July 8.   Matriculated Sizar([2]) (Quadrantarius).

1664    Feb. 19.   Observations on two halos about the Moon([3]).

        Thursday, Apr. 28.   Elected Scholar (44 vacancies).

1665    Jan.   Takes B.A. degree with 25 other Trinity men([4]).

        May 20.   Paper on fluxions([5]), in which the notation of *points* is used.

        Nov. 13.   "Discourse" on fluxions and their applications to tangents and curvature of curves([6]).

1666    In the beginning of this year (the year beginning March 25) "applies himself to the grinding of Optic glasses of other figures than spherical," and "procures a triangular glass prism to try therewith the celebrated Phænomena of Colours:" DISCOVERS THE UNEQUAL REFRANGIBILITY OF LIGHT([7]), and abandoning in consequence the idea of improving the refracting telescope, leaves off his "glassworks," and turns his attention to "Reflections," but while engaged thereon is "forced from Cambridge in {June} by the intervening plague([8]), and it was more than two years before he proceeded further."

        May 16.   Another paper on fluxions.

        First idea of gravity occurs to him from observing the fall of an apple([9]) in the garden at Woolsthorpe; proves (from Kepler's 3d law) that it must vary inversely as the square of the distance.

1666 Octob. Small tract on fluxions and fluents with their applications to a variety of problems on tangents, curvature, areas,
lengths, and centres of gravity of curves([10]).

Nov. Small tract similar to the preceding, but apparently more
comprehensive([11]). (Notation by *points* in first and second
fluxions. Basis of his larger tract of 1671).

1667 Oct. 1. Elected minor fellow([12]). Spiritual Chamber([13]).

1668 March 16. Admitted major fellow.

July 7. Created M.A.([14])

Makes a reflecting telescope([15]) (probably towards the end of the
year): is interrupted until the autumn of 1671.

1669 Feb. 23. Describes his Reflecting Telescope in a letter to a friend.

May 18. Letter of advice to his friend Francis Aston.

July 31. His *De Analysi* sent by Barrow to Collins.

Oct. 29. Appointed Lucasian Professor([16]).

Dec. Writes notes upon Kinkhuysen's Algebra sent by Collins
through Barrow.

1670 Jan. 19. Letter to Collins([17]). (Summation of harmonic series.
Solution of equations by tables. Is writing notes at his
leisure upon Kinkhuysen's Algebra).

Feb. 6. Letter to Collins. (Solution of annuity problem, given
all the other quantities, find the rate per cent. Kinkhuysen's
Algebra not worth the pains of a formal comment).

—— 18. Letter to Collins([18]). (Could give exacter solutions
of the annuity problem, but has no leisure for computations.
Sees also a way of summing a harmonic series by logarithms).

July 11. Letter to Collins (with his notes upon Kinkhuysen's
Algebra)([19]).

—— 16. Letter to Collins (proposing to make further additions
to Kinkhuysen's Algebra, which is accordingly sent back for
the purpose).

Sept. 27. Letter to Collins (two mean proportionals cannot be
found by trisecting an arc. General methods best adapted for
instruction. Kinkhuysen's Algebra not so imperfect as he
had thought).

1671 July 20. Letter to Collins. (Prevented by a sudden fit of sickness from visiting him at the Duke of Buckingham's installa-
tion as Chancellor. Will not, he fears, have time to return
to discourse of infinite series before winter. Approximate
sum of harmonic series).

Autumn. Makes his 2nd Reflecting Telescope (in its essential
parts like the former): it is sent up in December " for his
Majesty's perusal([20])."

1671 Dec. 21.  Proposed candidate at the Royal Society by Dr Seth Ward, Bishop of Salisbury.

(Towards the end of the year) occupied in enlarging his method of infinite series([21]), and preparing 20 Optical Lectures for the press.

1672 Jan. 6.  Letter to Oldenburg([22]), "altering and enlarging the {Latin} description([23]) of his instrument which had been sent him for his review before it should go abroad" to Huygens at Paris.

—— 11.  Elected Fellow of the Royal Society. His telescope the subject of conversation at the meeting: the revised description of it read([24]).

—— 18.  Letter to Oldenburg on "a fit metalline matter" for the specula: (announces his intention of sending to the Royal Society "an account of a philosophical discovery," "being the oddest, if not the most considerable detection, which hath hitherto been made in the operations of nature," viz. the composition of light).

—— 29.  Letter to Oldenburg on the proportions of arsenic and bell-metal for specula.

Feb. 6.  Letter to Oldenburg communicating his discovery of the unequal refrangibility of the rays of light (read to the Soc. Feb. 8: printed in the Trans. for Feb. 19).

—— 10.  Letter to Oldenburg, in acknowledgment of the flattering reception of his letter of Feb. 6, and acceding to the wish of the Society that it should be printed.

—— 20.  Letter to Oldenburg, "promising an answer to Mr Hooke's observations upon his new theory of light and colour," and acknowledging "the handsome and ingenious remarks" in Huygens's letter on his telescope (read to the Soc. Feb. 22).

March 16.  Letter to Oldenburg([25]).

—— 19.  Letter to Oldenburg, "containing several particulars relating to his new telescope([26])," (read to the Soc. March 21: printed in Trans. for March 25).

—— 26.  Letter to Oldenburg, "containing some more particulars relating to his new telescope([27])," (read to the Soc. March 28: printed in Trans. for Apr. 22).

—— 30.  Letter to Oldenburg, "containing his answer to the difficulties objected by M. Auzout against his reflecting telescope; as also the queries of M. Denys concerning it; together with his proposal of a way of using, instead of the little oval metal, a crystal figured like a $\Delta^r$ prism([28])." (Read to Soc. Apr. 4: extract printed in Trans. for Apr. 22).

1672 Apr. 13. Latin letter to Oldenburg, in answer to the objections of Pardies (professor in the college of Clermont, in Paris) against his theory of light and colours (read to the Soc. Apr. 18: printed in Trans. for June 17).

Same date. Letter to Oldenburg, " answering some experiments proposed by Sir Robert Moray, for the clearing of his theory of light and colours([29])" (read to the Soc. Apr. 18: extract printed in Trans. for May 20).

May 4. Letter to Oldenburg, "containing his judgment of M. Cassegraine's telescope" (read to the Soc. May 8: printed in Trans. for May 20).

——— 25. Letter to Collins (does not intend to publish his lectures)([30]).

June 11. Letter to Oldenburg([31]), accompanying (1) his 2nd answer to Pardies, who is satisfied by it, (printed in Trans. July 15), and (2) his answer to Hooke's " considerations upon his discourse on light and colours" (part of it read to the Soc. June 12: printed in the Trans. Nov. 18).

——— 19. Letter to Oldenburg from Woolsthorpe.

July 6. Letter to Oldenburg from Stoke, in Northamptonshire, in answer to an inquiry concerning refraction, and containing 8 queries to test his theory of light and colours (partly printed in English and Latin in the Trans. July 15).

——— 8. Letter to Oldenburg from Stoke (containing remarks upon Huygens's letter of July 1, N. S.)([32]).

——— 13. Letter to Collins from Stoke.

——— —   ——— Oldenburg ——— in which he repeats his inquiry about the 4 feet telescope, and desires to know the terms on which Cox will make one.

——— 30. Letter to Collins with a copy of his edition of Varenius's Geography([33]).

——— — Letter to Oldenburg([34]).

Sept. 21. Letter to Oldenburg, in answer to one from Oldenburg of the 17th, inquiring whether the duplicate of July 16 had come to hand: (had drawn up some experiments adapted for determining the queries in his letter of July 6, and had intended from them to prove various propositions relating to colours by means of definitions and axioms, but prevented by other business from carrying out his design. But if the answer to Hooke will conduce to the determination of any of the queries, it may be published).

1672 Dec. 10. Letter to Collins, containing (1) an account, requested by Collins in a letter received two days before, of his method of tangents([35]), and (2) "a long scribble" on James Gregory's observations upon his paper on Cassegrain's telescope. (Very glad to have Barrow again, especially as Master).

1673 March 5. Joins in a protest against the claim of the Heads of Houses to nominate for the Public Oratorship. Votes for Isaac Craven of Trin. Coll. (not nominated)([36]).

—— 8. Letter to Oldenburg (desires to withdraw from the Royal Society)([37]).

Apr. 3. Letter to Oldenburg, in answer to Huygens's letter of Jan. 14 (read to the Soc. Apr. 9: printed in Trans. Oct. 6).

—— 9. Letter to Collins (containing remarks upon Gregory's "candid reply").

May 20. Letter to Collins([38]).

June 23. Letter to Oldenburg, thanking Huygens for the present of his *Horologium Oscillatorium*, and replying to his remarks (in his letter of Jun. 10) upon Newton's letter of Apr. 3 (partly printed in Trans. July 21)([39]).

Sept. 17. Letter to Collins: (postpones further discussion of telescope until Gregory pays his expected visit to Cambridge).

1674 June 20. Letter to Collins: (horizontal velocity of a bullet not uniform. Value of $y$ in $y^3 + a^2y - b^3 = 0$).

Nov. 17. Letter to Collins: (mentions rules for solving incomplete equations by logarithms).

Dec. 5. Letter to Oldenburg: declines to take any notice of Linus's "conjecture:" however Oldenburg may direct him to the figure in the 2nd answer to Pardies, and signify "but not from me," that the experiment with the prism was made on clear days, with the prism close to the hole and the coloured image, not parallel but transverse to the axis of the prism. (A letter was written by Oldenburg accordingly, and printed without Newton's knowledge in the Trans. Jan. 25, 1675).

1675 Chemical pursuits([40]).

Jan. 22. Letter to Michael Dary (length of an elliptic arc).

—— 28. Excused the weekly payments to the Royal Society([41]).

Feb. 18. Admitted F.R.S.([42]).

Apr. 27. Obtains from the Crown a patent allowing the Lucasian Professor to hold a fellowship without being obliged to go into orders([43]).

May 8. Letter to John Smith (construction of tables of square, cube, &c. roots)([44]).

1675  July 24. ⎫
             ⎬ Letters to the same on extraction of roots.
     Aug. 27. ⎭

     Nov. 13.  Letter to Oldenburg, with minute directions for Linus
how to make the spectrum experiment: (communicated to
the Soc. Nov. 18: principal part of it printed in Trans. Jan.
24, 1676) [45].  Offers to send a paper on colours.

     —— 30.  Letter to Oldenburg (is adding "an hypothesis" con-
cerning light to his paper on colours.  Description of ear-
trumpet) [46].

     Dec. 1.  Gives a copy of Irenæus (Paris, 1675) to College
Library.

     Decemb.  Sends to the Roy. Soc. his papers, containing (1) his
Hypothesis explaining the properties of light, (2) his explanation of the colours of thin plates, and of natural bodies [47].

     Dec. 14.  Letter to Oldenburg (suggesting that the glass in the
electrical experiment should be nearer the table than he had
stated in his paper).

     —— 21.  Letter to Oldenburg, with (1) further directions
respecting the electrical experiment (read to the Soc. Dec. 30 [48],
and the experiment ordered to be made at next meeting), and
(2) remarks on Hooke's "insinuation."

Communicates to Mercator his explanation of the Moon's li-
bration [49].

1676  Jan. 10.  Letter to Oldenburg, containing (1) suggestions respecting the electrical experiment, (2) remarks upon Hooke's
"insinuations," (3) further directions for Gascoines how to
make the spectrum experiment.  Oldenburg (Jan. 18) sends
them to Gascoines, who requests Lucas (Linus's successor in
the mathematical chair at Liege) to make the experiment.
(Last part of the letter printed in Trans. Jan. 24, 1676).

     —— 13.  At the meeting of the Royal Society, the electrical
experiment being made according to Newton's "more particular directions succeeded very well."  "It was ordered that
Mr Newton should have the thanks of the Society for giving
himself the trouble of imparting to them such full instructions
for making the experiment."

     —— 20.  On the reading of the first 15 "observations" of
Newton's discourse, the Society were "so well pleased" with
them, that Oldenburg was ordered to desire him "to permit
them to be published together with the rest."

A passage was also read from his letter of Dec. 21, "stating the
difference between his hypothesis and that of Mr Hooke," in

allusion to what had fallen from Hooke at the meeting of Dec. 16.

1676  Jan. 25.  Letter to Oldenburg, in acknowledgment of the favourable reception of his papers([50]), with alterations to be made in them. (Read to the Soc. Jan. 27).

Feb. 3.  On the reading of Newton's observations on colours, a discussion arose as to whether the difference of colour in the rays of light was not to be attributed to the different velocities of the pulses rather than, as he thought, to a connate difference of refrangibility in the rays themselves. Hooke expressed himself in favour of the former explanation. See Newton's Letter of Feb. 15.

——— 15.  Letter to Oldenburg, answering the objection that had been raised at the meeting of Feb. 3.

——— 29.  Letter to Oldenburg, occasioned by his having read in the Trans. for Jan. 24, Linus's letter of Feb. 25, 1675: it contains a particular answer to that letter, followed by explanatory remarks for the behoof of Linus's friends. (Printed in Trans. March 25).

Apr. 26.  Letter to Oldenburg, thanking him for "motioning to get" the spectrum experiment tried before the Royal Society([51]). Remarks upon Boyle's paper on the incalescence of gold and mercury.

May 11.  Letter to Oldenburg, thanking him for getting the experiment tried: during the summer may possibly work at his long-projected discourse about the prismatic colours([52]).

June 8.  At a meeting of the Soc. a letter from Lucas to Oldenburg (Liege, May 27) was read, containing partly an account of the success of the spectrum experiment, partly some new objections against Newton's theory of light and colours. A copy of the letter ordered to be sent to Newton immediately: printed in Trans. for Sept. 25.

——— 13.  Letter to Oldenburg, containing a general answer to Lucas with a promise of a particular one, and also "some communications of an algebraical nature for M. Leibniz, who by an express letter to Mr. Oldenburg had desired them." (read to the Soc. June 15: the part for Leibniz([53]) was sent to him at Paris, July 26).

Aug. 22.  Letter to Oldenburg([54]) (accompanied by another dated Aug. 18, the latter being an answer to Lucas, printed in Trans. for Sept. 25).

Sept. 5.  Letter to Collins. (Infinite Series of no great use in the numerical solution of equations. The University press cannot print Kinkhuysen's Algebra : the book is in the hands of a Cambridge bookseller with a view to its being printed : shall add nothing to it. Will alter an expression or two in his paper about infinite series, if Collins thinks it should be printed).

1676    Oct. 24.   Latin letter to Oldenburg[55] for Leibniz, who desired explanation with reference to some points in the letter of June 13.

—— 26.   Letter to Oldenburg, with corrections for his letter of Oct. 24, &c.[56]

Nov. 8.   Letter to Collins, thanking him for copies of the letters of Leibniz and Tschirnhaus, with remarks shewing that Leibniz's method is not more general or easy than his own[57].

—— 14.   Letter to Oldenburg (cider-fruit-trees: Lucas's 2nd letter: further alterations of his letter of Oct. 24)[58].

—— 18.   Letter to Oldenburg (answer to Lucas will not be ready so soon as he intended. Will never publish anything more on philosophy, after he has got clear of this dispute. Letter to Boyle)[59].

—— 28.   Rejoinder to Lucas[60].

Subscribes £40 towards New Library.

1677    March 5.   Letter of Collins to him[61].

Sept.   Death of Oldenburg.

1679    Feb. 7.   Letter to Dr Maddock[62].

—— 11.   Sir Thomas Exton, Master of Trin. Hall, and James Vernon, of Trinity, (the Duke of Monmouth's Secretary,) elected M.P. for the University. Newton plumps for the former[63].

—— 28.   Letter to Boyle (physical qualities of bodies)[64].

Nov. 8.   Charles Montagu entered a fellow-commoner at Trinity College[65].

December.   Determines (in consequence of a letter from Hooke) the curve described by a body under the action of a central force, and applies his theorem to the case of an ellipse[66].

Gives copy of Huet's *Demonstratio Evangelica* to College Library.

1680    Jan. 21.   Collins offers to print Newton's Algebra (along with Wallis's and Baker's), if the Society would take 60 copies, which the Council two years and a half afterwards agreed to do (July 12, 1682), but the design was carried out only with respect to Baker and Wallis.

Lends the College £100. for the New Library (sometime between Dec. 1679 and Michaelmas 1680)([67]).

1680 Dec. 3. Letter to Hooke([68]).

Gives copy of Grew's *Musœum Regalis Societatis* to College Library.

1681 Jan. Promises to assist Adams (probably by advice and calculations) in a survey of England([69]).

Feb. 28. Letter of Flamsteed (through Crompton, Fellow of Jes. Coll.) about the Comet([70]).

Apr. 16. Letter to Flamsteed about the Comet([71]).

1682 Apr. 3. Testimonial to Edw. Paget, Fellow of Trin. Coll., candidate for the Mathematical Mastership of Christ's Hospital([71])*.—Letter to Flamsteed (introducing Paget).

June 20. } Letters to Briggs on Vision([72]).
Sept. 12. }

1683 Nov. 7. Votes for James Halman of Caius College, the successful candidate for the Registraryship.

—— 10. Death of Collins.

Dec. 22. Letter to Aubrey, who had offered some books for sale to Trinity College or the University([73]).

1684 Jan. 19. Votes for James Manfeild of Trinity, the successful candidate for the Librarianship.

August. Halley on a visit to him, "learns the good news that he had brought the demonstration" of "the laws of the celestial motions to perfection." Newton cannot lay his hands upon his papers, but works them over again, and sends them in November by Paget to Halley in the form of 4 theorems and 7 problems([74]). Halley "thereupon takes another journey to Cambridge, on purpose to confer with him about them([75])."

1685 Feb. 23. Letter to Aston (unsuccessful attempt to establish a philosophical society at Cambridge([76]). Thanks for registering at the Royal Society his "notions about motion").

Apr. 25. Letter to Briggs([77]).

DETERMINES THE ATTRACTIONS OF *MASSES* AND THUS COMPLETES THE DEMONSTRATION OF THE LAW OF UNIVERSAL GRAVITATION.

Summer. The 2nd book of the Principia finished.

Sept. 10. Certificate of approval of Mabbot's Tables for renewal of leases([78]).

—— 19. Letter to Flamsteed (is about to calculate the orbit of the comet of 1680 from 3 observations. Tides at solstices and equinoxes)([79]).

1685   Oct. 14.  Letter to Flamsteed (acknowledging the receipt of Flamsteed's two letters in answer to the preceding).

       Dec. 30.  Letter to Flamsteed (with thanks for information about comet of 1680 and Jupiter's satellites. Kepler makes Saturn's orbit too small. Requests the greatest elongations of any of Jupiter's satellites, and of Saturn's satellite)[80].

1686   Jan. 13.  Letter to Flamsteed (wishes to know the major axes of the orbits of Jupiter, Saturn and his satellite)[81].

      —— 22.  Votes for John Laughton, of Trinity, the successful candidate for the Librarianship.

      Apr. 28.  FIRST BOOK OF THE PRINCIPIA EXHIBITED AT THE ROYAL SOCIETY[82].

      May 19.  At a meeting of the Society it was ordered "that M^r Newton's *Philosophiæ Naturalis Principia Mathematica* be printed forthwith in 4to. in a fair letter; and that a letter be written to him to signify the Society's resolution, and to desire his opinion as to the print, volume, cuts, &c."[83].

      June 2.  Halley undertakes the publication of the *Principia* at his own expense[84].

      —— 20.  Letter to Halley (demolishing the claim set up by Hooke of having communicated to him the law of decrease of gravity according to the inverse square).

      —— 30.  At a meeting of the Council of the Royal Society, the President was desired to license the *Philosophiæ Naturalis Principia Mathematica*.

      July 14.  Letter to Halley (approves of the suggestion of having wood-cuts. Conciliatory remarks respecting Hooke).

      —— 27.  Letter to Halley (further remarks on Hooke's claim).

      Aug. 20.  Letter to Halley (with Cor. 2 and 3 of Prop. xci. Lib. i. of *Princip.* on the attraction of a spheroid on a point in its axis produced, and on an internal point)[85].

      Sept. 3.  Letter to Flamsteed (Cassinian satellites. Cassini's observation of Jupiter's oblateness).

      Autumn.  Second Book of Principia made ready for the press[86].

      Oct. 18.  Letter to Halley (corrections of Scholium to Prop. 31. Lib. i.: transformation of a trapezium into a parallelogram).

1687   Feb. 18.  Letter to Halley (may have the second book of Principia when he pleases: has the sheets up to M: thanks him for putting forward the press again)[87].

      March 1. Tuesday.  Letter to Halley, advising him that the 2nd book will arrive on Thursday night or Friday, by coach: obliged to him for pushing on the edition because of people's expectation, though otherwise he could be as well satisfied to let it rest a year or two longer (read to the Soc. March 2).

1687 March 11. Deputed with Billers, the Public Orator, to carry to the Vice-Chancellor the opinions of the Non-Regent House respecting King James's second mandate, requiring the University to confer upon Alban Francis, a Benedictine monk, the degree of M.A. without the usual oaths([88]).

Apr. 6. The 3rd book of the Principia "produced and presented" to the Royal Society ([89]).

—— 11. Appointed one of eight delegates to represent the Senate, in conjunction with the Vice-Chancellor, before the Ecclesiastical Commission([90]).

PUBLICATION OF THE PRINCIPIA (about Midsummer)([91]).

1688 Spring. Charles Montagu vacates his fellowship([92]).

Dec. 15. Votes for Archbishop Sancroft (for Chancellor of the University) who declines the office.

1689 Jan. 15. Elected one of the representatives of the University in the Convention Parliament([93]).

First acquaintance with Locke. Furnishes him (March) with an easy proof of elliptic motion about a centre of force in one of the foci([94]).

June 12. Huygens and Newton at the Royal Society([95]).

Aug. 20. Parliament prorogued.

—— —— Contemplated appointment to the Provostship of King's College([96]).

Oct. 19. Meeting of Parliament ([97]).

1690 Jan. 27. Parliament prorogued.

Feb. 6. Parliament dissolved.

—— 21. Sir Robert Sawyer, who had been expelled the House of Commons, Jan. 20, for having been, as Attorney-General, one of the prosecutors of Sir Thomas Armstrong in 1684, re-elected M.P. for the University. Newton votes for him.

Oct. 28. Letter to Locke: (will send, as desired, his " Historical Account of two notable corruptions of scripture." Acknowledgments to Lord and Lady Monmouth for their endeavours to procure him preferment)([98]).

Nov. 14. Letter to Locke, with the " Historical Account."

1691 Feb. 7. Letter to Locke (Daniel and Apocalypse).

June 30. Letter to Locke. (Locke's good offices in trying to get him the place of comptroller of the Mint. Effects of looking at the Sun's image in a mirror).

1691    July (London). Testimonial to David Gregory, recommending him for the vacant chair of Astronomy at Oxford([99]).

Directions to Bentley about reading the *Principia*, (p. 273).

Aug. 10. (London). Letter to Flamsteed (introducing David Gregory. Hopes Flamsteed will publish his catalogue of the fixed stars before long. Would willingly have his observations of Jupiter and Saturn for next 4 or 5 years at least, or rather for the next 12 or 15, before thinking further of their theory. Does the light of Jupiter's satellites, immediately before eclipse, incline either to red or blue, or become ruddier or paler than before ?)

Dec. 13. Letter to Locke. (Declines "making a bustle" for the Mastership of the Charter-House)([100]).

1692    Jan. 26. Letter to Locke. (Charles Montagu a false friend. Desires to have his "Historical Account" returned.)

Feb. 16. Letter to Locke. (Desires the translation and impression of the "Historical Account" to be stopped. Miracles).

May 3. Letter to Locke (glad of his intended visit. Miracles).

June. Observations on three halos about the Sun([101]).

July 7. Letter to Locke (Boyle's recipe for producing gold by means of red earth and mercury)([102]).

Aug. 2. Letter to Locke (Boyle's recipe. Discourages Locke from trying it).

Aug. 27.
Sept. 17.  } Letters to Wallis, with illustrations of the calculus of fluxions and fluents, sent at Wallis's request ([103]).

Nov. 21. Election of a Member for the University in the place of Sir Robt. Sawyer, deceased. Votes for the unsuccessful candidate, Dr Brookbank, of Trin. Hall([104]).

Dec. 10. First letter to Bentley.

Paper on Acids (exact date uncertain) communicated to a friend this year([105]).

1693    Jan. 17. Second letter to Bentley([106]).

Feb. 11. Third letter to Bentley.

—— 25. Fourth letter to Bentley([107]).

March 14. Letter to Fatio (proposing to make him such an allowance as might make his subsistence at Cambridge easy to him)([108]).

September. Bad state of health.

—— 13. Letter to Samuel Pepys (desiring to " withdraw from his acquaintance")([109]).

—— 16. Letter to Locke (begging his pardon for having had "hard thoughts" of him)([110]).

1693  Oct. 15.  Letter to Locke (explaining the circumstances under which the letter of Sept. 16 was written)([111]).

—— 16.  Letter to Leibniz, (p. 276).

Nov. 23. }
Dec. 16. }  Letters to Pepys on a problem in chances ([112]).

1694  May 7.  Haunted house ([113]).

—— 11.  Charles Montagu, Chancellor of the Exchequer.

—— 25.  Letter to Hawes (explaining his views relative to the old and new schedules of mathematical studies at Christ's Hospital)([114]).

—— 26.  Letter to Hawes (supplementary to preceding).

May.  David Gregory at Cambridge([115]).

July.  Requested by the Royal Society to publish his optical and other treatises([116]).

Sept. 1.  Visits Flamsteed at Greenwich, who shews him upwards of 150 lunar observations, and a comparison of them with the places as calculated from tables([117]). Consequent correspondence between them, extending from Oct. until Sept. of the following year([118]).

Oct. 7.  Letter to Flamsteed (describing what further observations he will want, with which he believes he can "set right the moon's theory this winter").

—— 24.  Letter t6 Flamsteed (thanking him for his letter of Oct. 11, and particularly for the table of the difference of refractions of Sun and Venus. Parallactic Equation)([119]).

Nov. 1.  Letter to Flamsteed (errors in some of his observations. Lunar inequalities. Sun's menstrual parallax)([120]).

—— 17.  Letter to Flamsteed (will send back the two synopses of the Moon's places the next day, together with a table of refractions. His method of proceeding in determining the Moon's motions. Requests to have the Moon's right ascensions and meridian altitudes just as they are observed without any correction: if Flamsteed will do him this favour, he desires them as Flamsteed had observed them for the last six months).

Dec. 18.  Letter to Flamsteed (Table of refractions not so accurate as it may be made: intends to correct it and send a new copy of it. Thanks Flamsteed for complying with his request of sending the Moon's right ascensions and meridian altitudes unreduced : begs her places on certain days which he names : observations in this and next month or two of great importance).

1694   Dec. 20.   Letter to Flamsteed (theorem upon which his table
           of refraction is founded.   Equations of the mean motions of
           Jupiter's satellites.   "What you say about my having a mean
           opinion of you is a great mistake ").

1695   Jan. 15.   Letter to Flamsteed (thinks he has discovered a new
           theorem in refractions, but intends to consider it a little
           further.   Thanks Flamsteed for two lunar observations sent
           him, and as Flamsteed has calculated the Moon's places in
           these and the other three observations of last month, will be
           glad to have a synopsis of the calculations.   But for the
           rest of the observations, he merely wants the observed places ;
           at the same time is obliged to Flamsteed for offering to be at
           the pains of calculating them.   Suggestions respecting the
           kind of *time* to be employed in taking the observations).

—— 26.   Letter to Flamsteed (answer to Flamsteed's childish
           question respecting a book which Flamsteed, two or three
           years before, had intended as a present to him.   Moon's hori-
           zontal parallax.   Has at last found out a new theorem in
           Refractions : is at present a little indisposed but hopes in a
           few days to be well enough again to finish the subject.   The
           two observations mentioned in the last letter([121]).   Promises
           to send a table of a small equation of Moon's parallax.   If
           Flamsteed would rather have the observations perfectly his own
           in all respects, by calculating them himself, will stay his time).

Feb. 16.   Letter to Flamsteed (with thanks for the observations
           of Dec. and Jan.   Has been engaged since he wrote last upon
           making a new table of refractions, and has not yet finished it.
           Manly answer to Flamsteed's ungenerous suspicions of his
           observations having been communicated to Halley).

March 15.   Letter to Flamsteed (Candidates for mathematical
           mastership at Christ's Hospital.   Encloses a copy of table of
           refractions now finished([122]).   Will send the other tables he
           promised in a few days).

Apr. 23.   Letter to Flamsteed (with the promised tables of
           Moon's horizontal parallax, equations of apogee and eccentri-
           cities).

—— 25([123]).   Letter to Flamsteed (in reply to some remarks
           on the tables sent with his last letter).

Jun. 14.   Letter to Hawes (with new scheme of mathematical
           reading for Christ's Hospital)([124]).

—— 29.   Letter to Flamsteed (with thanks for solar tables.   As
           Flamsteed's health and other business will not permit him to

calculate the Moon's places from observations, he proposes once more that Flamsteed should send the bare observations, and first of all those of 1692. If not, let him propose some other way of supplying the desired observations, or say plainly that he will not send any. Recommends equestrian exercise).

1695 July 9. Letter to Flamsteed (thankfully accepts the offer of the observations prior to 1690. Parallactic equation. Points out the kind of observations that he wants).

—— 20. Letter to Flamsteed (has written to contradict the report about Flamsteed's not communicating his observations. Thanks for the lunar observations. Has not yet compassed the small equations, and begs him not to be impatient for them. Forbears to take notice of some querulous expressions of Flamsteed's. " Pray take care of your health ").

—— 27. Letter to Flamsteed (is glad that all misunderstandings are composed. Describes the observations that he wants. Remuneration to Flamsteed's servant)[125].

Sept. 14. Letter to Flamsteed (Halley's calculated orbit of the comet of 1683 agrees with his own and Flamsteed's observations to a minute. Is going on a journey and will not therefore have time to consider the lunar theory for a month or above. Hopes he gets ground of his distemper).

Oct. 25. In the contest for the University plumps for the Hon. H. Boyle.

Nov. Rumour of his appointment to Mastership of Mint[126].

1696 Feb. 19. Votes for W. Ayloffe of Trin. successful candidate for the Public Oratorship.

March 14. Letter to Halley (is not engaged upon the longitude. Not a candidate for any place in the Mint, nor would accept the Comptroller's place, if offered)[127].

—— 19. Letter from Charles Montagu announcing his appointment to Wardenship of Mint.

1697 Jan. 30. Solution of John Bernoulli's two problems[128]: (read to the Soc. Feb. 24: printed, without his name, in Trans. for Jan.).

Feb. 11. Letter to Halley : (has proposed Halley as a fit person to teach the mathematical principles of engineering)[129].

End of June, or beginning of July. Examines boys at Christ's Hospital [130].

1698 May 30. Letter to Harington, p. 302.

July 25. Votes for Hon. H. Boyle (re-elected)[131].

Dec. 4. Visit to Flamsteed, in order to obtain 12 computed places of the Moon[132].

1699   Jan. 6.   Letter to Flamsteed (explaining why he did not wish
his name to be mentioned in the letter to Wallis, and stating,
that there may be cases in which "friends should not be published without their leave")([133]).

Feb. 11.   Made Associé-Étranger of the French Academy([134]).

Aug. 16.   Exhibits at the Royal Society an improved form of
his sextant (commonly called Hadley's)([135]).

Nov. 30.   Chosen member of Council of Royal Society([136]).

This year the great re-coinage of silver was completed, having
occupied the greater part of this and of the three preceding
years([137]).

Contributes towards the expenses of Lhuyd's Lithophylacii
Britannici Ichnographia([138]).

1700   Apr.   Paper on time of vernal equinox (p. 304).

July 24.   His opinion of the method proposed by an Italian
mathematician for trisecting an angle, doubling the cube, and
squaring the circle by means of a spiral line([139]).

1701   Jan. 27.   Whiston begins his Astronomical Lectures, as Newton's deputy, receiving "the full profits of the place."

May 28.   His scale of heat read to the Society([140]), (printed in
the Trans. for March-April).

Nov. 26.   Elected M.P. for the University([141]).

Dec. 10.   Resigns his Professorship, and his Fellowship shortly
after([142]).

1702   May 25.   Parliament prorogued.

(About June) his "Lunæ Theoria" published in Gregory's
Astronomy([143]).

July 2.   Parliament dissolved.

Autumn.   On a visit to Locke at Oates([144]).

1703   May 15.   Letter to Locke (giving his opinion of Locke's MS.
papers on the Epistles to the Corinthians, and criticising his
paraphrase on the 1st Ep. vii. 14).

Nov. 30.   Elected President of the Royal Society([145]).

1704   Jan. 20.   Mentions to the Royal Society his burning-glass([146]).

Feb.   Publication of Optics([147]).

Dec. 5.   Note to Sloane (desiring him to be in readiness on the
7th, the day fixed for their introduction to Prince George, for
the purpose of having the honour of his signature in the
Statute book of the Society, of which he was elected a member, Nov. 30).

—— 7.   Waits on the Prince, and takes the opportunity of
giving him a copy of Flamsteed's estimate of his Observations.

1704 Dec. 18. Letter to Flamsteed (inviting him to dinner to meet the gentlemen appointed by Prince George to inspect his papers, and requesting him to bring his papers, or specimens of them for the referees to examine).

—— 26. Letter to Flamsteed (begging him to bring his papers for the referees to examine).

1705 Jan. 1. (N. S.) Equivocal expressions in the review of his tract, De Quadraturâ Curvarum, in the Leipsic Acts([148]). (Origin of dispute on the priority of discovery of the new analysis).

—— 23. Report to Prince George recommending the publication of Flamsteed's Observations([149]).

March 2. Letter to Flamsteed (earnestly desiring him to attend a meeting of the referees, in order to agree about an amanuensis, calculators, and what else he has to propose for dispatching the work).

—— 7. Presents Royal Society with the 1st Vol. of Rymer's Fœdera, lately published([150]).

—— Visit to Cambridge([151])

Subscribes £60. towards the repairs of Trin. Coll. Chapel([152]).

April Returns to London (about the 5th).

—— 16. Knighted by Queen Anne at Trinity College.

—— 24 or 25. Goes to Cambridge to contest the University.

May 17. Defeated in the contest for the University([153]).

June 8. Note to Flamsteed (inviting him to meet the referees at dinner, "that we may set the press a going as soon as possible").

Sept. 14. Note to Sloane (begging him to get Hauksbee to bring his air-pump some evening to his house. " I can then get some philosophical friends to see his experiments, who will otherwise be difficultly got together ")([154]).

—— 17. Letter to Flamsteed (urging him to put his papers to press. "If you stick at anything, pray give Sir Chr. Wren and me a meeting as soon as you can conveniently, that what you stick at may be removed").

—— Note to Sloane (desiring Hauksbee's experiments to be put off for a while, as Lord Halifax, Archbishop of Dublin, and Robartes are out of town).

Nov. 14. Note to Flamsteed (inviting him to meet the referees at dinner, to finish the agreement and sign the articles about printing his book).

—— 20. Signature to pedigree([155]).

1706  Latin edition of Optics([156]).

Sept. 13.  Note to Sloane (thinks Bishop Wilkins's Legacy of £400 in 1672 should be defended at any cost)([157]).

1707  Jan. 14.  Date of statutes of recently founded Plumian Professorship, drawn up partly under his eye([158]).

Apr. 9.  Note to Flamsteed (requesting him to meet the referees, that all things may be now settled and adjusted, and to bring his bill of disbursements).

——  Letter to Sir John Newton (recommending a poor kinsman as undertaker to conduct the funeral of his cousin Coke)([159]).

1709  Jan. 12.  Gives the Royal Society £20 ([160]).

Oct. 11.  Commencement of his correspondence with Cotes relative to the 2nd ed. of the Principia, extending from this date to March 31, 1713([161]).

1710  Sept. 13.  Note to Sloane (glad that Sir Christopher and Mr Wren like the house in Crane Court, proposed to be purchased for the Royal Society, and hopes they will like the price also).

Dec. 14.  Promises to give £100 towards the easing of the debt for the house, besides the £20 mentioned Jan. 12, 1709.

1713  Midsummer.  Second edition of Principia([162]).

Nov.  Paper on the different kinds of years in use among the nations of antiquity ([163]).

1714  Apr. 2.  Letter to Keill (respecting an answer to be made to Leibniz's "charta volans" as reprinted with remarks in the Journal Literaire)([164]).

——  20.  Letter to Keill (on same subject).

May 11.  Letter to Keill (on same subject).

——  Letter to Chamberlayne in reply to one from Leibniz of Apr. 28, (if it can be pointed out where he has wronged Leibniz, he will endeavour to make satisfaction, but he cannot retract what he knows to be true, and believes the Committee of the Royal Society has not wronged Leibniz)([165]).

——  15.  Letter to Keill (in continuation of his letter of the 11th).

May—June.  One of Bishop Moore's Assessors at Bentley's trial([166]).

End of May or be- } Evidence before a Committee of the House
ginning of June. } of Commons, on the different methods of finding the longitude at sea ([167]).

Woodward's Classification of Fossils dedicated to him([168]).

1716   Feb. 26.   Letter to Conti in answer to one from Leibniz([169]).
       May 18.    Observations upon Leibniz's reply ([170]).
       June 5.    Death of Cotes([171]).
1717   May 16.    Presents his portrait to the Royal Society([172]).
       Sept. 21.  Report on the state of the Coin([173]).
       Nov. 23.   Another Report on the Coin([174])
1718   Second edition of Optics([175]).
       Jan. 21.   At the House of Lords with accounts relating to the
                  coin([176]).
       May 2.     Letter to Keill (will John Bernoulli's denial, in a pri-
                  vate letter, of the authorship of the *Epistola pro eminente
                  Mathematico*, satisfy him ([177]) ?)
       Oct. 22.   Observations on the state of the Coin([178]).
       Gift of £70 to the Royal Society([179]).
1719   July 13.   Present to Pound the Astronomer ([180]).
       Letter to Monmort, enclosing one to Bernoulli([180])*.
1721   Third edition of Optics([181]).
1722   Attack of stone.
       Oct. 22.   Letter to Arland the artist (thanking him for his pro-
                  fessional services in the matter of a plate in the French
                  translation of the Optics)([182]).
1723   Jan. 17.   Appoints (at a meeting of the Council of the Royal
                  Society) Martin Folkes his Deputy or Vice-President.
1724   Apr. 27.   Report on Wood's Halfpence and Farthings([183]).
       Jun. 25.   Imprimatur for new edition of Ray's *Synopsis Plan-
                  tarum Britannicarum.*
       Aug. ⎫ Delisle in England([184]).
       Sept. ⎭
       Aug. 25.   Letter to Lord Townshend (respecting a criminal
                  under sentence of death for coining: thinks the law should
                  take its course)([185]).
       Dec. 3.    Letter to Halley (requesting him to examine two of
                  the calculated places in the elliptic orbit of the Comet of
                  1680, and to calculate another place, supposing the orbit a
                  parabola)([186]).
1725   Jan.       Violent cough and inflammation of the lungs.  Prevailed
                  upon to take a house at Kensington.
       Feb.       Fit of the gout in both his feet (had had a slight attack
                  a few years before).  Improved health after it.
       Letter to Mason, Rector of Colsterworth, notifying his subscrip-
                  tion of £12. towards erecting a gallery in Colsterworth
                  church([187]).

1725  March 7.  Conversation with Conduitt on the formation of the
        planetary bodies ([188]).

—— 25.  Grant of rents (£25) for four years of the ancestral
        part of his estate at Woolsthorpe to his god-son Isaac Warner.

May 12.  Letter to Mason (very glad to understand that the
        gallery in Colsterworth church is finished.  The surplus in
        Mason's hands belonging to him to be applied " to the use of
        the young people of the parish that are learning to sing
        Psalms," according to Mason's desire).

May 12.  Letter to his tenant Percival of Woolsthorpe, agree-
        ing to a proposed distribution of the commons there and at
        Colsterworth ([189]).

—— 27.  Refuses his sanction to Freret's Translation of his
        Chronological Summary ([190]).

July 1.  Visit of Abbe Alari ([191]).

Date not given.  Letter to Maclaurin (glad that he has a pros-
        pect of being joined to James Gregory in the Professorship of
        Mathematics at Edinburgh, and heartily wishes him good
        success) ([192]).

Date not given.  Letter to Lord Provost of Edinburgh (is ready
        to contribute £20 per ann. towards a provision for Maclaurin,
        if he will act as assistant to Gregory).

Towards the end of the year.  Remarks upon Freret's observa-
        tions in his unauthorised translation of Newton's Chronologi-
        cal Summary ([193]).

1726  Third Edition of the Principia ([194]).

May 10.  Letter to Mason (with note for £3 for repair of the
        floor of Colsterworth church).

1727  Feb. 4.  Letter to Mason (has procured assays to be made of
        the pieces of ore left with him by a Woolsthorpe friend of
        Mason's, but they contain no metal).

Feb. 16.  Writes *Imprimatur* for Hales's Vegetable Statics.

March 2.  Present for the last time at a meeting of the Royal
        Society, at which he calls attention to the fact of the Astrono-
        mer-Royal (Halley) having omitted to send to the Society a
        copy of his annual observations, as required by the late Queen's
        letter ([195]).

—— 20.  Monday, between 1 and 2 A.M.  Dies ([196]).

# NOTES.

(¹) " Natus est Isaacus Neutonus...horâ primâ vel secundâ post mediam noctem, idque tempore ipso Plenilunii. Capillis effloruit sensim in summam canitiem versis, Annum ætatis inter trigesimum & quadragesimum." (Nicolas Fatio, in a printed copy of Latin Hexameters, entitled *Neutonus Ecloga*, inserted in his copy of the 3rd ed. of the *Principia* which is preserved in the Bodleian Library.)

For a description of his person and habits see his nephew Conduitt's account in Turnor's *Grantham* (pp. 163, 165), or Brewster's *Newton*, pp. 340—342.

According to Flamsteed he was short-sighted. "I happened once {during the year 1707} to visit the press while he was there, and took the opportunity to shew him how ill the compositor had placed the types of the figures {in Flamsteed's Observations}... He put his head a little nearer to the paper, but not near enough to see the fault, (for he is very near sighted,) and making a slighting motion with his hand, said, ' Methinks they are well enough.' " (Baily, p. 83.)

(²) This class of students were required to perform various menial services, which now seem to be considered degrading to a young man who is endeavouring by the force of his intellect to raise himself to his proper position in society. The following extract from the *Conclusion Book* of Trinity College, while it affords an example of one of their duties, will also serve to illustrate the rampant buoyancy of the Academic youth at the period of the Restoration. "Jan. 16. 1660-1. Ordered also that no bachelor of what condition soever, nor any undergraduate, come into the upper butteries, save only a Sizar that is sent to see his Tutor's quantum, and then to stay no longer than is requisite for that purpose, under penalty of 6d. for every time ; but if any shall leap over the hatch or strike a butler or his servant, upon this account of being hindered to come into the butteries, he shall undergo the censure of the Master and Seniors."

(³) *Optics*, Bk. ii. Part iv. Obs. 13.

(⁴) The persons appointed (in conjunction with the Proctors, John Slader of Cath. Hall and Benj. Pulleyn of Trin. Newton's tutor) to examine the Questionists, were John Eachard (the satirical author of *The Grounds...of the contempt of the Clergy...*) of Cath. Hall and Tho. Gipps of Trinity. I am sorry that I cannot gratify the curiosity of those who may expect to find here a notice of the Academical estimate formed of the acquirements of the most illustrious candidate that ever offered himself for a degree, as the " Ordo Senioritatis" of the Bachelors of Arts for this year is provokingly omitted in the Grace Book.

(⁵) Shewing how to take the fluxion of (or to differentiate) an equation connecting any number of variables. It is referred to in a paper which seems to be part of a draught of his observations on Leibniz's letter of Apr. 9, 1716. (Rigaud's *Appendix*, p. 23, compared with Raphson's *History of Fluxions*, p. 116).

(⁶) Rigaud and Raphson, *u. s.*

(⁷) The recipe described in the subjoined extract is at least as worthy of being recorded as Tasso's malmsey, or Blackstone's port. " I have been credibly informed that Sir Isaac Newton, when he applied himself to what is esteemed the greatest stretch of human invention and penetration (viz. the study, investigation and analysis of the theory of light and colours) to quicken his faculties and fix his attention, confined himself to a small quantity of bread, during all the time, with a little sack and water, of which, without any regulation, he took as he found a craving or failure of spirits." Cheyne's *Natural Method of curing diseases of the body and disorders of mind, &c.* Lond. 1742, p. 81.

(⁸) The College was "dismissed" June 22 on the reappearance of the plague. The Fellows and Scholars were allowed their commons during their absence. Newton received on this account 3s. 4d. weekly, for 13 weeks in the quarter ending Mich⁸ 1666.

........................ 12 .................................... Dec. 21......

........................ 5 .............................. Lady Day, 1667.

The College had been also dismissed the previous year, Aug. 8, on the breaking out of the plague, but Newton must have left Cambridge before that, as his name does not appear in the list of those who received *extra coes* for 6½ weeks on the occasion. "Aug. 7, 1665. A month's commons (beginning Aug. 8) allowed to all Fellows and scholars which now go into the country upon occasion of the pestilence." (*Conclusion Book*). On the continuance of the scourge we find him, with others, receiving the allowance for commons for 12 weeks in the quarter ending Dec. 21, 1665, and for 13 weeks in that ending Lady-Day, 1666.

(⁹) To the authorities for this anecdote (Biot, *Journal des Savans*, 1832, p. 265) may be added Green (*Philosophy of Expansive and Contractive Forces*, p. 972), whose information on the point was derived from a very good source : " quæ sententia...originem ducit, uti omnis, ut fertur, Cognitio nostra, a Pomo, id quod accepi ab...amicissimo Martino Folkes." For the sentiment, compare the following from the meditations of a modern speculatist : " plebi autem vis gravitatis cognita placuit...quia...corpora cœlestia in orbes revolvi præsertim per tritissimam illam pomi coram Newtone delapsi historiam edocta securitatem adversus cœlum hausit, oblita scilicet, universæ generis humani, deinde Trojæ miseriæ principiis pomum adfuisse, malum etiam scientiis philosophicis omen." Hegel's *Dissertatio Philosophica de Orbitis Planetarum*—an exercise written at the age of 31, *pro licentia docendi.* Werke, Band 16, p. 18. Berlin, 1834.

(¹⁰) In this tract his previous method of taking fluxions is extended to surds. The area of a curve, whose ordinate is *y*, is denoted by ▢ *y*. (Rigaud's Append. p. 23.)

(¹¹) Raphson, p. 116. Wilson's Appendix to Robins' Tracts (II. 351—356).

(¹²) There were nine fellowships vacant; among them those of Duport, Thorndike, and Cowley (the last by death in July, 1667). Two of the other vacancies were caused by the parties falling down staircases, one of which was that in which Newton subsequently "kept." All the nine successful candidates were in their last year. One of the middle bachelors had procured a King's letter for his election, but an order was passed by the Seniority putting him off until the following year. Besides Pearson, the Master, Babington and Lynnet were probably two of the examiners at this election. It is very improbable that Barrow examined : he was thirteenth on the list of fellows, and by the absence of one of the Seniors, and the exclusion of another (Barton) on the ground of mental aberration, he became temporarily the eleventh, but it is not likely that he would come within the first eight on so important an occasion, though in the preceding June he had sat upon the Seniority which ejected Barton from College.

In a MS. calendar, drawn up by Lynnet, of the routine events of an academical life, we find the following memorandum relative to the fellowship-examination ; it was written five-and-twenty years or more posterior to the period under consideration, but the practice had probably undergone little change in the interval. " The fellowes on the 3ᵈ day of their sitting must have a theme given them by the Master, wʰ the chappelclerk fetcheth for them : they sit 3 dayes being excused the 4ᵗʰ for their theme.

"They sit from 7 till 10, & from one to 4, each writing his name his age & his country ; as doe the scholars, & also yᵉ Masters of Arts, wᶜʰ papers are carried to yᵉ

Master & Vice-M$^r$, the first morning so soon as all have written...Octob. 1...by y$^e$ tolling of y$^e$ little bell at 8 in y$^e$ morning y$^e$ seniours are called & the day after at one o'clock to swear them y$^t$ are chosen......"

There was no election of fellows in the years 1665 and 1666, probably on account of the plague. At the election in 1664, there were seventeen fellows chosen, seven out of the middle year, and five out of each of the other years.

($^{13}$) It was usual, in Trinity College, as rooms fell vacant to distribute them among the fellows in the order of their seniority, and the chamber so assigned to a person was called his "seniority" or "fellowship chamber." A few of the papers containing a schedule of the succession to the various rooms at these periodical distributions are still preserved in the archives of the College, and among them is the one which was arranged on Sept. 30, 1667, with Pearson's signature, confirming the arrangement: "Oct. 5, 1667. I confirm this Succession of Chambers. Jo. Pearson Master." The last line on this paper runs thus: "to S$^r$ Newton — Spirituall chamber," a locality with respect to which the only conjecture that I have to offer (and it is not altogether free from objection) is that the apartment so designated may have been the ground-room next the Chapel, in the north-east corner of the great court. There is some reason for supposing that this room was, previously to 1640, the vestry, and that it is the same as that which is denominated the "vestry," or "vestry chamber," in the Junior Bursar's Books of 1648 and 1649. Though "spiritual chamber" is put down in the schedule as the habitation assigned to Newton, it does not follow that he actually dwelt there ; if he did not occupy the room himself, he would receive the rent of it from the person who was his tenant.

The rooms that he occupied before he was elected fellow—the scene of the experiments by which he analysed light—are not known. There is no mention of them in the Junior Bursar's books during that period. Neither is it known in what part of the College he lived from the epoch just mentioned to 1683. He himself states, that in June, 1673, John Wickins (a fellow, two years junior to him) was his chamber-fellow (Letter to Halley, July 27, 1686). But in the Junior Bursar's Book for the year ending at Michaelmas, 1673, we find the two entries "for seiling M$^r$ Newton's chamber," "for mending the slating...over M$^r$ Wickins," from which perhaps we may infer that one of them had changed his rooms in the interval between June and September*. In 1678 he had a sizar living with him: "for mending over M$^r$ Newton's sizar's chamber." (Junior Bursar's Book.) The first notice of Newton's rooms which fixes their position, occurs in the Junior Bursar's Book for the year ending at Michaelmas 1683, and we then find him inhabiting the rooms which well-informed tradition still points out to the stranger (the rooms on the first floor to the north of the Great Gateway): " For mending the wall betwixt Mr Newton's garden and St John's" (probably about the end of 1682). I am unable to determine satisfactorily the date of his taking these rooms, but the most probable supposition is that he went into them in the summer of 1679 †. Herbert Thorndike preceded him in the occupation of them (with one or two removes): when Newton left Cambridge in 1696, they seem to have come into the possession of

* If it was Newton that changed, we may find in that fact a foundation for the statement made by a grandson of Wickins, who, in making mention of a wooden pint flagon given to his grandfather by Newton, says: "This with the whole furniture of the chambers devolved upon my ancestor upon Sir Isaac's leaving the college, and hath with some other articles remained in the family ever since." (Gent. Mag. Apr. 1802.) Wickins vacated his fellowship in 1685 (eleven years before Newton left College), and had ceased to reside for several years. Yet, curiously enough, in Walker's account-book, quoted p. xLiv, in the statement of the "income" of his rooms, there is the following item (date 1716): "Paid D$^s$ Wickins a bill for repairing what M$^r$ Hanbury's brother took away, £1. 8$^s$." "D$^s$ Wickins" was a son of Newton's friend, and had just taken his bachelor's degree. Perhaps he had occupied part of the rooms jointly with Hanbury.

† A view of Newton's rooms from the east, with the garden attached, may be seen in Loggan's plate of the College. The following chronological notices, in conjunction with Loggan's plates,

Daniel Hopkins, whom Bentley describes as "a Fellow of Trinity College and a very useful person in it, having the greatest number of pupils of any one amongst us" (*Correspondence*, p. 185); Nat. Hanbury (see p. 192) took them in 1704, and was succeeded in 1715 by "Our hat" Walker, who continued in them until his death in 1764. Cumberland, who came up a freshman in 1747, speaking of the kindness shewn to him by Walker, who was Vice-Master, says: "He frequently invited me to his rooms, which I had so often visited as a child, and which had the further merit with me as having been the residence of Sir Isaac Newton, every relic of whose studies and experiments were respectfully preserved to the minutest particular, and pointed out to me by the good old Vice-Master with the most circumstantial precision. He had many little anecdotes of my grandfather {Bentley}, which to me at least were interesting, and an old servant Deborah, whom he made a kind of companion, and who was much in request for the many entertaining circumstances she could narrate of Sir Isaac Newton, when she waited upon him as his bedmaker, and also of Dr Bentley, with whom she lived for several years after Sir Isaac left college, and at the death of my grandfather was passed over to Dr Walker, in whose service she died." (*Memoirs*, p. 73.) What the "relics" alluded to were I cannot exactly say. It happens that Walker's private account-book has been preserved. It contains a statement of what is called the "income" of his rooms, and an inventory of the furniture and movables in them and in the garden. In the list there appears a "thermometer," "a bureau bought of Dr Smith {the Master}," a "violoncello (sold)," "a picture of Vandyke," "a barometer," and 10 pounds' worth of books, but there is nothing to indicate that any of these or the other articles ever belonged to Newton. In 1730 Walker made considerable alterations in the rooms. The same book contains his accounts with his bedmaker, Betty Baxter, and on her death, in Feb. 1744, with her sister "Deb." They seem to have been both women of thrift, and improved their capital by loans to their master. Deborah did not profit by her attendance upon Newton to learn the art of writing: in Walker's book, instead of her signature, she appends, like our early kings, her mark.

($^{14}$) He was 23rd on the list of 148 signed by the Sen. Proctor (Thomas Burnet, author of *Theoria Telluris Sacra*).

($^{15}$) It was 6 inches long, aperture something more than an inch, depth of plano-convex eye-glass, one-sixth or one-seventh of an inch, magnifying power about 40. (Letter of Feb. 23, 1669 in Macc. *Corr.* II. 289. Comp. Brewster's *Newton*, p. 27.)

($^{16}$) The Lucasian statutes, dated Dec. 19, 1663 (they are printed in the Appendix to Whiston's Account of his Prosecution, ed. 1718-9) require the Professor to lecture at least once a week during term-time, on some portion "Geometriæ, Arithmeticæ, Astronomiæ, Geographiæ, Opticæ, Staticæ aut alterius alicujus Mathematicæ Disciplinæ" ..."per unius circiter horæ spatium," and also two days in the week during term-time (and during vacation *one* day, if the Professor is in residence) "per duas horas...omni-

---

will enable the academical reader to picture to himself the College as it was when Newton walked to and fro within its courts :

1670-1 Gerrard's Hostle rebuilt at the expense of Bishop Hacket and thence called Bishop's Hostle.

1676 Feb. Foundation of new Library dug.

1678 Rooms over eight arches next the Library in north cloister finished, those next the library being built out of the subscriptions for the Library, those next to them to the east at the expense of Sir Thomas Sclater.

1681-2 Rooms over eight arches next the Library in south cloister built, those adjoining the library out of the library subscriptions, the others at the expense of Dr Humfrey Babington.

1681 May 7. Four statues on the top of the library by Cibber for which he received £80.

1685 Feb. New Library ceiled.

1386 Library floor laid down.

1687-8 Library paved.

1694 Ruinous part of King's Hostle pulled down.

1695 Books removed from the old library to the new.

bus illum consulturis vacare, liberum adeuntibus aperto cubiculo accessum præbere, circa propositas ipsi quæstiones & difficultates haud gravate respondere."......This last-mentioned part of the Professor's prescribed duties explains a passage in the Life of Henry Wharton (B.A. in 1684), who, we are told, "attained...no mean skill in mathematics. Which last was much increased by the kindness of Mʳ Isaac Newton, Fellow of Trinity College, the incomparable Lucas-Professor of Mathematics in the University, who was pleased to give him further instructions in that noble science, amongst a select company in his own private chamber." Life of Wharton, prefixed to his Sermons, 2nd ed. 1700.

The Letter of Charles II. (confirming the Lucasian statutes), dated Jan. 18, 1664, further ordered that all Undergraduates after their 2nd year, and all Bachelors of Arts " usque ad annum tertium," should attend the Professor's lectures: it also allowed the Professor to hold a Fellowship along with his Professorship, but forbad him " Decani, Thesaurarii, Seneschalli, aut Lectoris cujusvis in suo Collegio munus capessat, aut... inibi Tutorem se gerat (nisi Nobilium forte vel Generosorum Sociis Commensalium), vel...Procuratoris, Taxatoris, aut alterius cujuslibet Lectoris publicum in Academiâ Officium sustineat...Ab omnibus et singulis Muneribus istis prædictis liberatum volumus et exemptum." (Baker MSS. xxix. 403.) This prohibition will account for our not finding Newton's name at any time among the College or University Officers. He availed himself of the privilege of taking Fellow-Commoners as pupils in two instances only: viz. Mr George Markham (son of Sir Robt. Markham, of Sedgebroke, Notts.), afterwards Baronet and F.R.S., entered Jun. 26, 1680, and Mr Robt. Sacheverell, whose mother was daughter of the 2nd Sir John Newton, and sister of the 3rd Baronet of the name (to whom Letter No. XXXI. Appendix, is addressed) entered Sept. 16, 1687. We also find Mr St Leger Scroope (possibly connected afterwards by marriage with Sir John Newton's family) entered Fellow-Commoner under him Apr. 2, 1669, before he was appointed Lucasian Professor.

In 1675 Newton obtained a Royal Patent allowing the Professor to remain Fellow of a College without being obliged to go into orders, as the statutes of some Colleges require. See below, under that year.

In packet No. E. of the Lucasian MSS. there is a copy (with a few clerical errors) of the Statutes and the King's Confirmation of them in Newton's handwriting on a folio sheet doubled twice. On the last page he has written the following, as a help to his memory, the almanacs not having yet begun regularly to register the information:

Termini durant 1. a 10⁰ Octob. ad 16ᵘᵐ Decemb.

2. a 13⁰ Jan. ad 10 ante Pascha

3. ab 11⁰ post Pascha ad diem veneris Comitia sequentem.

(¹⁷) This, like most of Newton's letters, is in answer to questions proposed to him.

(¹⁸) In this letter he says: "That solution of the annuity problem {in letter of Feb. 6} ...you have my leave to insert it into the Philos. Trans. so it be without my name to it. For I see not what there is desirable in public esteem, were I able to acquire and maintain it. It would perhaps increase my acquaintance, the thing which I chiefly study to decline." Macc. Corr. II. 296.

(¹⁹) Newton wishes his name to be suppressed in connexion with the improvements made in the book, and suggests that in the title-page, after the words "Nunc e Belgico Latine versa," some such words as " et ab alio authore locupletata " should be added.

(²⁰) Collins, writing to Vernon at Paris, Dec. 26, says: "As to Mr Newton's Telescope, I suppose Mr Bernard { of Oxford } writ the same to you as he did to me upon the authority of one Mr Gale of Cambridge { Fellow of Trin. Coll. afterwards Dean of York } : since it hath been brought up for his Majesty's perusal, & I have seen an object in it," &c. He then proceeds to give a description of the instrument. (Royal Soc. MSS. lxxxi.) Compare Collins to Vernon, Dec. 14, in Macc. Corr. I. 176. This instrument is in the possession of the Royal Society. The instrument in

Trinity College Library, which is usually shewn to visitors as Newton's own telescope, I believe to have belonged to Robert Smith, and to be that which is described in his *Optics*, p. 304, note. The inscription upon it, " Sir Isaac Newton's Telescope," merely means " a Newtonian Telescope."

(²¹) It was never finished. It was published by Horsley, ɪ. 391—518, under the title of *Geometria Analytica*. It first appeared in 1736, in Colson's translation, with the title, " The Method of Fluxions and Infinite Series, with its Application to the Geometry of Curve-Lines. By the Inventor, Sir Isaac Newton...translated from the Author's Latin Original not yet made public..." Pemberton, in speaking of the treatise, tells us that he had prevailed upon Newton " to let it go abroad." " I had examined all the calculations and prepared part of the figures; but as the latter part of the treatise had never been finished, he was about letting me have other papers in order to supply what was wanting. But his death put a stop to that design." (Preface to *View of Newton's Philosophy*, Lond. 1728.)

(²²) In answer to Oldenburg's letter of Jan. 2, printed in the Appendix, No. I. The opening and concluding paragraphs are transcribed here, principally on account of the touching modesty of the closing words of the latter.

"At the reading of your letter I was surprised to see so much care taken about securing an invention to me, of which I have hitherto had so little value. And therefore since the Royal Society is pleased to think it worth the patronising, I must acknowledge it deserves much more of them for that, than of me, who, had not the communication of it been desired, might have let it still remain in private as it hath already done some years.

" I am very sensible of the honour done me by the Bishop of Sarum in proposing me candidate, and which I hope will be further conferred upon me by my election into the Society. And if so, I shall endeavour to testify my gratitude by communicating what my poor and solitary endeavours can effect towards the promoting your philosophical designs." Macc. *Corr.* ɪɪ. 311, 313.

(²³) A copy of this description, with Newton's alterations added by Oldenburg, is preserved at the Royal Society. *Orig. Lett. Bk.* N. ɪ. 37. It is printed in Horsley's *Newton*, ɪᴠ. 270.

Voltaire informs us that he had seen a little work by a German Jesuit, published about this time, " dans lequel, en parlant du télescope de Newton, on le prend pour un lunetier: *Artifex quidam Anglus nomine Newton*. La posterité l'a bien vengé." (*Dict. Philos.* and some editions of the *Lettres Philos.*)

(²⁴) " It was ordered that a letter should be written by the secretary to Mr Newton to acquaint him of his election into the Society, and to thank him for the communication of his telescope, and to assure him that the Society would take care that all right should be done him with respect to this invention." Birch, ɪɪɪ. 1. Picart's recent measure of the earth was also communicated at the same meeting in a letter from Vernon to Oldenburg, dated Paris, Jan. 9, but Oldenburg does not seem to have made any allusion to it in the letter which he was directed to write to Newton.

(²⁵) Appendix, No. II.

(²⁶) Appendix, No. III.

(²⁷) Appendix, No. IV.

(²⁸) Appendix, No. V.

(²⁹) Appendix, No. VI.

(³⁰) " Finding already, by that little use I have made of the press that I shall not enjoy my former serene liberty till I have done with it, which I hope will be so soon as I have made good what is already extant on my account." He adds that he may possibly complete his method of infinite series, "the better half of which was written last Christmas." Macc. *Corr.* ɪɪ. 322.

Under this date may be given the anecdote related in Nichols's *History of Hinckley* (p. 61, note), if, as is probable, it refers to the action between the English and Dutch fleets in Southwold bay on the 28th of May. "There is a traditional story at Cambridge ... { that } Sir Isaac Newton came into the hall of Trinity College and told the other fellows that there had been an action just then between the Dutch and English, and that the latter had the worst of it. Being asked how he came by his knowledge, he said that being in the observatory, he heard the report of a great firing of cannon, such as could only be between two great fleets, and that as the noise grew louder and louder he concluded that they drew nearer to our coasts and consequently that we had the worst of it, which the event verified." Jones, in his *Physiological Disquisitions*, p. 299 (quoted *ib.*), says that he had been informed "that the great engagement between the Dutch and English at sea in 1672 was heard by the people who were out at work in the fields to the very centre of England: Mr Derham says it was heard 200 miles." The "observatory" in the passage quoted above is a prolepsis for the "great gateway," which was not converted into an observatory until several years after Newton had left Cambridge.

(³¹) Appendix, No. VII.

(³²) He also says, "I should be glad to hear whether Mr Cox hath finished the 4 feet telescope and what its effects are...But I know not whether I shall make any further trials myself, being desirous to prosecute some other studies." Macc. *Corr.* II. 329.

(³³) For a character of this work see Humboldt's *Kosmos*, Vol. I. The edition of 1681 seems to be almost a reprint of the preceding one, in spite of the "auctior et emendatior" of the title-page.

(³⁴) Appendix, No. VIII.

(³⁵) This part of the letter is cited in the 3rd edition of the *Principia*, p. 246, instead of the letters to Leibniz referred to in the two first editions. Its contents were sent to Leibniz July 26, 1676, along with Newton's letter of June 13 of that year. There is a copy of it at the Royal Society (Miscell. MSS. LXXXI.) written in a tremulous hand, a consequence probably of the endeavour of the copyist to imitate Newton's writing. It has an address in Newton's hand, "These to his ever Honoured ffriend Mr John Collins...," and bears the post-mark of *May* 27 (probably 1676). This transcript may be conjectured to have been made at Collins's request for the purpose of accompanying the other papers which he was preparing to send through Oldenburg to Leibniz. See *Commerc. Epist.* p. 47. (128, 2nd ed.) Doubts have been expressed whether these papers were actually sent to Leibniz. We have however Collins's own testimony that they were sent as had been desired (*Comm. Epist.* p. 48, or 129, 2nd ed.), besides Leibniz's and Tschirnhaus's acknowledgments of the receipt of them. (*Ib.* pp. 58, 66, or 129, 142.) It may also be observed that the papers actually sent (in a letter dated July 26, 1676) to Leibniz by Oldenburg have been recently printed from the originals in the Royal Library at Hanover (Leibn. *Math. Schrift.* Berlin, 1849), and that in them, as in Collins's draught, which is preserved in the Royal Society ("To Leibnitz the 14th of June 1676 About Mr Gregories remains" MSS. LXXXI.), we find the contents of Newton's letter of Dec. 10, 1672, except that instead of the example of drawing a tangent to a curve, there is merely allusion made to the method. Collins's larger paper (called "Collectio" and "Historiola" in the *Commercium Epistolicum*), of which the paper just quoted "About Mr Gregories remains" is an abridgment, and which contains Newton's letter of Dec. 10 without curtailment, is stated in the second edition of the *Commercium* to have been sent to Leibniz, but whether that was the case may be fairly questioned. This paper was intended by Collins to be deposited in the archives of the Royal Society, where it is still preserved, with the title "Extracts from Mr Gregories Letter" (MSS. LXXXI.), consisting of thirteen sheets. A copy of Newton's letter was sent to Tschirnhaus in May, 1675, in Collins's paper "About Descartes" (14 folio leaves, Roy. Soc. MSS. LXXXI.)

(³⁶) On the Public Oratorship becoming vacant by the resignation of Ralph Widdrington, the mode of electing his successor became a subject of dispute between the

Masters of Colleges and the Senate. The Statutes of Elizabeth contain no express provision for the election of Orator, but the Heads (under the 40th Statute, which enacts that " Nominationes et electiones lectorum, bedellorum, stationariorum, gageatorum, vinopolarum et aliorum ministrorum seu officiariorum academiæ quorumcunque de quibus aliter a nobis non est provisum sequentur modum et formam in electione procancellarii præscriptam fientque intra xiv dies post vacationem nisi aliter statutis nostris aut fundatione cautum sit ") claimed, as had been usual, the right of nominating two persons, one of whom was to be elected by the Senate. The Senate, however, maintained that the proper mode of procedure was by an open election, as directed by the *Statuta Antiqua*, which they contended were still in force, except upon points where they were contrary to the Elizabethan code. The Chancellor (" great Villiers") endeavoured to effect an arrangement between the contending parties. "Being informed," he writes, " that there may be a contest between the Heads of the Colleges and the body of the University about the manner of electing an Orator,...he thinks it becomes his duty and affection to the University to communicate his thoughts :...he thinks that the election of Orator should be regulated by the statute of Henry VIII. made only for that purpose rather than by that of Queen Elizabeth." He suggests an expedient, which he says " I hope may for the present satisfy both sides. I propose that the Heads may for this time nominate and the Body comply, yet interposing (if they think fit) a Protestation concerning their plea that this election may not hereafter pass for a decisive Precedent in prejudice to their claim." And "whereas I understand that the whole University has chiefly a consideration for Dr Paman of St John's and Mr Craven of Trinity College I do recommend them both to be nominated. For it is very reasonable that in this nomination, before the difference be determined between you, the Heads should have regard to the inclination of the Body, especially seeing you all agree in two men that are very worthy and very fit for the place." (Letter read to the Senate, March 3. *Mandates in Registr. Office*, Vol. ii. p. 251*.) These conciliatory suggestions were not attended to. A majority of the Heads nominated Paman and a Mr Ralph Sanderson, likewise of St John's, on the day after the letter was read, and on the next day 121 Members of the Senate recorded their votes in favour of Craven and 98 for Paman. On the morning of the election, before the polling commenced, the following protest was read and entered in the Regent House: " Nos Antonius Marshall, Georgius Chamberlaine, Humfredus Babington, Gulielmus Lynnet,...Ioannes Hawkins, Isaacus Newton...aliique quorum nomina sunt infra scripta, coram Matthæo Whinn, Notario Publico, Protestamur de invaliditate et nullitate Nominationis et Notationis per puncta Præfectorum Collegiorum ad Officium Oratoris hujus Academiæ. Etiam et de nullitate omnis actus exin facti aut faciendi." The Vice-Chancellor admitted Paman the same morning; Craven, as " legitimè electus...per majorem partem suffragantium secundum statutum de electione Oratoris," gave in a protest against the validity of his competitor's election and admission, and there, so far as our information goes, the matter seems to have ended.

The reader who wishes to see what may be said on both sides of the question may consult an anonymous pamphlet, entitled *An Argument to prove that the* 39th *section of the* 50th *chapter of the statute, given by Queen Elizabeth...includes the Old Statutes* [by Mr Burford, fellow of King's]...*with an Answer to the Argument* [by Bentley] *and...* [Burford's] *Reply*. London, 1727. Comp. Monk's *Bentley*, pp. 524—6.

(³⁷) "Since I see I shall neither profit them, nor (by reason of this distance) can partake of the advantage of their assemblies." Macc. *Corr.* ii. 348.

(³⁸) It begins, "I received your two last letters with Heuret's Optics, which (not being so ready in the French tongue myself, as to read it without the continual use of a dictionary) I committed to the perusal of another..."

Here may be mentioned the myth respecting his not being elected into the law-fellowship, which became vacant Feb. 14, in this year, by the death of Dr Robert Crane. The story as told by a great-grandson of the person who was selected to fill the vacancy is, that Newton and Robert Uvedale (who was two years senior to Newton, and would, in the usual course of things, vacate his fellowship in a few months) were candidates for the fellowship in question; and that " Mr Barrow { who had been admitted Master on Feb. 27 } decided it in favour of Mr U. saying that Mr U. and Mr N. being (at that time) equal in literary attainments, he must give the fellowship to Mr U. as senior." (*Gentleman's Mag.* Supplement for 1799, p. 1186.) I apprehend the tenure of the law-fellowship of Trinity College was considered to be scarcely compatible with the efficient discharge of the duties of the Mathematical Professor, and I believe that it would argue much misconception of the characters of the two great men concerned to suppose them capable of being parties to a lax interpretation of the statutes which they had sworn to obey. The person who holds this fellowship is required "operam dare juri civili," and accordingly we find Uvedale, on receiving the appointment, excused by the University from appearing, according to an announcement made in April previous to his election, as Respondent in the Theological Schools on June 26 (the fellow next below him being called upon to perform the exercise), the ground assigned for the exemption being that "jam interea temporis Juris Civilis studio sese addixerit et ad ejusdem facultatis professionem virtute sodalitii sui prædicto collegio teneatur..." (*Grace Book*, June 11, 1673.) The turn given in the above story to the real facts of the case (viz. that Uvedale was appointed to a lay-fellowship, and that Newton would have been glad to have one) is a very natural family embellishment.

(39) Appendix, No. IX.

(40) We hear of these incidentally from a letter of Collins to James Gregory, dated Oct. 19, 1675. " Mr Newton.....I have not writ to or seen these 11 or 12 months, not troubling him as being intent upon chemical studies and practices, and both he and Dr Barrow beginning to think mathematical speculations to grow at least dry, if not somewhat barren." Macc. *Corr.* II. 280.

(41) Jan. 28. At a meeting of the council " Mr Oldenburg having mentioned, that Mr Newton had intimated his being now in such circumstances, that he desired to be excused from the weekly payments {1s.}, it was agreed to by the council, that he should be dispensed with, as several others were." It seems probable that the "intimation" respecting Newton's altered " circumstances" is to be referred to the expected vacating of his fellowship, which in the usual course of things would expire in the following autumn.

(42) On March 11, partly in consequence of Linus's second letter (Feb. 25. N. S.) "containing assertions directly opposite to those of Mr Newton," Hooke was ordered by the Royal Society to have the apparatus ready for the next meeting in order to make the spectrum experiment, but the day proved unfavourable. Newton was present at both meetings. While Newton was in London, Oldenburg shewed him Linus's letter, but upon reading it, he did not think it worth noticing. However, on the old man's writing again on the subject (Sept. 11), Newton was induced to send him in a letter to Oldenburg (Nov. 13) further directions for performing the controverted experiment. Linus's 3rd letter is preserved in the Royal Society Collection (L. 5. 89). The writer feeling the disadvantageous position in which the publication of his first letter with Oldenburg's rider left him, requests that his 2nd letter may be printed. It accordingly appeared in the *Trans.* for Jan. 24, 1676 in company with Newton's letter of Nov. 13.

(43) A draught of the patent (probably Newton's own composition) from a paper in his handwriting among the Lucasian MSS. (No. E.) is here subjoined.

"Carolus secundus Dei gratia Angliæ Scotiæ ffranciæ et Hiberniæ Rex, fidei Defensor, &c. : Omnibus et singulis has literas visuris salutem.—Cùm munus Professoris Mathematici in Academia nra Cantabrigiensi a Consulto Viro Henrico Lucas non ita pridem institutum authoritate nostra regia et Literis Patentibus stabiliverimus, et Ordinationes ad idem munus spectantes ratificaverimus, et ad petitionem executorum cum consilio Procancellnrii et Præfectorum privilegia insuper nonnulla eidem Professori Mathematico in perpetuum concesserimus : inter quæ statuimus ut dictus Professor eligi possit in Socium cujusvis Collegii non vetante Professione sua, et ne Is sodalitio suo, si quod ante susceptum hoc munus obtinuit aut postea obtinebit, vel ullis so{da}litii sui emolumentis aut privilegiis eo tantum nomine seu causâ privetur quovis cujuscunque collegii statuto non obstante. Quod privilegium ea intensione {sic} illi indulsimus ut eidem Professori liceret quodvis sodalitium capessere et retinere. Quod ut debitum sortiatur effectum nec restrictioni alicui in damnum aut præjudicium ejusdem Professoris pateat indulgentia nostra; Insuper volumus & statuimus ut verba nostra prædicta in favorem dicti Professoris semper accipiantur, ut non eo tantum sed nec alio quovis nomine aut causa sodalitio suo aut ejus emolumento privetur nisi quod quemlibet ejusdem Collegii Socium cujuscunque professionis & ordinis meritò privare debeat. Et speciatim volumus et ordinamus ut ordines sacros non nisi ipse voluerit, suscipiat, nec ob defectum sacrorum ordinum sodalitio cedere ipse teneatur aut ab aliis quibuscunque cogatur, sed ea immunitate quamdiu suo munere fungitur gaudeat et fruatur quo quilibet socius Medicinæ aut Juri Civili vel Canonico dicatus frui solet quovis cujuscunque Collegii Statuto aut consuetudine vel interpretatione quacunque non obstante. In cujus rei Testimonium has Literas nostras fieri fecimus patentes. Teste meipso apud Westmonasterium vicesimo septimo die Aprilis, Anno Regni nostri vicesimo septimo.

Per Breve de Privato Sigillo

Pigott."

After the above comes the following, also in Newton's hand :

" Whitehall, March 2, 1674 {O. S.}.

His Ma<sup>ty</sup> being willing to give all just encouragement to learned men who are & shall be elected into y<sup>e</sup> said Professorship, is graciously pleased to refer this draught of a Patent unto M<sup>r</sup> Atturney Generall to consider y<sup>e</sup> same, & to report his opinion what his Ma<sup>ty</sup> may lawfully do in favour of y<sup>e</sup> said Professors as to y<sup>e</sup> indulgence & dispensation proposed & desired. And then his Ma<sup>ty</sup> will declare his further pleasure.

A. COVENTRY."

The above draught was adopted: the actual instrument, (coinciding with the draught except in two unimportant particulars), with the broad seal attached, is in the Registrary's office (Box 21. G. 1. 2*) :

" A grant to the Mathematical Professor in Cambridge.

Pigott."

A transcript of it will be found in a large folio copy of the Elizabethan statutes of Trinity College, preserved in the College Archives, with the heading "Indulgentia Regia Professori Mathematico concessa, dignissimo Viro M<sup>ro</sup> Isaaco Newton, hujus Collegii Socio, istud munus tunc temporis obeunte."

Newton's visit to London in February may have been connected with his application to the Crown.

Towards the end of the preceding year, Francis Aston endeavoured to obtain a similar dispensation on his own individual account, and was backed by the interest of Sir Joseph Williamson, Principal Secretary of State. There is extant in the State Paper Office, (Domestic, No. 102), a characteristic letter from Barrow to Williamson on the subject (Dec. 4, 1674), in which he gives his reasons for resisting the application. One short extract from it may be given here : "Indeed a Fellowship with us is now so poor, that I cannot think it worth holding by an ingenuous person upon terms liable to so much scruple."

(44) Letter CIX (bis) in this work.

(45) Appendix, No. XI.

(46) Appendix, No. XII.

(47) "Dec. 9. There was produced a MS. of Mr Newton, touching his theory of light & colours, containing partly an hypothesis to explain the properties of light discoursed of by him in his former papers, partly the principal phænomena of the various colours exhibited by thin plates or bubbles, esteemed by him to be of a more

difficult consideration ; yet to depend also on the said properties of light." See Birch III. 247, seqq. One experiment mentioned in the " hypothesis" relative to the effects of glass electrised by friction particularly struck some of the members, and it was ordered to be tried at the next meeting. The paper was read by instalments, the "hypothesis" on Dec. 9 and 16, the "observations" respecting colours on Jan. 20, Feb. 3, and 10. The "observations" afterwards formed part of the 2nd Book of his Optics. The "hypothesis" has been lately reprinted in the *Phil. Mag.* for Sept. 1846, pp. 187—213.

"Dec. 16. Mr Newton's experiment of glass rubbed to cause various motions in bits of paper underneath was tried, but did not succeed......This trial was made upon the reading of a letter of his to Mr Oldenburg (Dec. 14) in which he gives some more particular directions about that experiment." Oldenburg was ordered to write to him again upon the subject " & desire him to send his own apparatus, as also to enquire whether he had secured the papers from being moved by the air, that might somewhere steal in."

On the second part of Newton's hypothesis being read, Hooke, according to his wont, said that the main of it was contained in his *Micrographia*.

(⁴⁸) At the meeting on Dec. 30, there was also read a letter from John Gascoines (Liege, 15 Dec. 1675) to Oldenburg, acquainting him with the death of Linus from the prevailing epidemic, "and with the resolution of Mr Linus's disciples to try Mr Newton's experiment concerning light and colours, more clearly and carefully"..... according to his directions of Nov. 13 : " intimating withal that if the said experiment be made before the Royal Society, and be attested by them to succeed, as Mr Newton affirmed, they would rest satisfied. It was ordered that when the sun should serve, the experiment should be made before the Society."

(⁴⁹) Harum....librationum causas Hypothesi elegantissimâ explicavit nobis vir Cl. Isaac Newton, cujus Humanitati hoc et aliis nominibus plurimum debere me lubens profiteor. Mercator's *Institutiones Astronomicæ* (p. 286) published in the beginning of 1676. See Princip. (3d ed.) Lib. 3. Prop. 17. *Mécan. Célest.* Tom. v. p. 279. Newton seems to have been in possession of his explanation in 1673. See his letter to Oldenburg, June 23 of that year, Horsley IV. 343. Rigaud, Append. 42.

(⁵⁰) He returns his hearty thanks for "the favour of the Society in their kind acceptance of his late papers ;" "that he knew not how to deny any thing which they desired should be done, but he requested that the printing of his observations about colours might be suspended for a time, because he had some thoughts of writing such another set of observations...which ought to precede those now in the Society's possession." Macc. *Corr.* II. 388.

(⁵¹) We find the following notices in the Journal Book upon this subject. On March 2, Oldenburg reminded the Society that the sky was favourable for making the experiment. Hooke said that he had an apparatus ready whenever it should be called for. March 16. The experiment ordered to be made at next meeting if the weather should prove favourable. Apr. 6. A committee appointed to try the experiment and repeat it before the Society. Apr. 27. The experiment tried with success, of which Oldenburg sends an account to Gascoines (May 4).

(⁵²) Appendix, No. XIII.

(⁵³) It was afterwards printed in Wallis's *Opp.* III. 622—629. (Oxf. 1699), and, from that work, in the *Commercium Epistolicum*, where the typographical error of 26 *Junii* for *Julii*, which is corrected in Wallis's *errata*, is also copied in the heading of the letter.

(⁵⁴) Appendix, No. XIV.

(⁵⁵) The original letter extending over 14 folio pages is in the British Museum (MSS. Birch 4294). It was accompanied by a note to Oldenburg (Macc. *Corr.* II. 400) in a postscript to which he observes: "I hope that this will so far satisfy M. Leibnitz that it will not be necessary for me to write any more about this subject ; for having other things in my head, it proves an unwelcome interruption to me to be at this time put upon considering these things." Newton sent some corrections by the next post (Appendix, p. 257).

A copy of the letter so corrected was not despatched to Leibniz until May 2 of the following year, the delay arising from Oldenburg's anxiety to send this "Thesaurus Newtonianus" by a safe hand. Leibn. *Mathem. Schrift.* i. 1. 151 (Berlin, 1849).

On Nov. 14 he desired Oldenburg to make some further corrections, (Appendix, No. XVII.) which, however, were not introduced into the copy sent to Leibniz, which was made ten days before.

This letter, like its predecessor of June 13, was printed in the 3rd Volume of Wallis's *Opera*, from which it was copied into the *Commercium Epistolicum*. Wallis says that he obtained his copies of the two letters from Oldenburg.

Leibniz wrote two letters in answer (June 21, July 12, 1677) in the former of which he gives examples in differentiation. Oldenburg acknowledged the receipt of these Aug. 9, observing, "Non est quod dicti Newtoni vel etiam Collinii nostri responsum tam cito ad eas expectes, cum et urbe absint, et variis aliis negotiis distineantur." (Leibn. *Math. Schrift.* i. i. 167, Berlin, 1849). Oldenburg died the following month, but there is no reason to think that, if that event had not taken place, Newton would have departed from his intention of not continuing the correspondence. Leibniz's answers will be found in Wallis's 3rd volume, the *Commercium Epistolicum* and his Works.

(⁵⁶) Appendix, No. XVI.

(⁵⁷) Macc. *Corr.* ii. 403.

(⁵⁸) Appendix, No. XVII.

(⁵⁹) Macc. *Corr.* ii. 405. See next note.

(⁶⁰) Lucas replied to Newton's letter of Aug. 18 in a letter of four pages closely written, dated Oct. 23, "containing further objections and experiments against Mr Newton's theory of light and colours with an examination of his experimentum crucis:" among other things he professes to prove that the red rays suffer the same refraction as the blue ones. Newton sent an answer to this (Nov. 28), but with a determination that it should close the controversy. In a letter to Oldenburg (Nov. 18), he writes : "1 see I have made myself a slave to philosophy, but if 1 get free of Mr Linus's business, I will resolutely bid adieu to it eternally; excepting what I do for my private satisfaction, or leave to come out after me; for I see a man must either resolve to put out nothing new, or to become a slave to defend it." Macc. *Corr.* ii. 405.

His opponent, however, was not satisfied with the answer, and indited another letter (Feb. 2, 1677 N. S.), the sole value of which to us consists in its preserving for us a few words out of Newton's letter of Nov. 28. "In his last of Nov. 28," writes the Liege professor, "I still meet with new demurs....He is pleased to quarrel with my examining his Experimentum Crucis, representing it 'a jostling out of the point in dispute by a new attempted digression,' or as he is pleased to term it 'a running from one thing to another.' He tells us 'that he intends to take into consideration one or two of my experiments, which I shall recommend for the best: and when there appears to be no weight in them, let others judge what there may be in the number of the rest'." Lucas closes his epistle with a desire that the whole of his previous letter of Oct. 23 may be printed, but the request was not attended to. The matter does not seem to have altogether dropt here, for in Oldenburg's letter to Leibniz of May 2, accompanying Newton's letter of Octob. 24 preceding, we read, "Ad alia nunc distrahitur Newtonus ab iis, qui Leodii, Francisco Lino succenturiati, novam ipsius de Lumine et Coloribus Theoriam vehementer insectantur: qua de re brevi plura accipies, ni rationes meas male subduxi," but our information extends no further.

Goethe, in his "Geschichte der Farbenlehre" (Werke, Band 55. Stuttg. 1833) gives an account of the reception of Newton's discovery of the composition of light, which does not indicate a very intimate acquaintance with the circumstances of the history.

For example, he does not know that the three persons whose suggestions or objections accompanied by Newton's answers are printed without their names in the *Philosophical Transactions* were Moray, Hooke and Huygens. One of them, indeed, he conjectures rightly enough to be Hooke, the loss of whose paper of "considerations," he says, is greatly to be regretted. It will, however, be found in a book which he himself quotes not many pages before, viz. Birch's *History of the Royal Society*, III. 10—15. In p. 56 he confounds John Gascoines, Linus's pupil, with William Gascoigne, the inventor of telescopic sights, who fell at the age of 23 at Marston Moor fighting on the Royalist side. Again, Newton, in his answer to Lucas (*Phil. Trans.* Sept. 1676, p. 703) says that the principal experiments which Lucas had sent him were detailed in a "tractate" which he had written upon light. Goethe, in quoting the passage, for the word "tractate" writes "Optical Lectures," and adds that the statement "keineswegs der Wahrheit gemäss ist" (p. 64). It is true that the treatise in question consisted in the main of the Optical Lectures, but it would not have been amiss to have ascertained the perfect identity of the two works before using language like that which has just been quoted. For Goethe's speculations on colours, see Whewell's *Hist. Ind. Sci.* II. Wilde's *Geschichte der Optik. Theil.* II. p. 153 sqq. (Berlin, 1843), and the works referred to by him.

(⁶¹) Printed in Wallis's *Works*, III. 646 (extracts from it in the *Commercium Epist.*). At the end of the letter Collins says : " Narrat mihi D. Loggan (Chalcographus) quod Effigiem tuam delineavit ille, in ordine ad Sculpturam ; Quæ præfigenda sit libro tuo de *Lumine, Coloribus, Dioptricis,* &c. quem edendum intendis. Qua de re desideramus esse certiores." Nothing further is known of the "effigies" here spoken of.

We may mention here Loggan's Dedication of his Plate of St Mary's Church. Its date is uncertain, as, though Loggan's *Cantabrigia Illustrata* was published in 1690, the dates of the separate plates range over a period of several years. " Clarisso. Viro Dᵒ. *Isaaco Newton* Matheseos apud *Cantabrigienses Professori Lucasiano SSⁱᵃ. Trinitatis Collⁱⁱ.* ibidem, et *Regiæ Societ*ᵗ. *Socio, Mathematico, Philosopho, Chymico* consummatissᵒ. Nec minus suavitate Morum et Candore Animi, Cum rerum Humanarum Divinarumq : Peritiâ spectabili, *Hanc Tabulam* Observantiæ ergo D. D. C. Q. Dav. Loggan." Loggan had the use of a room in Trinity College for his press.

(⁶²) Appendix, No. XVII.

(⁶³) In this and other instances where Newton is mentioned as voting at University elections of Members of Parliament or Officers, our information is derived from the actual slips of paper on which each voter recorded his suffrage, and which are still preserved in the Registrary's office. A copy of Newton's voting paper on this occasion is given as a specimen. " Isaacus Newton eligit Thomam Exton Militem in Burgensem hujus Academiæ in Regni Comitiis."

(⁶⁴) *Boyle's Life* (by Birch) prefixed to his Works, p. 70. Macc. *Corr.* II. 407. and elsewhere.

(⁶⁵) A very pretty story is told of him by his biographer—how that in 1682 when his schoolfellow George Stepney was elected scholar from Westminster to Trinity College, Montagu, unable to bear the thoughts of being separated from his "dearest friend," went to College a year before the proper time—but, like many other pretty stories, it will not stand the test of dates. Montagu was matriculated Dec. 18, 1679, the "chamber" in which he "kept" in 1680 and following years is known, being the same, in fact, in which these lines are written, and on Oct. 6, 1681, he was made M.A. by Royal Mandate.

(⁶⁶) Newton seems to have been requested to give his opinion on a wild hypothesis of the heavens, which a Frenchman of the name of Mallemont had sent to the Royal Society. His judgment was given briefly, and with some reluctance, in a letter to Hooke, one of the Secretaries, (Nov. 28, read to the Soc. Dec. 4), in which, to make amends for the curtness of his answer, he suggested "an experiment whereby to try whether the earth moves with a diurnal motion or not, viz. by the falling of a body from a considerable height, which, he alleged, must fall to the eastward of the perpendicular, if the earth moved. This proposal was highly approved of by the Society, & it was desired that it might be tried as soon as could be with convenience."

At the meeting of the Soc. Dec. 11, Hooke read his answer to Newton's letter, in which he shewed that the path of the falling body would not be a spiral "as Mr. Newton seemed to suppose," and that it would fall " not directly east, but to the south-east & more to the south than the east. It was desired that what was tryable in this experiment might be done with the first opportunity."

At the meeting on Dec. 18, Hooke read his answer again, and also a reply to it from Newton, "containing his farther thoughts and examinations of what had been propounded by Mr Hooke." He also gave an account of three trials that he had made of the experiment.

At the meeting on Jan. 8, 1680, Hooke read another letter of his to Newton, giving " a further account of his theory of circular motion & attraction, as also several observations & deductions from it." Newton declined answering this letter. At the same meeting Hooke " was desired to make his trials " of Newton's experiment as soon as possible.

($^{67}$) Library Account Bk. for year from Dec. 22, 1679, to Dec. 22, 1680. The charge for the bond appears in the Sen. Bursar's Bk. for year ending Mich. 1680. The money seems to have been repaid Nov. 12, 1688. Conclus. Bk. Feb. 5, 1689.

($^{68}$) Appendix, No. XVIII.

($^{69}$) Birch, IV. 65. A letter of his to his kinsman Sir John Newton, introducing Adams, is printed in Turnor's *Grantham*, p. 85, note.

($^{70}$) *Gen. Dict.* VII. 788. The originals of this and the other letters to Flamsteed down to 1698, are preserved in the Library of Corpus Christi College, Oxford.

($^{71}$) *Gen. Dict.* VII. 791.

($^{71}$)*           " Cambridge April y$^e$ 3$^d$ 1682

These are to signify y$^t$ M$^r$ Ellis advising w$^{th}$ me ab$^t$ a person fit to be intrusted w$^{th}$ y$^e$ Charge of teaching Navigation to y$^e$ Boys of y$^e$ King's late foundation, I propounded M$^r$ Edw$^d$ Paget Master of Arts & ffellow of Trinity College in this University, as y$^e$ most promising person for this end I could think of; and that upon these considerations. He is of a temper very sober & industrious, as I am confident all that know him are ready to testify. He understands y$^e$ several parts of Mathematics, Arithmetic, Geometry, Algebra, Trigonometry, Geography, Astronomy, Navigation, & w$^h$ is y$^e$ surest character of a true Mathematical Genius, learned these of his own inclination & by his own Industry without a Teacher: And to make him y$^e$ readier in practicall Matters, his hand is very steady & accurate, as well as his fancy & apprehension, good; as may be seen by his writing & drawing w$^{th}$ his Pencil very well: Perfections w$^{ch}$ I conceive considerable for making y$^e$ Boys accurate & curious in their Draughts of Charts, Mapps & Prospects from Sea, w$^{ch}$ joyn'd w$^{th}$ his knowledge in perspective and projections of y$^e$ Sphere will enable him to contrive & draw schemes after y$^e$ best manner for y$^e$ Boys apprehension, & perswades me y$^t$ he will not only be dexterous & nice in y$^e$ use of Instrum$^{ts}$ but improve them: His long acquaintance also w$^{th}$ variety of Learning here, will help him to be methodical & clear in his teaching; w$^{ch}$ much conduces to y$^e$ Boys ready & distinct apprehension of what they are taught. So y$^t$ tho it may be easy to find persons valuable for some of these Qualifications, yet considering him in all respects as I could not think of any other person in this University so fit in my Opinion to be intrusted w$^{th}$ a place of so great concerne as that of preparing Boys to make more skilful Navigators than formerly, so I believe it will be difficult to meet w$^{th}$ fitter persons abroad for that purpose. These things made me forward to propound him to y$^e$ Electors; but to compare him w$^{th}$ other Competitors & chuse y$^e$ best I leave wholly to their judgment.

Is. NEWTON, Profess. Math. Luc."<br>( Pepysian MSS. 2612. p. 536).

Newton also wrote to his friend Collins requesting him to use his interest in behalf of Paget. There is in the same MS. volume from which the above is taken, a copy of the letter which Collins wrote in consequence, enumerating from Newton's letter to him Paget's qualifications, and dwelling upon the weight which the recommendation of the greatest mathematician of the age ought to have with the electors.

(⁷²) Appendix, Nos. XIX. XX.

(⁷³) "The charge of building" the College Library, "disables us from buying books at present.......We know not yet whether the University will purchase them, their chest being at present very low." *Gentleman's Magazine*, LXI. 504.

(⁷⁴) The propositions here mentioned as sent to Halley, have been printed by Rigaud from the copy in the Register Book of the Royal Society, vi. 218. (Appendix to Essay on Publication of Principia, No. I.) It is to be observed, however, that the title which Rigaud gives to the Paper (*Newtoni Propositiones de Motu*) is not to be found in the MS.

(⁷⁵) At the Meeting of the Royal Society, Dec. 10, "Mr Halley gave an account that he had lately seen Mr Newton at Cambridge, who had shewed him a curious treatise, *De Motu* { drawn up since August } ; which, upon Mr Halley's desire, was, he said, promised to be sent to the Society, to be entered upon their Register. Mr Halley was desired to put Mr Newton in mind of his promise for the securing his invention to himself till such time as he could be at leisure to publish it. Mr Paget was desired to join with Mr Halley." Birch, iv. 347.

The treatise *De Motu*, mentioned here, was probably the same as that of which a copy is preserved in the University Library (D d. IX. 46,) beginning "De motu corporum Liber primus, Definitiones," &c. consisting of the Lectures which he delivered as Lucasian Professor, (the first of them is dated Octob. 1684), and forming, to a certain extent, the first draught of the Principia. (See Letter CIV.). The paper which Newton sent up to Halley, in Nov. 1684, was the germ of this treatise. It is probable that Halley produced the paper at the meeting on Dec. 10, though the fact is not recorded in the Journal Book. The treatise was never registered, but the paper *was*, apparently in February 1685, with the date Dec. 10, 1684.

Rigaud's idea that the paper which he has printed from the Register of the Royal Society (consisting of 4 theorems and 7 problems) is different from the paper which Newton sent to Halley, and that it was sent to the Society in Feb. 1685, is founded upon what I conceive to be a misapprehension of a passage in Newton's letter to Aston, (Feb. 23, 1685). The words are as follow : " I thank you for entering in your Register my notions about motion. I designed them for you before now, but the examining several things has taken a greater part of my time than I expected, and a great deal of it to no purpose. And now I am to go into Lincolnshire for a month or six weeks. Afterwards I intend to finish it as soon as I can conveniently," &c. We possess only a part of the letter, and that in a copy. We cannot therefore be sure that the grammar is Newton's. It seems clear to me that what he " designed " for the Society " before now," was not yet finished and sent to the Society : that he was in fact working at his Treatise *De Motu* with a view to fulfil the promise which he had made to Halley, that he would " send it to the Society to be entered upon their register."

That the paper sent to Halley is identical with that which we find in the Register of the Royal Society, is evident from the whole tenor of our information on the subject : it is sufficient to refer to Halley's own statement ( Rigaud, Appendix to Essay, p. 37), and a letter of his to Wallis, dated Dec. 11, 1686, in which he says: " Mr Is. Newton about two years since gave me the inclosed propositions, touching the opposition of the medium to a direct impressed motion and to falling bodies, upon supposition that the opposition is as the velocity ; which tis possible is not true ; howe er, I thought any thing of his might not be unacceptable to you, and I beg your opinic thereupon, if it might not be (especially the 7th problem) somewhat better illustrated." (The original of this letter is in the collection of Dawson Turner, Esq. Compare Birch, iv. 514. Rigaud, 77.)

The probability is that Halley saw no immediate prospect of obtaining the treatise *De Motu*, and determined to secure the author's rights by at once registering the 11 propositions which he had received in November.

(⁷⁶)   Birch, iv. 370.   Rigaud (Appendix to Essay, p. 24).   Newton observes that "that which chiefly dashed the business was the want of persons willing to try experiments, he whom we chiefly relied on refusing to concern himself in that kind.......I should be very ready to concur with any persons for promoting such a design, so far as I can do it without engaging the loss of my own time in those things."

(⁷⁷)   Appendix, No. XXI.

(⁷⁸)   *Tables for renewing and purchasing of the leases of Cathedral Churches and Colleges, &c.*, Cambridge, 1686.   Newton's certificate prefixed to this work, the author of which was manciple (*mancipium*) or caterer of King's College, runs as follows: "Methodus hujus Libri recte se habet, numerique ut ex quibusdam ad calculum revocatis judico, satis exacte computantur.   Is. Newton, Math. Prof. Luc."   The later editions on the strength of this testimonial were published under the title of "Newton's Tables."

In the treasury of Trinity College in a book labelled "Notitia E," which belonged to Humfrey Babington, as Bursar (1674—1678), containing "a true particular of the rents and leases belonging to Trin. Coll. 1674-5," there is a table and an explanation of it in Newton's handwriting, of the fines to be paid for renewing any number of years lapsed in a lease for 20 years.   It is entitled *Tabula redemptionalis ad reditus Collegii SS. Trinitatis accommodata*.   It is constructed on the hypothesis that a lease for 20 years is worth 7 years' purchase, and that for the renewal of 7 years lapsed, one year's purchase must be paid.   (This is equivalent to allowing the lessee between 12 and 13 per cent. for his money).   This table which was apparently drawn up by Newton for Babington's official use, continued to be employed by the College until 1700, when Bentley, on his appointment to the Mastership, introduced the 10 per cent. tables.   The innovation however, according to Vice-Master Walker, was unpalatable to the Seniors and Officers, whose "greediness for present sealing money" superadded to "quarrels in the College," compelled a return to the old system, and occasionally the granting of terms still more favourable to the tenant.   On Dr Robert Smith's succeeding to the Mastership in 1742, the 10 per cent. tables were introduced, and these were replaced in 1750, by 9 per cent. tables.

(⁷⁹)   *Gen. Dict.* vii. 793, where also the next four letters to Flamsteed will be found.

(⁸⁰)   "You seem to insinuate as if Saturn had not yet any more satellites than one discovered by Hugenius.   I should be glad to know if it be so."   If Flamsteed returned an answer to this question, it seems to have been still in the negative.   Writing to him on Sept. 3, of the following year, Newton says: "He [Mr Philips] tells me he apprehended by some of your discourses, that you had seen two of Cassini's new planets about Saturn.   Hugenius with' a sixty foot glass could see none of them.   Mr Halley (who was lately here) I find still suspicious of them, notwithstanding what Cassini has lately published of two more.   I was glad to hear two of them confirmed by your observation."   Mr Philips' information does not appear to have been correct, for in a paper in Cotes's handwriting (Trin. Coll. Newtonian MSS. No. 382) which is apparently a memorandum of a conversation which he had had with Flamsteed some time between 1706 and 1716) it is stated "that he (Flamsteed) thought there were but 3 satellites of Saturn, himself had never seen above one."

The first discovered satellite of Saturn (now the 6th, reckoning outwards) was observed by Huygens March 25, 1655.   In 1671, 2, 3 Cassini discovered what is now the 8th, in 1672, 3 (while in pursuit of the last-mentioned one) the 5th, (see *Phil. Trans.* March 25, 1673), and in 1684 the 3d and 4th: (an account of this last discovery, given in the *Journal des Savans* for April 1686, was mentioned at the Royal Society April 28, communicated at their next meeting, and printed in the Transactions for May 25: a letter from Cassini to Halley, dated Oct. 10, giving more correct elements of the then known 5 satellites was read to the Society Nov. 3, and published in the Transactions for Apr.—June of the following year.)

In the first edition of the *Principia* Newton mentioned only the Hugenian satellite, but in the second he introduced the others, availing himself of Cassini's paper in the *Mémoires* of the Academy for 1705, published in 1706 (comp. p. 49 of this work). Pound (in 1718) was the first English astronomer who succeeded in observing the Cassinian satellites: this he did by means of corrected elements supplied by the younger Cassini, in the *Mémoires* for 1714 (published in 1717), and a telescope with an object-glass of 123 feet focal length, which Huygens had presented to the Royal Society in 1691. (See *Phil. Trans.* Jan.—Apr. 1718. Delisle's "Seconde Lettre sur les Tables Astronomiques de M. Halley..." *Journal des Savans*, June, 1750). Flamsteed, however, was not convinced. (See his letter to A. Sharp, Sept. 13, 1718, Baily, p. 331).

(81) The date is taken from the post-mark, which is Jan. 14.

(82) "Dr Vincent, { Fellow of Clare Hall } presented to the Society a manuscript treatise intitled, *Philosophiæ Naturalis principia mathematica*, and dedicated to the Society by Mr Isaac Newton, wherein he gives a mathematical demonstration of the Copernican hypothesis as proposed by Kepler, and makes out all the phenomena of the celestial motions by the only supposition of a gravitation towards the center of sun decreasing as the squares of the distances therefrom reciprocally.

It was ordered that a letter of thanks be written to Mr Newton; and that the printing of his book be referred to the consideration of the Council: and that in the mean time the book be put into the hands of Mr Halley, to make a report thereof to the council." Birch, iv. 479.

For some account of Dr Vincent, see Whiston's *Memoirs*, who was his sizar. It may perhaps prevent further currency being given to the supposition of his being the husband of the lady to whom in early life Newton is said to have been attached, if I state that he was a Senior Fellow of Clare Hall at the time of his death (March 1722).

(83) See Birch, iv. 484.

(84) At a meeting of the *Council* of the Royal Society "it was ordered that Mr Newton's book be printed, & that Mr Halley undertake the business of looking after it, & printing it at his own charge, which he engaged to do." Birch, iv. 486.

(85) My knowledge of this letter is derived from a memorandum by Halley, on the back of Newton's letter of July 14, mentioning among Newton's letters one of this date. The contents as stated above are purely conjectural, and founded upon a sentence in Newton's letter of Febr. 18, 1686-7, ("I hope you received a letter with two corollaries I sent you in autumn,") coupled with the fact that the two corollaries abovementioned are not found in Newton's MS.

(86) It had been finished in the summer of the preceding year. Writing to Halley June 20, 1686, he says that it "only wants transcribing and drawing the cuts fairly."

(87) "I think I have the solution of your problem about the sun's parallax, but through other occasions shall scarce have time to think further on these things: and besides, I want something of observation." The "occasions" may refer to the anticipated effects of James's mandate, which had been received in Cambridge nine days before. See under March 11.

(88) The first mandate was dated Febr. 7, received by the Vice-Chancellor on the 9th, and read to the Senate on the 21st, the second was dated Febr. 24, and read March 11.

(89) "It contained the whole system of celestial motions, as well of the secondary as primary planets, with the theory of comets, which is illustrated by the example of the great comet of 1680-1, proving that that which appeared in the morning in Nov......to have been the same that was observed in Dec. and Jan. in the evening." Birch, iv. 530.

The MS. sheets of the Principia (without the preface) have been bound up into a Volume which is preserved at the Royal Society. It is from no wish to detract from the value of this treasure that I state that I do not think the MS. to be in Newton's auto-graph. I believe it to be written by the same hand as the first draught of the Principia in the University Library. The author's own hand is easily recognised in both MSS. in additions and alterations.

The Preface in the first edition has no date. The date "Dabam Cantabrigiæ e Collegio S. Trinitatis, Maii 8, 1686," first appeared in the second edition in 1713. See note to Febr. 1704.

(⁹⁰) The following are the dates of the proceedings connected with this affair. Apr. 21. Vice-Chancellor and delegates appear before the Commissioners. Apr. 27. Give in their plea. May 7. Plea discussed. Vice-Chancellor sentenced to be deprived of his office, and suspended from his Mastership. May 12. The delegates reprimanded. Jeffreys wound up his address to them with the words: "Therefore I shall say to you what the scripture says, and rather because most of you are divines; 'Go your way and sin no more, lest a worse thing come unto you.'" See *State Trials*, or Cooper's *Annals of Cambridge*. Newton does not appear at all as a speaker during the proceedings. The Chancellor alludes twice to his having himself formerly been a member of the Univer-sity. Until some other College can establish a claim to him, Trinity College is liable to the suspicion of having had him for an *alumnus*. A "Georgius Jeffrys" was admitted pensioner there March 15, 1661-2, under Mr Hill, and he would therefore be a year junior to Newton.

Under this date may be given the following entry in the College Account Book of the building of the New Library, which probably refers to our philosopher. "May 28, 1687. Pᵈ...for erecting a scaffold for Mr Newton to measure the fret work of the stair-case: 4s. 6d."

We may also notice under this year an elegant method given by him of finding (by infinitesimals) the volume of a segment of a parabolic conoid cut off by a plane per-pendicular to the axis. "Construction and Demonstration as I received it from M. Isaac Newton, Prof. of the Mathematics, in Cambridge." *Guager's Magazine*, by Wm. Hunt, Lond. 1687.

(⁹¹) Rigaud, 81, 82. The copy which he gave to the College Library does not contain his autograph. In a copy in Emmanuel College Library is written, "Ex dono Authoris sumè docti Iulii 13ᵗⁱᵒ. 1687." The copy in Keill's catalogue of his books is priced at 10s., as also is a copy in Clare Hall Library, given by Cornelius Crownfield to Cotes's friend Morgan, of which however the price at the time of the gift is put 5s. There is in the same Library a copy of the Theses printed at Edinburgh, in the first half of 1690, by James Gregory, of St Andrew's, containing a compend of the *Principia*, alluded to in the *Museum Criticum*, ii. 518, note, and Brewster's *Newton*, p. 174, note.

The following anecdote of Demoivre's first introduction to the *Principia* may not be altogether out of place here. The scene is probably to be laid in the year after its publication, when Newton is known to have been out of College. (See Table of Exits and Redits). Demoivre, then about 21, was earning a livelihood in London by teaching mathematics, in which he thought himself a perfect master. "Il en fut bientôt et bien singulièrement desabusé. Le hasard le conduisit chez Mylord Devonshire dans le moment où M. Newton venoit de laisser chez ce Seigneur un exemplaire de ses Principes. Le jeune Mathématicien ouvrit le livre, et, séduit par la simplicité appa-rente de l'ouvrage, se persuada qu'il alloit l'entendre sans difficulté; mais il fut bien surpris de le trouver hors de la portée de ses connoissances, et de se voir obligé de convenir que ce qu'il avoit pris pour le faîte des Mathématiques n'étoit que l'entrée d'une longue et pénible carrière qui lui restoit à parcourir. Il se procura cependant le livre, et comme les leçons qu'il étoit obligé de donner l'engageoient à des courses pres-que continuelles, il en déchira les feuillets pour les porter dans sa poche et les étudier dans les intervalles de ses travaux." *Eloge, Hist. de l'Académie*, 1754.

(⁹²) He did not give up his rooms until Midsummer. On Sept. 14, a donatiou of £50 towards the New Library was received from him.

(93) In many of the voting papers his name is preceded by the words "præclarum virum," in some the adjective is "doctissimum," "integerrimum," "venerabilem," "reverendum." Pulleyn, his old tutor, calls him "summum virum." Thirteen letters from Newton to the Vice-Chancellor, written between February and May 1689, on matters connected with the University as affected by the new order of things, have been recently printed by Dawson Turner, Esq. from the originals in his possession.

Laplace, in speaking of the publication of the *Principia*, observes: " Les principes du système social furent posés dans l'année suivant, et Newton concourut à leur établissement." *Syst. du Mond.* p. 372, Paris, 1824.

(94) See Lord King's *Life of Locke*, i. 389 (2nd. ed.)

(95) " Mr Huygens of Zulichem being present gave an account that he himself was now about publishing a Treatise concerning the cause of gravity, and another about Refractions giving amongst other things the reasons of the double refracting Island Crystal.

Mr Newton considering a piece of the Island Crystal did observe that of the two species wherewith things do appear through that body, the one suffered no refraction when the visual ray came parallel to the oblique sides of the parallelepiped; the other, as is usual in all other transparent bodies, suffered none, when the beam came perpendicular to the planes through which the object appeared." *Journ. Bk.*

The first mentioned observation of Newton is due to Erasmus Bartholinus, but was found by Huygens not to be rigorously true, (*Traité de la Lumière*, 1690, p. 57).

I take this opportunity of offering my grateful acknowledgments to the President and Council of the Royal Society for their liberality in granting me access to their Archives. Perhaps I may be permitted in this place to express my opinion of the obligation which that illustrious body would confer upon the world by the continuation of Birch's History of the Society, at least down to the close of Newton's Presidentship. Independently of the value, great or small, of such a work to the historian of science, it would give us an opportunity of meeting our philosopher once or twice a week for the twenty three last years of his life. The following extracts from the Journal Books of the period are given not as specimens of their contents, but are selected solely for the local allusions. " March 31, 1720. The President... mentioned a remarkable experiment he made formerly in Trinity College kitchin at Cambridge, upon the heart of an eel which he cut into three pieces, and observed every one of them beat at the same instant and interval: putting spittle upon any of the sections had no effect, but a drop of vinegar utterly extinguished its motion." (He had mentioned the same experiment more briefly at the meeting on Nov. 13, 1712). " Febr. 20, 1723-4. The President upon reading this { a letter containing an account of the effects of a violent thunderstorm } made mention of an accident much like it which he once saw at Trinity College in Cambridge. He was suddenly surprized with a violent strong flash of lightning which was so exceeding bright that he was forced immediately to guard his eyes with his hands. And at the same instant a violent clap of thunder broke down the window in the next room, and forced some splinters out of the floor which darted against the cieling, and there being another window opposite to that which was broke down they observed it to be bowed outwards by the violence of the shock."

(96) " Aug. 29, 1689. Before the King & Council was heard the matter of King's College about Mr Isaac Newton, why he or any other not of that foundation should be Provost, & after the reasons shewed & argued Mr Newton was laid aside." (Alderman Newton's Diary among Bowtell MSS. at Downing College.) The Statutes of King's College require the Provost to be in Priest's Orders and to be chosen from the existing or former fellows of the Society, Newton therefore was disqualified for the post.

(97) The following entry among the gratuities given by the College in the course of the year ending at Michaelmas 1690, is probably to be referred to the end of 1689, or beginning of 1690, when Newton was in London in attendance on his parliamentary

duties. "To Mr John Lamb, commended by Mr Newton, lately an operator to the Royal Society. 10s."

(⁹⁸) This and the other letters to Locke, except that of July 7, 1692, will be found in Lord King's *Life of Locke.* This letter is dated " Sept." by mistake, the London post mark being " Oc. 29."

(⁹⁹) Nichols's *Illustrat. Lit. Hist.* xiii. 49.

(¹⁰⁰) " Besides a coach which I consider not, it is but 200*l.* per annum, with a confinement to the London air, & to such a way of living as I am not in love with."

(¹⁰¹) *Optics,* Bk. 2. Part iv. Obs. 13.

(¹⁰²) Appendix, No. XXIII.

(¹⁰³) Wallis, *Opp.* ii. 391. seqq.

(¹⁰⁴) Brookbank was originally of Trinity College. The successful candidate was the Hon. H. Boyle, " a near relation" of the Chancellor, (Duke of Somerset) who wrote a letter (Sept. 6) recommending him to the University. (Baker MSS. xxx. 355).

(¹⁰⁵) It was read at a meeting of the Royal Society, Febr. 15, 1710, and ordered to be printed in the Transactions. It was printed in the Introduction to Vol. ii. of Harris's *Lex. Techn.* 1710.

(¹⁰⁶) It may be doubted whether this letter is in Newton's handwriting. The conclusion " Sʳ I am" &c., and the address, are evidently in his hand.

(¹⁰⁷) The four letters to Bentley were given to the College by Cumberland. They were printed in 1756, and reviewed by Johnson in the Literary Magazine. See Monk's Bentley, p. 33 ; Brewster's Newton, p. 286. They first appeared in their correct order in Bentley's Correspondence (Lond. 1842), the third and fourth having previously changed places.

(¹⁰⁸) " I have now received the box of rulers, with your receipt of £14. I sent you that money because I thought it was just ; & therefore you compliment me if you reckon it an obligation. The chamber next me is disposed of ; but that which I was contriving was ... to make you such an allowance, &c." *Gentleman's Magazine,* lxxxiv. 3.

(¹⁰⁹) Brewster's *Life of Newton,* p. 232. In this letter he says: " I have neither ate nor slept well this twelvemonth, nor have my former consistency of mind." A fortnight afterwards he apologized through a common friend for having written such " a very odd letter," saying, " that it was in a distemper that much seized his head, & that kept him awake for above five nights together." *Ib.* p. 234.

(¹¹⁰) Dated " At the Bull, in Shoreditch." When he wrote this letter, he " had not slept an hour a night for a fortnight together, & for five nights together not a wink." See his letter of Oct. 15, in which he explains the cause of this state of his health. " The last winter, by sleeping too often by my fire, I got an ill habit of sleeping ; & a distemper, which this summer has been epidemical, put me farther out of order." Lord King's *Life of Locke,* i. 420, Brewster's *Life of Newton,* p. 240, where the date is printed by mistake, Oct. 5.

Intelligence of his being out of health was conveyed in a very exaggerated form to Huygens in May of the following year by a Scotchman, of whom we know nothing whatever except that his name was Colm, (M. Biot's *Colin*): this person's information as recorded in a sort of journal by Huygens, who was himself troubled at the time with symptoms which in little more than a year afterwards terminated fatally, and would drink in with a morbid sympathy the tale of the affliction of a kindred spirit, is in the following terms : " 29 Maj. 1694. Narravit mihi D. Colm Scotus virum celeberrimum ac summum geometram Is. Neutonum in phrenesin incidisse, abhinc anno et 6 mensibus. An ex nimia studii assiduitate an dolore infortunii, quod incendio laboratorium chymicum et scripta quædam amiserat ? Cum ad Archiepiscopum Cantabrigiensem venisset, ea locutum, quæ alienationem mentis indicarent. Deinde ab amicis curam ejus susceptam, domoque clauso remedia volenti nolenti adhibita, quibus jam sanitatem recuperavit, ut jam rursus librum suum Principiorum Philosophiæ Mathematicorum

NOTES.

intelligere incipiat." (Hugenii *Exercitationes*....Uylenbroek, Fascic. II. p. 171. Hag. Com. 1833). This extract was first published by M. Biot in the *Biographie Universelle* (art. Newton, p. 168). Sir David Brewster has pointed out the improbability of the story and shewn the impossibility of reconciling it with known facts, (*Life of Newton*, p. 230 foll.) but not to M. Biot's satisfaction. We will first quote at length an anecdote which has been brought to bear upon the question, which, however, I think an attentive perusal will prove to refer to a period some years antecedent to the epoch under consideration. It is found in a MS. diary written by a member of St. John's College, who, at the date of the entry about to be quoted, was in his second year of residence at Cambridge. He seems to have heard the anecdote in company, and immediately chronicled it in his journal. He does not tell us who was his informant, and therefore we do not know the precise correction to be applied in this instance to an undergraduate's story. We shall not, however, probably err much in believing in the substantial truth of the narrative. It runs as follows:—

"1692. Feb. 3d. What I heard to-day I must relate. There is one Mr Newton (whom I have very oft seen) Fellow of Trinity College, that is mighty famous for his learning, being a most excellent Mathematician, Philosopher, Divine, &c. He has been fellow of the Royal Society these many years, & amongst other very learned Books & Tracts he's written one upon yᵉ mathematical principles of Philosophy, which has got him a mighty name, he having received especially from Scotland abundance of congratulatory letters for the same : but of all the Books that he ever wrote there was one of colours & light established upon thousands of Experiments which he had been 20 years of making, & which had cost him many hundred of pounds. This Book which he valued so much, & which was so much talked of, had the ill luck to perish, & be utterly lost just when the learned Author was almost at putting a conclusion at the same, after this manner :—

In a winter's morning leaving it amongst his other Papers, on his Study table whilst he went to Chapel, the Candle which he had unfortunately left burning there too, catched hold by some means of other papers, & they fired the aforesaid Book, & utterly consumed it, & several other valuable writings, & which is most wonderful did no further mischief.

But when Mr Newton came from Chapel and had seen what was done, every one thought he would have run mad, he was so troubled thereat that he was not himself for a Month after. A long account of this his system of light & colours you may find in the Transactions of the Royal Society which he had sent up to them long before this sad mischance happened unto him." (Abraham de la Pryme's *Diary*, in the possession of Prof. Pryme).

The foregoing narrative is shewn by Sir David Brewster to be irreconcileable with Huygens's memorandum, on the supposition that they both refer to the same circumstance. But, as I have stated, I believe De la Pryme's anecdote to refer to an earlier period not exactly known but admitting of being fixed within certain limits, as I will hereafter endeavour to point out. The discrepancy between the two statements is adverted to here solely for the purpose of noticing the singular hold which a traveller's gossip has acquired over M. Biot. "Nous trouvons au contraire," observes that distinguished philosopher, "entre ces dates un parfait accord," and twits Sir David Brewster with having overlooked the difference of calendar (*Journal des Savans* 1832, p. 325). M. Biot tells us that in English documents, previous to the change of style in the middle of last century, we are to add 1 to the year of our Lord for dates between January 1 and March 25, in order to find the year according to the present reckoning, and that therefore 1692 in the above extract is what would now be written 1693. It does not require a very extensive acquaintance with the literature of our diaries and correspondence to know that this rule is by no means a safe one to follow. In the case before us it is a matter of fact that the author of the diary commences the year in

January\*: (*ex. gr.* the death of Charles II. is placed in Febr. 1685; under Jan. 1692, which follows 1691, the writer laments the loss of Robert Boyle who died Dec. 31, 1691; Dec. 1692 is followed by "1693 Jan. 1. This year begins very ill, &c."; Dec. 1693 is followed by "1694 Jan. This month we sat for our degrees, &c.")

Sir David Brewster points to the fact that Newton wrote his four celebrated letters to Bentley during the time when Colm's gossiping statement represents him as having fallen into "phrenesis." Upon which M. Biot says, "nous admettrons volontiers maintenant qu'il {the fire which consumed Newton's papers} est postérieur à la première lettre" (*Journ. de Savans,* p. 332), and proposes to place the catastrophe between the 10th and 30th of Dec. 1692. " C'est à cela sans doute," he remarks, " que se rapporte le passage suivant des œuvres de Wallis imprimées en 1693....'Quam (methodum) speraverim Neutonum ipsum aliquando fusius traditurum; et quidem *audio* illum hujusmodi aliquid prelo paratum habuisse anno 1671, sed quod (infortunio quodam) flammis periit.' Wallis, Tom. II. p. 390. Le temps présent du verbe audio, écrit en 1693, ne peut s'appliquer qu'à un accident récent, tel que celui que les autres docu- mens nous attestent." Now the extract here quoted from Wallis is merely a trans- lation of what had originally appeared in English in his Algebra some years before. (Wallis's *Algebra* bears the booksellers' date of 1685. The bulk of the work was sent to London to be printed in 1676 or 7, but the printing was not proceeded with until about the beginning of Aug. 1683, some additions having been made to it in the mean time. The Preface is dated Nov. 20, 1684.) The passage alluded to is as follows: " But I here only give some *specimen* of what we hope Mr Newton will himself publish in due time. And it was, I hear, near ready for the press in 1671. But most of those papers have since (by a mischance) been unhappily burned" (p. 347). It is the more remarkable that M. Biot should have fallen into such an error, as nine lines below in the same page from which he has taken the above extract, Wallis goes on to say, "Atque hæc sunt quæ, ex memoratis Newtoni literis excerpta, inserueram in editione Anglicana 1685."

M. Biot makes another application of his chronological rule to Newton's fourth letter to Bentley, dated Febr. 11, 1693, which he affirms to mean our 1694, and that " les propres expressions de celleci et sa relation avec les autres" shew that it was written a long time after the third, dated Febr. 25, 169⅘. Now the letters here called the third and fourth, though printed in that order until the appearance of Bentley's *Correspondence* in 1842, are wrongly placed. The four letters are endorsed by Bentley in the order in which they were received: on the back of the letter of Febr. 11, 1693 he has written " A 3ᵈ Letter from Mʳ Newton," and on that of Febr. 25, 169⅘ he has written " A 4ᵗʰ Letter from Mr Newton." Besides, it can be shewn, I think satisfac- torily, that Bentley's two last sermons were printed in 1693, and as Newton must have known that, his words in his letter of Febr. 11, " if this come not too late for your use" · would have no meaning if they were written in 1694.

By way of supplement to Sir David Brewster's refutation of the statement in Huygens's journal, it may be observed that the words " Archiepiscopum Cantabri- giensem" (probably a mistake for *Cantuariensem*) imply that the crisis of Newton's "phrenesis" took place in London. A glance at the Table in p. LXXXIX. will shew that he was not absent from College for more than a fortnight at a time in 1692 and 1693, and therefore if the calamity which M. Biot first made known to the world really occurred, Newton must have been brought down to Cambridge very soon. Now if this had been the case, we should, almost to a certainty, have found Newton's name among the invalids in the Steward's Books, where a record is kept of the "commons" allowed to sick fellows in their own rooms. For example, in the year in question, ending at

\* I am enabled, by the kindness of the family in whose possession the diary now is, to state this distinctly.

Michaelmas 1693, we find one valetudinarian fellow allowed his commons in his rooms ("ex. co.") for 8 weeks, another for 1: in 1694 one for 6 weeks, another for 2; in 1692 one for 19 weeks, a second for 15 and a third for 20½; in 1691 one for 9 weeks, another for 1½, three others for half a week each, and another for 3.

But probably the most elaborate and complete refutation will have less weight with the majority of persons than the testimony of a trustworthy contemporary witness. I will therefore lay before the reader an extract from a letter of Dr Wallis to Waller, the Secretary of the Royal Society, dated May 31, 1695, from which by the way it will be observed what "strength" Colm's story had "acquired" in the course of its circulation to this country. Wallis had sent a copy of the second Volume of his Works as a present to Sturm a Professor at Altorf. Sturm wrote to thank Wallis for the present, and it is this letter of thanks which Wallis alludes to in the beginning of the following extract: "I have, since, one from Sturmius, which signifies that he had, some weeks before, received the Book I sent him. He sends me word of a Rumor amongst them concerning Mr Newton as if his House & Books & all his Goods were Burnt, & himself so disturbed in mind thereupon, as to be reduced to very ill circumstances. Which being all false, I thought fit presently to rectify that groundless mistake" {in a letter which he desires Waller to forward}. (Lett. Bk. Roy. Soc. W. 2. 50.)

I may observe that I should not have devoted so large a space to so transparent a piece of exaggeration but for the remarkable fact of its adoption by M. Biot, whose veneration for the creator of Natural Philosophy will not, I hope, suffer diminution by this exposure of an idle traveller's tale. ("Et si le sort eût voulu le frapper aussi cruellement, quel sentiment devrait faire naître en nous son infortune, sinon de plaindre et de vénérer davantage cet autre Tirésias, dont l'intelligence se serait ainsi aveuglée pour avoir vu de trop près les secrets des dieux? Toute autre pensée serait un sacrilége." Biot in *Jour. des Sav.* Apr. 1836, p. 216).

A word may be added on the probable date of the fire in Newton's rooms. The notice which we have given above respecting the publication of Wallis's *Algebra* shews that the accident happened before Aug. 1683. The superior limit is the winter of 1677, 1678 as Wallis believed copies of Leibniz's letters, the last of which was dated June 21, 1677, to have perished in the flames. (Letter to Leibn. Dec. 1, 1696). One of the winters therefore from 1677 to 1682 (excluding perhaps that of 1680, 1681 during which we know a little more of Newton's movements than in the others) may be fixed upon as the probable date of the occurrence.

The version of the story in which "Diamond" is made to play a prominent part, and according to which the scene is laid in Newton's latter years, and consequently in London, may perhaps deserve a place here. "His temper was so mild and equal, that scarce any accidents disturbed it. One instance in particular, which is authenticated by a person now living, [1780,] brings this assertion to a proof. Sir Isaac being called out of his study to a contiguous room, a little dog, called Diamond, the constant but incurious attendant of his master's researches, happened to be left among the papers, and by a fatality not to be retrieved, as it was in the latter part of Sir Isaac's days, threw down a lighted candle, which consumed the almost finished labours of some years. Sir Isaac returning too late, but to behold the dreadful wreck, rebuked the author of it with an exclamation (*ad sidera palmas*) 'Oh Diamond! Diamond! thou little knowest the mischief done!'—without adding a single stripe." (Notes to Maude's *Wensleydale*, p. 102. 4th ed. 1816.)

(¹¹¹) See under Sept. 16.

(¹¹²) A Mr Smith "took a journey" to Cambridge for the purpose of consulting Newton on a problem in chances which had its origin in a lottery recently drawn, and brought with him a letter of introduction from Pepys. The 1st of Newton's letters is principally occupied with settling the meaning of the question (What are the chances of throwing 1 six with 6 dice, 2 sixes with 12 dice, and 3 sixes with 18 dice?). The 2nd contains his "easy computation." See Pepys's *Correspondence*.

([113]) "On {the} Monday {night} likewise there being a great number of people at the door {of the haunted house,—it was a house opposite St. John's College in the occupation of Valentine Austin} there chanced to come by Mr Newton, fellow of Trinity College, a very learned man, and perceiving our fellows to have gone in {three fellows of St John's with a fellow-commoner of that college had rushed in armed with pistols} , and seeing several scholars about the door, Oh ye fools! says he, will you never have any wit? Know you not that all such things are mere cheats and impostures? Fie! fie! go home for shame. And so he left them, scorning to go in." (De la Pryme's MS. Diary, where there is a full account of the proceedings of the "spirit" which the writer of the diary had received in a letter from Cambridge.)

([114])    Appendix, No. XXIV.

([115]) " Quoniam varii errores in Prop. 37 & 38 (Lib. II.) irrepsere, illos omnes restitutos hic apponam, prout in autoris exemplari inveni, ineunte Maio 1694, dum Cantabrigiæ hærerem, consulendi divini autoris gratiâ." MS. of Dav. Gregory (Rigaud. p. 100).

([116]) "July 4. Ordered that a letter be written to Mr Isaac Newton praying that he will please to communicate to the Society in order to be published his Treatise of light & colours & what other Mathematical or Physical Treatises he has ready by him." *Journ. Bk.*

([117]) "Mr Newton coming to see me Sept. 1, 1694, and discoursing of the theory of the moon, to let him see what I had done in order to restore her motion, I produced and shewed him these 3 sheets { or synopses } of her observed and calculated places compared." Flamsteed *ap.* Baily, p. 191. Shortly afterwards Flamsteed lent him copies of two of the synopses, of which Newton made transcripts at Cambridge. A copy of the 3d was sent Oct. 29.

([118]) The whole of the known correspondence is printed in Baily's *Flamsteed,* pp. 133—160. Newton's letters are preserved in the library of Corpus Christi College, Oxford, to which Society they were given in 1764 by S. Adee, M.D., formerly Scholar of the College.

Mr Baily has attempted from this correspondence to shew, in opposition to a prevailing opinion, that Flamsteed manifested no unwillingness to furnish Newton with the observations necessary to enable him to complete the lunar theory, but, on the contrary, freely communicated every observation that Newton required. (*Supplement to Flamsteed's History,* pp. 708—720.) I regret that I cannot concur in Mr Baily's conclusion. Assuming, what is far from clear, that up to December, 1694, Flamsteed sent Newton all the observations that he asked for, I think that in the following month, and afterwards, we discover traces of a feeling which is scarcely compatible with Mr Baily's hypothesis. The following particulars are gleaned from Newton's letters, and Flamsteed's rough draughts or notes ; additional light will be thrown upon the subject when the correspondence between them is made complete by the discovery of Flamsteed's actual letters, which it is hoped may be found among the Portsmouth papers :—

1694 Dec. 6. Flamsteed promises to send Newton the observations that he wants after the Christmas holidays.

1695 Jan. 15. Newton acknowledges the receipt of two observations uncalculated, and as Flamsteed had calculated these and the other three of last month, he desires a synopsis of the calculations, merely to save himself the trouble of doing what was already done. But as regards the rest of Flamsteed's observations, he repeats what he had said in his letter of Nov. 17, that he desires only the naked observations.

—— 19. Flamsteed wrote back, " but no observations imparted...I have not time to send the synopsis now ; may do it hereafter : but would gladly see what places you have derived from the given Right Ascensions first. Shall give more hereafter."

—— 26. Newton replies : " Since I perceive you have a mind to see whether we can compute correctly, if you please to send me the latitude of Greenwich, I'll send you what you desire.''...'' I told you in autumn that it would be necessary to have about half of the observations in your synopses set right by the correct places of the fixt stars. If you please to do it at your leisure, I'll send you a catalogue of the observations.'' This request is again alluded to by Newton in his letters of Apr. 23 and July 9, but was never complied with.

'' One thing,'' he continues, and we now come to an important part of the correspondence as affecting the question under discussion, '' I did not consider. The observations being yours, perhaps you had rather have them perfectly your own in all respects, by determining the moon's longitude and latitude from them all yourself. If so (for that's what you have a very just right unto) I will stay your time. And when I have got a little further in the theory...I'll make a new table of the moon's eccentricities and equations of her apogee for finding her mean anomaly, and send you a copy of it......Chuse you therefore whether you will compute the moon's places from the observations or leave that work to me.''

This was answered in haste on the day on which it was received, but we do not know in what terms. Flamsteed sent a fuller answer, Feb. 7, with some lunar observations calculated and reduced, (among them the three mentioned by Newton Jan. 15, but not the two others.) In his draught of this answer he says : '' I shall mind my business of the fixt stars and give him an account of my progress, whilst he is employed on the moon : and shall be very well pleased with an account of his success.'' Flamsteed accepted Newton's proposal with respect to the observations, hinting, at the same time, that he should devote himself to his catalogue of the fixt stars. At this point therefore Newton's labours upon the lunar theory are suspended while he is '' staying the time '' of the Astronomer Royal.

March 2. Flamsteed, in a draught of an answer to Newton's letter of Febr. 16, has these words : '' Vindication of myself for not imparting my observations, and an account of my northern correspondence.''

Apr. 23. Newton writes : '' When I have your materials, I reckon it { the moon's theory } will prove a work of about three or four months : and when I have done it once I would have done with it for ever.''

June 29. Newton, who is still staying the Astronomer's time, thanks him for sending his solar tables (which Newton does not seem to have wanted) : '' But these, and almost all your communications will be useless to me, unless you can propose some practicable way or other of supplying me with observations. For as your health and other business will not permit you to calculate the moon's places from your observations, so it was never my inclination to put you upon such a task, knowing that the tediousness of such a design will make me as weary with expectation as you with drudgery...I will therefore once more propose it to you { as he had done Nov. 17 and Jan. 15 } to send me your naked observations of the moon's right ascensions and meridional altitudes ; and leave it to me to get her places calculated from them. If you like this proposal, then pray send me first your observations for the year 1692, and I will get them calculated, and send you a copy of the calculated places. But if you like it not, then I desire you would propose some other practicable method of supplying me with observations ; or else let me know plainly that I must be content to lose all the time and pains I have hitherto taken about the moon's theory and about the table of refractions.''

July 2. Flamsteed, stung to the quick, offers not the mural arc observations of 1692, but the sextant observations from 1677 to 1690. It would also seem, from a statement written by Flamsteed on the back of Newton's letter, as if he had sent at the same time the 30 observations which he had made from Febr. 8

to June 25 in the current year.   But as Newton makes no mention of having received them, merely saying, "when you have computed your 30 observations, you will know no more of it { the parallactic equation } than at present," I suspect that there is some mistake in Flamsteed's memorandum.

1695   July 9.   Newton writes: "After I had helped you where you stuck... { he particularly mentions the table of refractions, which he says } cost me above two months' hard labour which I should never have undertaken but upon your account, and which I told you I undertook that I might have something to return you for the observations you then gave me hopes of, and yet, when I had done, saw no prospect of obtaining them* or of getting your synopses rectified, I despaired of compassing the moon's theory, and had thoughts of giving it over as impracticable, and occasionally told a friend so who then made me a visit.  But now you offer me those observations which you made before the year 1690, I thankfully accept of your offer, and will get as many of them computed as are sufficient for my purpose."

——— 13.   Flamsteed sends his observations from Jan. to July 1677.

——— 20.   Newton says, "The report you mention { which was current in London about Flamsteed's not furnishing Newton with observations } was much against my mind, and I have written to put a stop to it.  I thank you for...your lunar observations."

——— 27.   Newton says, "The other day I had an excuse sent me for what was said at London about your not communicating, and that the report should proceed no further.  I am glad all misunderstandings are composed."  He then specifies the further observations (out of the sextant stock) that he wants.

Sept. 14.   Newton returned to Cambridge on Sept. 10, and went away again on the 14th: before leaving, he writes, "I have not yet got any time to think of the theory of the moon nor shall have leisure for it this month or above: which I thought fit to give you notice of, that you may not wonder at my silence." He however returned in a fortnight, but had sublunary matters to attend to, was named by rumour shortly afterwards as Master of the Mint, and in the March of next year was actually appointed Warden.

——— 17.   In Flamsteed's draught, written on Newton's letter, we read, "My exercise will devour no small part of my time, and therefore I shall desire my friends to excuse me if I answer not their letters so fully nor readily as formerly; however, when you want more of my lunar observations { i.e. those made before 1690 with the sextant, not those which he had made or was making with the mural arc } I shall cause them to be transcribed and it will be no trouble."  Mr Baily has printed the words "however........no trouble" in italics; the preceding part of the sentence is not however destitute of significance.

Here the correspondence terminates.   There are several allusions to it in Flamsteed's extant memoranda, two of which are produced here as evidence in the question we are examining:  " { Mr Newton } ceased not to importune me (though he was informed of my illness) for more observations, and with that earnestness that looked as if he thought he had a right to command them, and had about 50 more imparted to him.  But I did not think myself obliged to employ my pains to serve a person that was so inconsiderate as to presume he had a right to that which was only a courtesy.  And I therefore went on with my business of the fixed stars; leaving Mr Newton to examine the lunar observations over again: which had he done, he had found that he needed not be so importunate for new,—the old would have been sufficient for the purpose and design for which

---

* Flamsteed has written on the letter "My sickness has hindered."  But we shall see by and by from his own statement that that was not the sole cause.

I had imparted them to him. I was therefore forced to leave off my correspondence with him at that time." (Baily, p. 63.) Again: " I continued since furnishing him with lunar observations, as I gained them, until Midsummer 1695, when being troubled with a distemper......I was forced to intermit my correspondence with him." (Ib. p. 191.)

Upon the whole, I think, we may conclude that the combined action of Flamsteed's bad temper and bad health, for which great allowance must be made, coupled with his professional jealousy of Halley * and his exaggerated opinion of the value of his own astronomical labours, has robbed us of the lunar theory in the form that its creator would have given it, and that the following words contain more truth than is sometimes to be met with in epistolary statements : " Flamsteedius suas de Luna observationes Newtono negaverat. Inde factum aiunt quod hic quædam in motu Lunari adhuc indeterminata reliquit." (Leibniz to Roemer, Oct. 4, 1706. *Opp.* Tom. IV. Pars II. p. 126.)

(119) This inequality in the Moon's longitude is proportional to $\dfrac{\sin \odot\text{'s parallax}}{\sin \mathbb{D}\text{'s parallax}}$, its argument being $\mathbb{D}$'s mean angular distance from $\odot$. " On la considérer...avec raison comme une des applications les plus délicates de l'analyse moderne." (Biot, *Journ. des Sav.* Apr. 1836, p. 218.) In his letter of July 9, 1695, Newton says that its maximum value scarce exceeds 2 or 3, or at most 4 minutes. Bürg ( *Mécan. Cel.* Tom. III. p. 282) gives it 2′, 2″, 38. Compare Pontécoulant, IV. 605, who (ib. XIV. note) does not seem to be aware that this equation was known to Newton. M. Biot says that this equation is omitted in the second edition of the *Principia*, and suggests reasons to account for the omission. But see p. 120 of this work, where the " Variatio secunda " is described.

(120) This is now called the lunar equation of the Sun, " et l'on avait tout lieu de la considérer comme une des corrections les plus délicates des tables modernes." (Biot, *Journ. des Sav.* Apr. 1836, p. 220.) It $= \dfrac{\mathbb{D}\text{'s mass}}{\oplus\text{'s mass}} \cdot \dfrac{\text{dist. of } \mathbb{D} \text{ from } \oplus}{\text{dist. of } \odot \text{ from } \oplus} \cdot \sin$ difference of longitudes of $\mathbb{D}$ and $\odot$. The coefficient is given 8″,83 in the *Mécan. Cel.* Tom. III. p. 108. Newton in the above letter says that he had not yet ascertained its magnitude, but that it may be assumed 16″ or 20″ until it be determined more exactly. Comp. Pontécoulant, IV. 653.

(121) Flamsteed's coquetry about his two observations draws from Newton a little playful irony—an indulgence extremely rare with him : " The places of the moon from your two observations I have not yet computed : for I thought it superfluous to do what you had done to my hands ; and desired a copy of your computations only to save myself that labour. But since I perceive you have a mind to see whether we can compute exactly, if you please to send me the latitude of Greenwich I'll send you what you desire." (Baily, p. 149.)

(122) This is the table afterwards published by Halley in the *Phil. Trans.* May—Aug. 1721, " such as I long since received it from its Great Author." See Biot's third article on Baily's *Flamsteed* in the *Journal des Savans* for Nov. 1836, which he commences by observing that he is in arrear with the article, " et pourtant, depuis environ neuf mois que mon second article a paru, je n'ai pas été occupé d'autre chose que de sa continuation. Mais, pendant tout ce temps, je puis dire en vérité, comme Jacob, que j'ai lutté avec L'ESPRIT." For the results of the struggle see that article, and his paper " Analyse des Tables de réfraction construites par Newton, avec l'indication des procédés numériques par lesquels il a pu les calculer." (Ib. pp. 735—754.)

---

* The torrents of vituperation poured by Flamsteed upon this illustrious man are, I believe, to be explained on the principle alluded to: (κεραμεὺς κεραμεῖ...) At the meeting of the Royal Society, June 1, 1692, Halley read a paper vindicating his St Helena Observations "from some groundless exceptions" of Flamsteed's.

([123]) Some delay occurred in sending this letter. Flamsteed did not receive it until May 6.

([124]) Appendix, No. xxvi.

([125]) " I shall order Will Martin...to pay him two guineas, if you please to let him call for them, or to pay it to his or your order in London if you please to let me know where." The words in this extract which follow "pay him" are crossed out in the MS. and the word "guineas" altered into "shillings" apparently by Flamsteed. The words after "for them," to the end of the passage, are conjectural, the original writing being most skilfully blotted out. I believe however that it might be made out on a bright day, if it were thought worth the trouble. What motive Flamsteed could have had for disguising any part of the above sentence I do not pretend to divine. It is curious that Mr Rigaud, who, at Mr Baily's request, examined the MS. with reference to this very point, should have overlooked the original "guineas." (Baily, p. 159, note.)

([126]) Wallis, writing to Halley from Oxford. Nov. 26, says: "We are told here that he is made Master of the Mint, which if so, I doe congratulate to him and am his & your &c." *Orig. Lett. Bk.* Roy. Soc. W. 2. 56. See Appendix, p. 302.

([127]) Macc. *Corr.* II. 419.

([128]) The original MS. with the address, "For the Right Honourable Charles Montagu Esq. Chancellour of the Exchequer," is preserved at the Royal Society, *Orig. Lett. Bk.* N. 1. 61[b]. The problems are (1) To determine the brachistochron between two given points not in the same vertical line: (2) $APP'$ is a straight line passing through a fixed point $A$, and meeting a curve in $P, P'$: to find the curve such that $AP^m + AP'^m =$ constant. One of the two identical papers (a printed folio half-sheet) which were sent to Newton by Bernoulli, containing the problems, still exists in the Archives of the Royal Society, (Volume lettered "Arithmetic, Algebra," &c. 13). At the bottom, in Newton's hand, are the words "Chartam hanc ex Gallia missam accepi Jan. 29, 169$\frac{6}{7}$."

([129]) Macc. *Corr.* II. 420.

([130]) See Appendix, p. 299.

([131]) "Isaac Newton chuseth the Hon[ble] Henry Boyle Esq[re], Burghess of this University." The votes were given in English on account of the election occurring during the vacation.

([132]) James Hodgson had calculated these 12 places for Newton by Flamsteed's orders, during the absence of the latter in Derbyshire, and sent them to him Sept. 8. Flamsteed on examining them Nov. 11, "found them all false," and computed them afresh. The results of these last calculations were communicated to Newton on his visit to Greenwich, Dec. 4.

On December 29 Flamsteed sent him a correction of the time of one of the observations, and afterwards found that his results required further modification. "I acquainted him," he says, "there was a further fault in them, when I was last with him. He is reserved to me, contrary to his promise. I lie under no obligation to be open to him." (Baily, p. 166). Flamsteed was in London on Dec. 30 and 31, (Friday and Saturday), and the words "when I was last with him," probably refer to one of those days. Newton was then aware of the liberty which Flamsteed had taken, in mentioning his name in connexion with the Lunar Theory, in the Letter to Dr Wallis. Hence we may explain the "reserve" of which Flamsteed complains, and to which Mr Baily has attached a different meaning, (p. 710, note).

([133]) In a letter to Dr Wallis on annual parallax, which was to appear in the 3d volume of the Doctor's Works, Flamsteed alluded to his having supplied Newton with lunar observations. On being informed by David Gregory of the fact, Newton desired him to request Dr Wallis not to print the paragraph containing the allusion in question. When Flamsteed, who does not seem to have anticipated that there could be any objection to his making public use of Newton's name without previously obtaining permission to do so, received intimation of this from Wallis, he wrote to Newton on the subject, (Monday, Jan. 2,) and again on the 5th. Newton in his answer, dated Jan. 6, states his reasons for having requested the suppression of the paragraph. " I was concerned," he says, " to be publicly brought upon the stage about what, perhaps, will never be fitted for the public, and thereby the world put into an expectation of what, perhaps, they are never like to have. I do not love to be printed on every occasion, much less to be dunned and teased by foreigners about mathematical things, or to be thought by our own people to be trifling away my time about them, when I should be about the King's business." (The great re-coinage of silver was not yet completed).

(¹³⁴) The eight foreign Associates created on the re-modelling of the Academy in 1699, were

| | |
|---|---|
| 1. Leibniz, | |
| 2. Guglielmini, | Febr. 4. |
| 3. Hartsoeker, | |
| 4. Tschirnhausen, | |
| 5. James Bernoulli, | Febr. 14. |
| 6. John Bernoulli, | |
| 7. Newton, | Febr. 21. |
| 8. Roemer, | |

The first four seem to have been nominated by the King, the rest by the Academy.

(¹³⁵) " Mr Newton shewed a new instrument contrived by him for observing the moon, stars and { so finding the } longitude at sea, being the old instrument mended of some faults, with which notwithstanding Mr Halley had found the longitude better than the seamen by other methods." *Journal Book.* (Hooke, as usual, at the next meeting of the Society, Oct. 25, laid claim to the discovery). A paper, in Newton's hand, describing the instrument, headed " An instrument for observing the distance of the moon from the fixt stars at sea," is preserved in No. LXXXI. MSS. Roy. Soc. It was found among Halley's papers after his death, and was published in the Transactions for Oct.—Nov. 1742.

The following extract from a letter of Charles Montagu to Sloane, dated Aug. 7, 1699, refers to the "mending" of the "faults" of the "old instrument." After stating that he was to have waited on the Lord Chancellor (Somers) at Gresham College, next Wednesday, he says : " But I understand that Mr Newton's experiment will not be ready by that time......I hear the engine will not be made within 10 days, and then I believe my Lord will wait upon you." (Sloane MSS. Brit. Mus. 4053).

(¹³⁶) With Aston and Flamsteed. Lord Chancellor Somers was re-elected President.

(¹³⁷) Ruding's *Annals of the Coinage*, ii. 427.

(¹³⁸) 120 copies of the work were printed "impensis illustrissimorum...Somers... Dorset...Car. Montagu...Newton..." and five others, including Sloane and Aston.

(¹³⁹) The method was sent by a M. du Verger, in a letter from Rome, with a description of an instrument for solving the three problems. (*Regist. Bk.* ix. 12.) At a meeting of the Royal Society, Apr. 8, Sloane was "ordered to give the letter and demonstrations to Mr Newton, to have his opinion and answer." At the next meeting, Apr. 15, Sloane "promised to take care to deliver" them. On July 24, was read a letter from Sloane to du Verger, containing Newton's report concerning his papers. The following is an extract from it : " Ipsissimo quo chartas accepit momento exami- nandas commisit uni e Sociis in hisce rebus versatissimo, qui nuper opinionem suam Societati retulit modum nimirum describendi *volutam* accuratum satis videri et in rebus mechanicis usui futurum, nec tamen geometrice demonstratum esse existimat ; et proinde *anguli trisectionem, duplicationem cubi* et *quadraturam circuli* non esse mathematice investigata." *Letter Bk.* xii. 328.

(¹⁴⁰) " Tabula quantitatum et graduum Caloris." *Orig. Lett. Bk.* Roy. Soc. N. l. 62. Comp. Brewster's *Newton*, 297.

([141]) The poll stood as follows : Right Hon. H. Boyle, (*Trin.*)    180
          Mr Newton, (*Trin.*)    161
          Mr Hammond, (*Joh.*)    64
Dr Bainbrigg, Vice-Master of Trinity, in voting for Newton, calls him " virum optimum," Dr Stubbe, one of the Seniors, and afterwards Vice-Master, terms him " clarissimum virum :" in some of the voting papers the epithet is " dignissimus " or " doctissimus." Bishop Monk, (*Life of Bentley*, p. 122,) says that Bentley " had the satisfaction of assisting in the return of his illustrious friend Sir Isaac Newton." Bentley' voting-paper however is not found among those of any of the three candidates. Newton himself voted for Boyle.

([142]) His resignation of the Professorship in his own handwriting, is preserved in the Registrary's office. With respect to the resignation of his fellowship, see p. lxxxii. note §.

([143]) It appeared in English, separately, the following August, also in Harris's *Lexicon Technicum*, 1704, (a work to which Newton was a subscriber), and, with a few corrections by Newton in the table of *Errata*, in the *Miscellanea Curiosa*, 1705, (this is the date of the 1st ed., not 1708 as stated by Mr Baily in his *Supplement to Flamsteed's History*, p. 688,) with the title of " The Famous Mr Isaac Newton's Theory of the Moon." With respect to Mr Baily's renewed assertion (*ib.* p. 735) that " in the *Theoria Lunæ* there is not a single allusion made to Flamsteed," it may be observed that in the three above mentioned English reprints the mention of Flamsteed's name comes *after* the title of the tract, not *before* it as in Gregory's *Astronomy*. Not that this is a point of any great consequence, for the acknowledgment of Flamsteed's services in supplying Observations is much the same in either case. It is extremely improbable that the essay was communicated to Gregory in the naked form in which it stands within inverted commas in his *Astronomy*: it must have been accompanied by some notice of Flamsteed's Observations and their near agreement with the results derived from the Theory, the substance of which Gregory chose to embody in an introductory paragraph, then prefixing the title " Lunæ Theoria Newtoniana," and finally giving us the actual Theory in its author's own words—a bare numerical statement of facts and rules, in which complimentary phrases would scarcely find an appropriate place.

([144]) During this visit Locke shewed him his *Essay upon the Corinthians*, with which " he seemed very well pleased, but had not time to look it all over." Locke sent it to him before Christmas for his more careful perusal, and not hearing anything from him, towards the end of March, 1703, sent him a further communication. Receiving no answer, Locke, who was now old and infirm, became impatient to learn something of the fate of his papers, and in a letter dated Apr. 30, commissioned his nephew Peter King (afterwards Lord Chancellor) to wait upon the Master of the Mint, with a letter which he had written for the purpose. " He lives in German St. You must not go on a Wednesday, for that is his day for being at the Tower. The reason why I desire you to deliver it to him yourself is that I would fain discover the reason of his so long silence. I have several reasons to think him truly my friend, but he is a nice man to deal with, and a little too apt to raise in himself suspicions where there is no ground ; therefore when you talk to him of my papers, and of his opinion of them, pray do it with all the tenderness in the world, and discover, if you can, why he kept them so long and was so silent. But this you must do without asking why he did so, or discovering in the least that you are desirous to know....Acquaint him that you intend to see me at Whitsuntide, and shall be glad to bring a letter to me from him, or any thing else he will please to send....Mr Newton is really a very valuable man, not only for his wonderful skill in mathematics, but in divinity too, and his great knowledge in the Scriptures, wherein I know few his equals. And therefore pray manage the whole matter, so as not only to preserve me in his good opinion, but to increase me in it ; and be sure to press him to nothing, but what he is forward in himself to do." Lord King's *Life of Locke*, ii. 38.

Newton accordingly sent an answer, apparently in the manner suggested, (it is dated May 15, the day before Whitsunday), the first clause of which shews that the groundless suspicions were on the part of Locke. " Upon my first receiving your papers, I read over those concerning the first Epistle of the Corinthians, but by so many intermissions, that I resolved to go over them again, so soon as I could get leisure to do it with more attention. I have now read it over a second time, and gone over also your papers on the second Epistle." *Ib.* 1. 420.

(145) He succeeded Lord Somers, who had held the office five years. He was re-elected annually during the remainder of his life.

(146) "The President said he had thought of a contrivance for burning-glasses, by uniting several, { probably *apropos* of a paper by Lowthorp on the subject } ....The President was desired to give directions to make such glasses as he shall think proper.

May 17. The President shewed a piece of silver money and iron wire, part of which were melted in the focus of a metallic speculum, &c. &c.

—— 24. The President said that he had tried the addition of a reflecting speculum, and he thought the focus of the burning-glass too near to produce the desired effect.

—— 31. The President shewed a piece of red tile { vitrified by the burning-glass }, &c.

June 21. The President tried some new experiments with his speculum.

July 12. The President gave the speculum lately contrived by him to the Society.

Nov. 15. Mr Halley was desired to draw up an account of Mr Newton's burning-spe-culum." (*Journal Bk.*)

The burning-glass given by Newton to the Society is described by Harris (*Lexicon Technicum*, Vol. II.), as consisting of 7 concave glasses (each about 11½ inches in diam.), with their foci coincident, 6 of them being placed round the 7th and in contact with it, and forming a sort of segment of a sphere, whose subtense is about 34½ inches. The central glass lies about an inch lower or farther in than the rest. The common focus is about 22½ inches distant, and about ½ inch in diam. It vitrifies brick or tile in a mo-ment, and melts gold in about ½ a minute. Comp. Hutton's *Math. Dict.*

Under the date of Febr. 2, may be mentioned the examination of the pseudo-For-mosan, George Psalmanazar, at the Royal Society. In the British Museum there is a letter from John Chamberlayne to Newton, dated Febr. 2, 1703-4, reminding him of " the famous conference appointed to take place this afternoon at Gresham College, be-tween Mr George, the Formosan, the bearer hereof, and Le Pere Fontenay, a Jesuit, lately come from China. I have engaged Mr George, and am to carry him thither this afternoon in my coach, but without telling him the reason. I beg therefore the same caution and security on your side." (MSS. Birch, 4292). Newton does not seem to have attended the meeting. Psalmanazar gives an account of the conference in the Preface to his Description of Formosa. (Lond. 1704. p. vii.). The impostor quailed under the searching scepticism of Halley, Mead and Woodward. (*Memoirs*, p. 196. Lond. 1764). For a brief account of this singular person, who at 32 repented of his ways, and in after life became a large contributor to the Universal History, and won the respect of Johnson, see Chalmers's *Biogr. Dict.*

(147) " Febr. 16, the President presented his book of Optics to the Society ; Mr Halley was desired to peruse it, and to give an abstract of it ; and the Society gave the President thanks for the book and for being pleased to publish it." (*Journ. Bk.*)

The Preface in the first edition bears no date. In the second edition (1718) the date "April 1, 1704," was added. There is a similar peculiarity about the Preface to the *Principia.* (See p. LVIII.) The dispute with Leibniz had probably taught our philosopher the importance of dates.

(148) The words are: " Pro differentiis igitur Leibnitianis D. Newtonus adhibet semperque adhibuit *fluxiones*...iisque tum in suis Principiis Naturæ Mathematicis, tum in aliis postea editis eleganter est usus, quemadmodum et Honoratus Fabrius in sua Sy-nopsi Geometrica motuum progressus Cavallerianæ Methodo substituit." (p. 35). Ludo-vici (*Historie der Leibnizischen Philosophie*, quoted by Guhrauer), and Guhrauer

(*Biographie of Leibniz*, I. 311, Breslau, 1846,) inform us that no other person than Leibniz himself was the writer of the review in question, for that in the Pauline Library at Leipsic there is a copy of the Acts in which Leibniz's name is added in writing to several of his anonymous articles, and to this among others. Keill, in a paper on central forces, (*Philos. Trans.* Sept. Oct. 1708, p. 185,) took occasion to retort in the following terms. " Hæc omnia sequuntur ex celebratissimâ nunc dierum Fluxionum Arithmeticâ, quam sine omni dubio Primus Invenit Dominus Newtonus, ut cui libet ejus Epistolas à Wallisio editas legenti, facile constabit, eadem tamen Arithmetica postea mutatis nomine et notationis modo à Domino Leibnitio in Actis Eruditorum edita est."
On receiving from Sloane, Secretary of the Royal Society, the Volume containing Keill's article (the Volume for 1708 and 1709, published in 1710), Leibniz, who was at Berlin, wrote to Sloane (March 4, 1711, N. S.) complaining of the imputation cast upon him, and begging the Society to interfere. " Nempe æquum esse vos ipsi credo judicabitis, ut D. Keillius testetur publice, non fuisse sibi animum imputandi mihi quod verba in-sinuare videntur, quasi ab alio hoc quicquid est Inventi didicerim et mihi attribuerim." A synopsis of the proceedings of the Society in relation to this affair is subjoined. 1711, March 22. President in the chair. Part of Leibniz's letter was read, and Sloane ordered to write an answer to him. Newton, before the article in the Acts was shewn to him, was annoyed at what Keill had said, but at the meeting on Apr. 5, Keill drew attention to the " unfair account" of Newton's tract. " Upon which the President gave a short account of that matter, with the particular time of his first mentioning or discovering his invention, referring to some letters published by Dr Wallis ; upon which Mr Keill was desired to draw up an account of the matter in dispute and set it in a just light." Apr. 12. " The former minutes being read gave occasion to further discourse of the matter mentioned in the Leipsic Acts. The President was pleased to mention his letters many years ago to Mr Collins about his method of treating Curves, &c., and Mr Keill being present was again desired to draw up a paper to assert the President's right in this matter." May 24. Keill's reply read, and a copy of it ordered to be sent to Leibniz, and to be printed in the Transactions on the receipt of Leibniz's answer to it. At the next meeting, May 31, at which Newton was not present, Sloane read his letter to Leibniz, which was approved of. 1712 Jan. 31. Leibniz's answer (Dec. 29, 1711) read and delivered to Newton. (See p. 55). Febr. 7. " The President not coming there was no account given of M. Leibniz's letter to Dr Sloane." March 6. In consequence of Leibniz's letter a committee was appointed consisting of Arbuthnot, Hill, Halley, Jones, Machin and Burnet, to inspect the letters and papers relating to the dispute, and make a report to the Society. On March 20, Francis Robartes, March 27, Bonet the Prussian Minister, and on Apr. 17, Demoivre, Aston and Brook Taylor were added to the Com-mittee. Apr. 24. The Report of the Committee read. (See *Commerc. Epistol.* p. 120, p. 241, 2d ed. Turnor's *Grantham*, p. 185. Brewster's *Newton*, p. 207. Weld's *Royal Soc.* I. 410.) The Committee conclude their Report as follows : " For which reasons we reckon Mr Newton the first Inventor, and are of opinion that Mr Keill, in asserting the same, has been noways injurious to Mr Leibniz. And we submit to the judgment of the Society, whether the extracts of Letters and Papers now presented, together with what is extant to the same purpose in Dr Wallis's 3rd Volume, may not deserve to be made public." The Report was unanimously adopted, and it was " ordered that the whole of the matter from the beginning, with the extracts of all the letters relating thereto, and Mr Keill's and Mr Leibniz's letters, be published with all convenient speed that may be, together with the Report of the said Committee." (*Journ. Bk.* Roy. Soc.) The collec-tion accordingly appeared early in 1713, under the title of " Commercium Epistolicum D. Johannis Collins et aliorum de Analysi promota : jussu Societatis Regiæ in lucem editum." The printing of the work was entrusted to Halley, Jones and Machin. " 1713 Jan. 8. Some copies of a book entitled *Commercium Epistolicum*, &c....being brought, the President ordered one to be delivered to each person of the Committee, appointed for that purpose, to examine it before its publication." (*Journ. Bk.*) It appears from the

Minutes of the Council, that on Jan. 29, it was "ordered by balloting that the Treasurer pay the charges of printing the *Commercium Epistolicum*," and that on June 11, the sum of £22. 2s. 6d., was ordered to be disbursed to Halley, "being money he had paid for printing" it. Only a few copies of the book were printed, and they were principally distributed as presents to Universities or distinguished scientific men, (see p. 221) but not entirely so, as is shewn by the following extract from the Journal Book. "1714 June 17. The President in the chair. Dr Keill acquainted the Society that Mr Johnson, Bookseller at the Hague, desired a parcel of the *Commercium Epistolicum* at a certain price, and that he would return the money upon the receipt of the books. Ordered that 25 complete books be delivered by Mr Thomas to Dr Keill to be transmitted to Mr Johnson accordingly, at 3s. per book."

At the meeting of the Society on Apr. 24, Keill "said he would draw up an answer to Mr Leibniz's last letter, it relating chiefly to himself, which he was also desired to do, and that it should be read at a meeting of the Royal Society." We hear no more of this contemplated answer of Keill's.

([149]) Signed by Robartes, Wren, Newton, &c. On this recommendation Prince George most liberally offered to defray the expenses of the work. Flamsteed instead of feeling grateful for Newton's intervention in his behalf, was annoyed at the thought of any other opinion than his own being taken on the propriety of publishing his Observations, and when the referees proceeded in the discharge of their trust, to take steps with reference to the publication, he naturally enough wished to have his own way in the management of it, and by his perverseness in this respect, gave them (to use their own language) "a great deal of trouble."

It is not necessary to enter further into this question here : the reader will find in Mr Baily's *Account of Flamsteed* a multiplicity of details upon the subject, through which the clue just given will guide him with tolerable safety. But I may remark that among the documents that are still wanting to complete our knowledge of the circumstances that attended the passage of Flamsteed's work through the press, there is one which it is hoped will yet be discovered—the paper of Articles actually agreed upon preliminary to the printing of the book. And yet Mr Baily (p. xlii. note) has ventured to assert in contradiction to Halley, that it was *not* agreed that the Catalogue should be prefixed to the first volume or book. It is true that we have a private memorandum of Flamsteed's (Baily, p. 253) stating that he "signed the Articles, but covenanted that the Catalogue of the fixed stars mentioned to make a part of the first volume should not be printed, but with the last;" but this implies that the point "covenanted" about did not form one of the Articles, and we have no proof that the "covenant" was accepted by the referees. Flamsteed uses the same phrase on a similar occasion. (*Ib.* p. 86).

([150]) On the 11th of July following Lord Halifax gave to the Society the 2nd Vol. of the work.

([151]) Probably on business connected with the approaching election. Parliament would expire under the triennial Act the following August, but that event was anticipated (after a prorogation on March 14) by dissolution on April 5. Flamsteed in a letter written on the last-mentioned day, which I think there can be no doubt was intended for Newton, though Mr Baily (p. 238) describes it as "probably addressed to Mr Hodgson," says : "Good success in your affairs: health and a happy return is heartily wished you by, Sir, your obliged and humble servant."...

([152]) In the Senior Bursar's Book for the year 1707 in a "particular account of several Benefactions received for the use of the Chapel and Library, by the R^d Mr Nicholas Spencer .... {who was Bursar from December 1701, to June 1705,} never yet accounted for to the College from the Sen^r Bursar's Office" we find, "R^d the Gift of Mr Isaac Newton £60." I have ventured to assume that this donation was intended for the Chapel, as he had already in 1676 subscribed liberally to the fund for building the Library. The date of the subscription may probably be assigned to his electioneering visit to Cambridge.

(¹⁵³)    The numbers were "Hon. A. Annesley,    (*Magd.*)    182

Hon. D. Windsor,    (*Trin.*)    170

Hon. Fra. Godolphin,    (*King's*)    162

Sir I. Newton,    (*Trin.*)    117."

Bentley voted for him.

In a letter to A. Sharp, Apr. 24, Flamsteéd writes : "Mr Newton is knighted : stands for parliament man at Cambridge ; and is going down thither, this day or tomorrow, in order to his election. 'Tis something doubtful whether he will succeed or no, by reason he put in too late." The Tory election cry was "the Church in danger." In the debate in the House of Lords on the subject of this alleged danger the following December, Patrick, Bishop of Ely, is reported as moving that the Judges "might be consulted what power the Queen had in visiting the Universities, complaining of the heat and passion of the gentlemen there, which they inculcated into their pupils ;... that at the election at Cambridge, it was shameful to see a hundred or more young students, encouraged in hollowing like schoolboys and porters, and crying, No Fanatic, No occasional Conformity, against two worthy gentlemen that stood candidates." Cobbett's *Parl. Hist.* vi. 496.

(¹⁵⁴)    The originals of this and five other notes to Sloane are in the British Museum.

(¹⁵⁵)    Turnor's *Grantham*, p. 169.

(¹⁵⁶)    With alterations and additions (among others, seven new queries).

The translation was made by Samuel Clarke, who was rewarded by the author with a present of £500. A second edition of it appeared in 1719.

Demoivre is stated to have "revû et conduit la traduction latine de l'optique de Newton, pour laquelle il n'épargna ni soins ni peines; aussi ce grand homme lui avoit-il accordé toute sa confiance. Il alloit tous les soirs l'attendre dans un café {probably Slaughter's Coffee House in St Martin's Lane } où M. Moivre se rendoit dès qu'il avoit fini ses leçons, et d'où il l'emmenoit chez lui pour y passer la soirée dans des tête-à-tête philosophique." (*Eloge*, 1754).

(¹⁵⁷)    "I thank you for giving me timely notice of the caveat, and think we should stick at no charge for defending the legacy. What money shall be wanting for this purpose I'll advance till the Council shall be called. If you see Dr Harwood before me, pray desire him to have an eye upon this matter. I do not know the method of proceeding in these cases ; but he can tell us. I will take the first opportunity to inform myself of what is to be done." (Sloane MSS. Brit. Mus. 4054; printed without the date in Nichols's *Illustrations of Lit. Hist.* xiii. 59). The note in the same volume, dated Thursday night, ("Lady Betty Gayer being engaged for tomorrow, and at liberty on Monday or Tuesday, I beg the favour we may wait on you on either of those days at three o'clock, and that you will let us know which of those two days you are most at leisure,") is recommended to the attention of those who are versed in the "fashionable arrangements" of Anne's reign.

(¹⁵⁸)    The trustees appointed under Plume's will (Covel, Bentley, Whiston, Fra. Thompson of Caius) were directed to frame statutes for the regulation of the Professorship, "with the advice of Sir John Ellis, (Master of Caius), Sir Isaac Newton and Dr { *sic* } Flamsteed." Cotes, the first professor, was elected Oct. 16, 1707. Flamsteed wrote to Whiston Febr. 13, 1705-6, (compare Baily, p. 258,) recommending his assistant Mr John Witty for the Professorship. (Flamsteed's MSS. at Greenwich, xxxiii. 65). In Vol. lxix. of the same Collection, there is a long letter, dated Dec. 31, 1706,) from Ellis to Thompson, on the subject of the Professorship, in which Cotes is spoken of in very high terms, and in Vol. xxxiii. p. 74, there is an answer to it, in which Flamsteed is reported as saying that "Trinity Gatehouse is not fit for" an observatory, (see p. 200) "and that that of St John's is preferable, and that the Virtutis Gateway at Caius is better than either." Flamsteed wished a separate building to be devoted to the purpose.

The substance of a note written by Prof. Smith on the fly-leaf of his copy of

Huygens's *Cosmotheoros* (*Hag. Com.* 1698) and dated 1764, is worth preserving. " I have been well informed that Dr Plume, Archdeacon of Rochester, was so pleased with this book, which the celebrated Mr Flamsteed had recommended to him, as to leave by his will £1800 to found the Plumian Professorship of Astronomy and Experimental Philosophy, which I held many years after Mr Cotes's decease."

(159) Appendix, No. XXXI.

(160) "Instead of the like sum he intended after his death. It was ordered to be put up by itself and to be subject to such end or benefaction as the President shall direct." This no doubt is the foundation for Thomas Hearne's scandal, " he promised to become a benefactor to the Royal Society, but failed." See under Dec. 14, of the following year.

(161) It fills pp. 4—157 of the present volume.

(162) Mr De Morgan, in his sketch of the life of Newton, says that in the 2nd edition Flamsteed's name was "erased in all the passages in which it appeared (we have verified, for this occasion, eight or nine places ourselves)." The name however will still be found in pages 441, 443, 455, 458, 465, 478 and 479 : the last two references occur in some additional matter on comets, which was put into Cotes's hands in October 1712. (See p. 141 of this work.) I question very much whether the suppression of Flamsteed's name in several places where it had appeared in the 1st edition was not such as was necessary in the process of improving the work. Newton's own experiments on the old echo in Trinity College cloister give way in the 2nd edition to more accurate researches.

(163) The original of this paper is in the British Museum, Add. MSS. 6489. fol. 67. ("ex dono Dnæ Sharp"). It is printed in the *Gentleman's Mag.* for Jan. 1755, pp. 3—5. (Compare his Chronology, p. 71, sqq.) In the same MS. volume (fol. 69) is an abstract of the paper in Newton's hand, (printed in the Appendix to this Work, No. XXXIII.), which was embodied in a letter to Bishop Lloyd by an unknown writer, dated Nov. 7, 1713, of which the draught is preserved in the volume referred to (fol. 65, 66), beginning " I had the honor to receive and the pleasure to read the papers your Lordship directed to the Dean of Norwich { Prideaux } : and before I sent them forward I communicated them to Sir Isaac Newton, according to your Lordship's order by Mr Archdeacon : when Sir Isaac brought them back, he told me that he found many excellent observations in them about the ancient year, and at the same time acquainted me that he had formerly discoursed with your Lordship about that year of 360 days, and represented" &c. (See Appendix, p. 314). Trimnell, Bishop of Norwich, may possibly have been the writer of this letter, as, three years before, he was the organ of communication between Lloyd and Prideaux, conveying to the latter Lloyd's scheme of Daniel's 70 weeks. (*Prideaux's Life*, p. 237). It would appear that Newton's abstract, and not the paper itself, was sent to Lloyd, but it does not seem very clear why the abstract was drawn up at all.

(164) This and four other letters to Keill are printed in this volume, p. 169, foll.

(165) John Chamberlayne was endeavouring to reconcile the two philosophers. He sent Newton's letter to Leibniz, who replied in a letter dated Vienna, Aug. 25, (Leibn. *Opp.* III. 491) part of which was read by Chamberlayne at the meeting of the Royal Society on Nov. 11. In it Leibniz " desires that some letters and papers of Mr Oldenburg and Mr Collins which he supposes to be in the custody of the Royal Society may be communicated to him in order to his publishing a Commercium Epistolicum in defence of himself at his return from Vienna to Hanover. The Society was of opinion that Mr Leibniz ought either to make good his charge against Dr Keill or to ask pardon of the Society for suspecting their judgment and integrity in the Commercium Epistolicum already published by their order and approbation. But Mr Chamberlayne saying that Mr Leibniz designed in a short time to be in England, the farther consideration of this affair was referred to some other opportunity." *Journ. Bk.* There is in the British Museum (MSS. Birch, 4284) a copy in Newton's hand of Leibniz's letter of Aug. 25.

(166) The other assessors were Sir James Montagu, Dr Cannon, Prebendary of Ely,

Dr Samuel Clarke, Dr Henry Newton, Chancellor of the Diocese of London, and Dr Johnson, Chancellor of the Diocese of Ely. (Colbatch's MSS.) The trial after continuing about six weeks, the Court holding its sittings two evenings in the week, ended on June 14. See Monk's *Bentley*, pp. 281—286.

(¹⁶⁷) Commons' Journals, xvii. 677, 716. I do not consider M. Biot's abstract of the proceedings on this occasion (*Biog. Univ.* art. Newton, pp. 192, 193) as a model of accurate condensation : I will therefore exhibit Whiston's statement as nearly as may be in his own words. In 1714 Whiston and Ditton communicated to Newton their method of discovering the longitude at sea by signals, and at his desire to Halley, as also to Sam. Clarke and Cotes, and soon had their approbation so far as to encourage them to apply to the House of Commons for a reward to such as should discover the Longitude. A Committee was appointed to examine into the matter, and the four persons just mentioned were summoned to attend. " As soon as the Committee was set, which was a very large one, Newton, Halley, Clarke and Cotes appeared. A chair was placed for Sir I. Newton near the Chairman { Mr Clayton, M.P. for Liverpool }, and I stood at the back of it. What the rest had to say they delivered by word of mouth, but Sir I. Newton delivered what he had to say in a paper { referred to above }. Upon the reading of this paper, the Committee were at a loss, as not well understanding its contents : Sir I. Newton sitting still and saying nothing by way of explication. This gave the chairman an opportunity which it was perceived he wanted of trying to drop the bill ; which he did by declaring his own opinion to be that ' Unless Sir I. Newton would say that the method now proposed was likely to be useful for the discovery of the Longitude, he was against making a bill in general for a reward for such a discovery' ; as Dr Clarke had particularly proposed to the Committee. Upon this opinion of his, not contradicted by any other of the Committee ; and upon Sir I. Newton's silence all the while, I saw the whole design was in the utmost danger of miscarrying. I thought it therefore absolutely necessary to speak myself : which I did nearly in these words, 'Mr Chairman, the occasion of the puzzle you are now in is nothing but Sir I. Newton's caution. He knows the usefulness of the present method near the shores' [which are the places of greatest danger]. Whereupon Sir Isaac stood up and said that ' He thought this bill ought to pass, because of the present method's usefulness near the shores.' Which declaration of his was much the same with what he had said in his own paper, but which was not understood by the Committee, and determined them unanimously to agree to such a bill." Historical Preface, date probably 1742, inserted in some copies of his " Longitude discovered...Lond. 1738," p. v.

I will now leave it to the reader, who will of course make the requisite allowance for the forwardness and vanity of the reporter, to judge whether M. Biot's term " presque puérile " be a proper epithet to apply to the part that Newton took on the occasion.

(¹⁶⁸) " Redit nunc demum Tibi, Vir illustris ! quod sane, si non omnino Tuum sit, Ortum saltem suum Tibi debet ; nempe Opusculum de Methodo Fossilium, te assiduè hortante, inceptum, provectum, absolutum," etc. *Naturalis Historia Telluris,* &c. Lond. 1714. The letter is given in English in the same author's " Fossils of all kinds," &c. Lond. 1728.

(¹⁶⁹) Raphson's *Hist. of Fluxions,* pp. 100—103. Des Maizeaux's *Recueil...*Tom. ii. Amsterd. 1720. Leibn. *Opp.* iii. 451—455.

(¹⁷⁰) Raphson's *Hist. of Fluxions,* pp. 111—123. Des Maizeaux's *Recueil.* Leibn. *Opp.* iii. 474—488. The French Translation of Newton's letters of Febr. 26 and May 18, as given by Des Maizeaux, had the benefit of Newton's supervision. His correc tions of the press (in his own hand) are preserved in the British Museum, MSS. Birch, 4284. fol. 235.

(¹⁷¹) " Mr Roger Cotes Astronomy Professor & Fell. dyed upon a Relapse into a Fever attended with a violent Diarrhœa and constant Delirium. He was bury'd on yᵉ 9ᵗʰ. yʳᵉ wʳᵉ 20 rings of 20s. each & 30 at 10s. each." (Rud's *Diary.*) Cotes "tout-à-la-fois géomètre, astronome et physicien" (see Delambre's *Hist. Astron.* 18 *siecle,*

p. 449, Mathieu's note) was born at Burbage in Leicestershire, July 10, 1682. He was entered pensioner at Trin. Coll. Apr. 6, 1699, from St Paul's School. His name stands in the admission book immediately after that of Conyers Middleton. They were elected scholars together in May, 1701, took their B.A. degree in 1703, and were sworn in minor fellows of the College on Octob. 3, 1705. An accurate life of Cotes is given in the *General Dictionary*, partly from materials supplied by his cousin Robert Smith. See also Knight's *Life of Colet*, (Lond. 1724) who says (p. 430) "I could run out many pages in the just character of this extraordinary man, being very intimate with him, and having the opportunity of knowing him perfectly, by being his chamber-fellow many years in Trinity College in Cambridge, but am obliged not to exceed the bounds of a short account"...and Monk's *Bentley* (p. 314 and elsewhere). Bentley's inscription on his monument has been frequently printed. The authority for the well-known saying attributed to Newton on the premature death of this promising mathematician is Robert Smith, who in his copy of the *Harmonia Mensurarum*, under Cotes's epitaph, has written the words "S$^r$ Isaac Newton, speaking of Mr Cotes, said, ' If He had lived we might have known something '." In his *Optics*, (Vol. ii. art. 465, *remarks*, p. 76) he gives the saying in exactly the same words, where in allusion to a theorem on the image of an object seen through a number of lenses he says : "That noble and beautiful theorem...was the last invention of that great Mathematician Mr Cotes, just before his death at the age of 32: upon which occasion I am told Sir Isaac Newton said "...The author of Cotes's Life in the *Biographie Universelle*, who has been followed by Delambre (p. 457), seems to have misunderstood this passage, taking Newton's remark to apply to the discovery of the optical theorem. Parne, who was six years junior to Smith, in his Collections for *Hist. of Trin. Coll.* p. 351, gives the saying with the single variation of "had" for "might have :" "On the death of Mr Cotes Sir Isaac Newton is said to have expressed himself in these honourable and remarkable words...."

(¹⁷²) "The President in the chair. The President gave the Society his picture drawn by Mr Jervase for which he had their thanks." *Journ. Bk.*

(¹⁷³) In pursuance of an Address to the king it was laid before the House of Lords on Jan. 21, 1718.

In consequence of this Report a Proclamation was issued in December 1717, reducing guineas from 21s. 6d. to 21s.

(¹⁷⁴) This Report was accompanied with an Account of the Gold and Silver coined from Jan. 1, 1702, to Nov. 20, 1717, and with the Report of Sept. 21, was laid before the House of Commons on Dec. 21, in pursuance of an address to the King. Both Reports will be found in the *Commons' Journals*, xviii. 664—6. That of Sept. 21, was printed in the *Daily Courant*, Dec. 30, 1717, and may also be seen in *The Political State of Great Britain*, Tindal's *Continuation of Rapin*, and *Macc. Corr.* ii. 424.

(¹⁷⁵) With additions (among others, eight new queries, from the 17th to the 24th.) The Advertisement is dated July 16, 1717.

(¹⁷⁶) "The House being informed ' That Sir Isaac Newton attended at the Door,' he was called in; and delivered at the Bar pursuant to the Address of this House to his Majesty of Thursday last : 'An Account of the Silver Monies coined in the four years ending at Christmas 1699, by weight.' Also 'An Account of the Gold and Silver Monies coined yearly from Christmas 1699 to Christmas 1716, by weight." And then he withdrew." (*Lords' Journals.*) The last "Account" is printed in *Macc. Corr.* ii. 434.

(¹⁷⁷) p. 185.

(¹⁷⁸) *Macc. Corr.* ii. 430.

(¹⁷⁹) Nov. 6, 1718. "The Treasurer acquainted the Council that Sir Isaac Newton {who was present} had lately paid him as a gift to the Society £70." (*Council Minutes.*)

(¹⁸⁰) "1719 July 13, to a free gift rec$^d$. from Sir I. Newton £52 10s. 1720 Apr. 28, to a gift rec$^d$. of Sir I. Newton £52 10s." Pound's *Account Book*, quoted by Rigaud (*Bradley*, p. iii.) These instances of Newton's liberality were probably in acknowledgment of astronomical observations supplied by Pound (*ex. gr.*

the magnitude of Jupiter's diameters, *Princip.* ed. 3. p. 416). Pound was Bradley's uncle.

([180])* It is written in a most peace-loving spirit. See p. 187, note †.

([181]) A fourth edition was published in 1730 from a copy, it is said, of the third corrected by the author's own hand, and left before his death with the bookseller.

([182]) p. 188.

([183]) Made jointly with E. Southwell and J. Scroope, Esqrs. Printed in *Hibernian Patriot*, " being a Collection of the Drapier's Letters," &c. 1730, p. 244. Comp. Scott's *Swift*, vi. 392, ed. 2.

([184]) Newton was then occupied with the 3rd edition of the *Principia*. Delisle tells us that Newton assured him that "si M. Halley avoit eu égard {in constructing his Lunar Tables} aux moindres équations dont il a fait mention dans sa Théorie, et qu'il eût ajoûté une minute et demie à la longitude de la Lune pour son accélération physique dans notre temps, il n'auroit trouvé aucune différence sensible entre ses observations et le calcul." *Journal des Savans*, June 1750, p. 428.

([185]) Appendix, No. XXXIV.

([186]) *Macc. Corr.* ii. 435. Newton wanted the calculations for the 3rd ed. of the *Principia.* If Halley re-examined the two calculations, the examination led to no new result, and·if he performed the calculation for the place in the parabolic orbit, no use was made of it in the 3rd ed. as had been intended.

([187]) *Gentleman's Mag.* lix. 775, (with three other letters to the same person)    It begins " A bad state of health makes me averse from minding business."

([188]) Turnor's *Grantham*, p. 172. Brewster's *Newton*, p. 363. " Just after he was come out of a fit of the gout...; he was better after it and his head clearer and his memory stronger than I had known them for some time."

([189]) Turnor's *Grantham*, p. 158.

([190]) *Phil. Trans.* for 1725, p. 315. Brewster's *Newton*, p. 262. The summary is entitled " A Short Chronicle from the first memory of things in Europe to the conquest of Persia by Alexander the Great," and was afterwards published in his *Chronology*. It was drawn up in a few days at the request of the Princess of Wales. Conti, at her desire, was allowed to have a copy of it, from which when he went to France other transcripts were made.

Newton's *Chronology*, (Lond. 1728) appeared towards the end of 1727. Conduitt's Advertisement states that it " was writ by the author many years since ; yet he lately revised it, and was actually preparing it for the press at the time of his death." Martin Folkes writing to Morgan, Master of Clare Hall, Jan. 6, 1727-8, says : " I am glad you have been so well entertained with Sir Isaac's book, and at the same time to find my own opinion of it so entirely confirmed...but indeed I have had that satisfaction from several hands, and I even hear your Neighbour of the great College { Bentley } who spoke very slightingly of the performance before it appeared begins not to talk so magisterially as he did before, but W. W. { Whiston } continues in the same way, and declares he shall overturn it so easily that he shall not be able to extend the whole confutation to a sheet of paper."

([191]) " Pendant les deux mois que l'abbé Alari passa à Londres { 1725 } , il visita l'université de Cambridge, et le grand Newton, qui jouissait alors dans la capitale de l'Angleterre, de l'estime générale de l'Europe et de cinquante mille livres de rente, en qualité d'intendant des monnaies. L'abbé étant allé chez lui à neuf heures du matin, l'Anglais débuta par lui apprendre qu'il avait quatre-vingt-trois ans. On voyait dans sa chambre le portrait du Lord Halifax, son protecteur et celui de l'abbé Varignon dont il estimait les ouvrages de géométrie. *Varignon et le père Sébastien carme, sont*, dit-il, *ceux qui ont le mieux entendu mon système sur les couleurs.* La conversation tomba ensuite sur l'histoire ancienne, dont Newton s'occupait alors. L'abbé, qui était plein de la lecture des auteurs grecs et latins, l'ayant satisfait, il le pria à dîner. Le repas fut détestable ; Newton était avare, et il ne fit boire à son convive que des vins de Palme ou de Madère, qu'il recevait en présens. Après le dîner, il mena l'abbé à la Société royale de Londres, dont il était président, et le fit asseoir à sa droìte. La séance commença et Newton s'endormit. A la fin de la séance, tout le monde signa le registre, et l'abbé comme les autres. Newton le ramena ensuite chez lui, où il le garda jusqu'à neuf heures du soir." ( *Essai Historique sur Bolingbroke*, compiled by General Grimoard, in *Lettres Historiques...de...Bolingbroke*, i. 155, Paris. 1808).

Alari was born in 1689 ; he was a friend, at least for some time, of Bolingbroke's, and instructor of Louis XV. The intelligent reader will make allowance for the spice added to give pungency to the story. The following is the simple record in the *Journal Book* of Alari's visit, " Mr Mildmay had leave to be present, as also Mr Petre Joseph Alary, a French Gentleman."

(¹⁹²)   Life of Maclaurin, prefixed to his *Account of Sir Isaac Newton's Philosophical Discoveries.*

(¹⁹³)   Printed in the *Phil. Trans.* for 1725, pp. 315—321. Comp. Brewster's *Newton*, pp. 261—265. The MS. written in a fine copper-plate hand is preserved in the Archives of the Royal Society, and is endorsed " read about the latter end of 1725." In this paper he incidentally informs us that when he lived at Cambridge he used sometimes to refresh himself with History and Chronology for a while, when he was weary with other studies.

(¹⁹⁴)   The Preface is dated Jan. 12, 1725-6. Twelve copies are stated to have been printed on large paper, (Rigaud's *Bradley*, p. xi.), of which there is one in Trinity College Library, another in that of Queens' College, (a presentation copy from the author to his friend J. F. Fauquier,) and a third in the Library of the Royal Society, of which we find the following *naïve* notice in the *Journal Book*. " March 31, 1726. Mr Folkes in the name of the President gave the Society a Book richly bound in morocco leather as a present for the Library, entitled *Philosophiæ Naturalis Principia Mathematica*, printed at London 1726. The Society ordered thanks to be rendered to the President for this invaluable present." It is to be hoped that the correspondence which passed between Newton and his editor (Henry Pemberton, M.D.) during the progress of the work through the press will yet be discovered. See Rigaud's *Essay*, p. 107. *Philos. Mag.* May 1836, p. 441.

We may give here the anecdote quoted by Mr De Morgan from Maty's *Memoirs of Demoivre* (*Phil. Trans.* 1846, p. 109.) " Comme tout ce qui regarde les grands hommes peut être intéressant, on sera peut-être bien aise de savoir que Newton a souvent dit à Mr. de Moivre que s'il avoit été moins vieux il auroit été tenté de revoir sur les dernières observations sa théorie de la Lune, ou comme il s'exprimoit *de l'attaquer de nouveau* (*to have another pull at the moon*). Je tiens ceci de Mr. de Moivre lui-même."

(¹⁹⁵)   Baily, *Memoirs of Astron. Soc.* viii. 188.

(¹⁹⁶)   " March 23. The chair being vacant by the death of Sir Isaac Newton there was no meeting this day." (*Journal Bk.*)   For the reflections which his death suggested to some minds, see Boyer's *Political State of Great Britain* for March 1727, (Vol. xxxiii. pp. 327—330). In Mist's *Weekly Journal* for March 25, the obituary opens with " Sir Isaac Newton, the greatest Mathematician that the World ever knew." Thomson's " Poem sacred to the Memory of Sir Isaac Newton," (dedicated to Walpole) seems to have had a large circulation. I have a copy before me of the 5th edition, dated 1727.

I wish that I had been able to contribute more local information respecting Sir Isaac Newton than it has been my fortune to meet with. But the age of " conversations with " and " reminiscences of " had not yet arrived, and we do not know that any fellow of his College kept a diary. Thomas Parne, who took his B.A. degree in 1718, collected materials for the history of Trinity College, and had opportunities of conversing with men who had been contemporaries of Newton (for example, George Modd who was two years junior to Newton, and lived in College until his death in 1722). He has given us many particulars of more or less interest relating to Ray, Thorndike, Pearson, Barrow, Duport and other members of the College, but the only allusions to its chief pride and boast that I have found in his MSS. are the following : under the head of

" Writers " the name " Newton " stands first in the list ; the dates of his return as
M.P. for the University and of his unsuccessful contest, (in the latter of which the
majority against him is erroneously stated) are given, and an anecdote is preserved
of his absence of mind in these terms : " Newton hath come into the Hall without his
Band, and went towards St. Maries in his surplice ;" for which Parne quotes as his
authority a " Mr Burwell," (perhaps Alexander Burrell, eleven years senior to Parne,
who may have been a connexion of the Alexander Burrell who took his B.A. degree in
1670, and was chaplain of the College from Oct. 1673 to June 1681.) I do not know
that I can find a more appropriate place for a similar anecdote which has already
appeared in English. It was told to the Swedish Professor Björnståhl at Basle by John
Bernoulli, son of the famous John, on Nov. 5, 1773 : " sagte uns, Newton sey eben-
falls sehr zerstreut gewesen, und habe einmahl den Finger eines Frauenzimmers genom-
men, um seine Tabakspfeife nachzustopfen." ( Briefe auf...Reisen. Leips. und Rostock.
1777—1783. v. 46). On Dec. 8, in the following year Björnståhl paid a visit at Amsterdam to the " gelehrten Herrn Fontein," an Anabaptist preacher and scholar of
Hemsterhuis and Albert Schultens. " Im Jahr 1738 hat er eine Reise nach England
gemacht und mit dem grossen Bentley Bekanntschaft unterhalten. Zu Cambridge hat
er verschiedne Anekdoten von Newton, welcher berühmte Mann neun oder zehn Jahr
vorher gestorben war, gehört, unter andern : Newton habe geglaubt, dass Mahomed von
Gott gesandt worden sey, um die Araber von der Finsterniss zurück, und zum Glauben
an einen Gott zu führen u. s. w. (Dies haben ihm wenigstens die Professoren oder
Fellows zu Cambridge als eine besondre Merkwürdigkeit aus Newtons Geschichte
erzählt ;) die im Koran und Mahomeds Leben vorkommenden Fabelu und Wunder
jedoch habe dieser aufgeklärte Mann nicht geglaubt. Er sagte mir, Newton habe eine
Abhandlung herausgegeben, um zu beweisen, die Stelle 1 Johann. v. 7. sey nicht
ächt, und der Text habe ohne diesen Vers einen weit bessern Zusammenhang." ( Ib.
462).
    The Professor was in England from April 1775 to March 1776. Writing from Oxford
Oct. 24, 1775, after saying that he passes over many remarkable objects, such as the
Marmora Oxoniensia, Cromwell's scull, Guy Faux's lantern, Blenheim, Stowe, &c. he
proceeds: " Dagegen aber will ich einen Umstand melden, der, wie ich mit Ueberzeugung
weiss, bisher in keinem Buche vorkommt : diesen, dass wir unter andern in der hiesigen
Nachbarschaft ausdrücklich zu dem Ende eine Reise gethan haben, um die eigne Bücher-
sammlung des grossen und unsterblichen Ritters Newton zu sehen. Jetzt besitzt sie Herr
Doctor Musgrave...Rector zu Chinnor, achtzehn...Meilen von Oxford. Sie hat ihm un-
gefehr vier hundert Pfund sterling gekostet. Hier findet man alle Ausgaben von Newton's
Werken, und, welches das merkwürdigste ist, am Rande mit seinen eigenhändigen
Anmerkungen angefüllt, und bisweilen mehrere Blätter am Schlusse der Bücher von
ihm ganz vollgeschrieben. Ich zweifle nicht, dass ein Newtonianer hier nicht viel
Vergnügen und manche Erläuterung antreffen würde. Hier sah ich auch das seltne
Buch von Herr Jones Vater, wovon ich oben angemerkt habe, das der Sohn selbst es
nicht einmahl besitze. Der Titel ist: Epitome of the Art of Practical Navigation...Lon-
don, 1706. Noch ein sehr seltnes Buch von eben diesem Jones: (dies ist ganz ausser-
ordentlich rar:) Synopsis Palmariorum Matheseos...London 1706...Uebrigens sieht
man, dass Newton eine vortrefliche Bibliothek gehabt hat. Alle griechischen und latein-
ischen classischen Schriftsteller finden sich daselbst. Sonst habe ich verschiedne
eigenhändige Briefe von Newton an Flamsteed gesehen, die in der Corpus-Christi-Bibli-
othek zu Oxford aufbewahrt werden. Zu Cambridge werden noch mehr Handschriften
von ihm angetroffen." (III. 288.)
    I have no means of confirming or impugning the accuracy of the account given by
the simple-hearted Swede of the disposal of Newton's Library. A statement of its
magnitude will be found at the end of the subjoined extract from Maude's Wensleydale (p. 106.)

Newton's nephew, Benjamin Smith, "left a small ivory bust { of his uncle } of admirable workmanship by that celebrated artist, Marchand, which from its elegance, similitude and placid expression is truly valuable. It is said to have cost Sir Isaac 100 guineas and is specified in an authentic inventory of his effects, taken by virtue of a commission of appraisement in April 1727, now in my possession. It appears that his personal estate amounted to £31,821 16s. 10d. which was distributed among eight relations, Sir Isaac dying intestate:...as a proof of his benevolence...at his death there was owing him by one tenant £60 for 3 years rent, and by another for 2½ years a smaller sum.... { His } wardrobe and cellar...in the valuation stand thus. Item, wearing apparel, woollen and linen, one silver hilted sword, and two canes, £8. 3s. Item, in the wine vault, a parcel of wine and cider in bottles, £14. 16s. 6d. The furniture and luxuries of his house bearing nearly the like proportion, his library excepted, which consisted of 2000 volumes and 100 weight of pamphlets."

It does not fall within the scope of our Chronological Synopsis of Newton's life to notice the great political events of his time, and I am therefore compelled to place here an extract from an ingenious French writer which might otherwise have been given under a more convenient head. I leave it to future inquirers to ascertain the precise embarrassment alluded to in it, and to determine the probable extent to which we are indebted for the story to the play of a lively imagination.

" Pour faire voir que l'universalité des talents est une chimère, je ne veux pas chercher mes autorités dans la classe commune des esprits ; montons jusqu'à la sphère de ces génies rares qui, en faisant honneur à l'humanité, humilient les hommes par la comparaison. Newton, qui a deviné le système de l'univers, du moins pour quelque temps, n'étoit pas regardé comme capable de tout par ceux mêmes qui s'honoroient de l'avoir pour compatriote.

Guillaume III, qui se connoissoit en hommes, étoit embarrassé sur une affaire politique ; on lui conseilla de consulter Newton : Newton, dit-il, n'est qu'un grand philosophe. Ce titre étoit sans doute un éloge rare ; mais enfin, dans cette occasion-là, Newton n'étoit pas ce qu'il falloit, il en étoit incapable, et n'étoit qu'un grand philosophe. Il est vraisemblable, mais non pas démontré, que, s'il eût appliqué à la science du gouvernement les travaux qu'il avoit consacrés à la connaissance de l'univers, le roi Guillaume n'eût pas dédaigné ses conseils.

Dans combien de circonstances, sur combien de questions le philosophe n'eût-il pas répondu à ceux qui lui auroient conseillé de consulter le monarque : Guillaume n'est qu'un politique, qu'un grand roi ?"

(Duclos's *Considérations sur les Moeurs, Œuvres*, I. 160. Paris, 1820.)

## DIVIDENDS RECEIVED BY NEWTON, AND NUMBER OF WEEKS HE RESIDED EACH YEAR WHEN FELLOW OF TRINITY COLLEGE.

| Year ending Michaelmas. | His Dividend. | Weeks resided by him. | Year ending Michaelmas. | His Dividend. | Weeks resided by him. |
|---|---|---|---|---|---|
| 1668 * | £15 | 4 {(in long vacation.) | 1685 | £25 | 51 |
| 9 | 25 | 52 | 6 | 12. 10 | 52 |
| 70 | 20 | 49½ | 7 | 12. 10 | 45 |
| 1 | 16. 13s. 4d. | 48 | 8 | Nil. | 45 |
| 2 | 16. 13s. 4d. | 48½ | 9† | Nil. | 19 |
| 3 | Nil. | 49½ | 90 | Nil. | 29½ |
| 4 | 25 | 51 | 1 | 12. 10 | 44½ |
| 5 | 25 | 46 | 2 | 12. 10 | 49 |
| 6 | 25 | 50½ | 3 | 25 | 49½ |
| 7 | 12. 10s. | 43½ | 4 | Nil. | 49 |
| 8 | 25 | 49 | 5 ‡ | 34 | 50 |
| 9 | 25 | 38 | 6 | 34 | 27½ |
| 80 | 25 | 36½ | 7 | Nil. | 0 |
| 1 | 25 | 49 | 8 | 34 | ½ |
| 2 | 12. 10 | 46½ | 9 | 34 | 0 |
| 3 | 12. 10 | 46 | 1700 | 37 | 0 |
| 4 | 25 | 52 | 1 § | 40 | 0 |
| | | | 2 | | 2 |

The dividend was voted at the annual audit in December, and paid by the Bursar "as money came to his hands," generally at the end of 6 or 12 months, but sometimes the payment was still further delayed.

Newton's own receipts for his dividend, livery and stipend for the four years of Humfrey Babington's Bursarship are to be found in Babington's Day-Book, which is one of two or three that are still pre-

---

* Steward's bill unpaid 19s. 7¼d.

As an illustration of the scrupulous exactness and regularity which characterised Newton in all matters of business, it may be mentioned, that in two instances only was he in arrear with his Steward's bill, viz. the one before us when he had just become Master of Arts, and probably did not know the proper mode of paying the bill until after the accounts for the year were made up, the other when he was absent in London as a member of the Convention Parliament.

† Steward's bill unpaid £5 12s.

‡ The augmented dividend of this and subsequent years is in consequence of Newton's increased standing in the College.

§ He must have resigned his fellowship before Dec. 21, 1701, otherwise the Bursar's Book would have contained a record of his receiving dividend for the quarter ending then. At the time of his resignation he stood 10th on the list: had he remained fellow until August of the year next but one following, he would have been elected a senior.

served in the Muniment Room.  The first of these receipts we give at
full length : the others are added for the sake of the dates :—

Oct. 11, 1675.    Rec<sup>d</sup> then my wages as fellow for the ⎫ £2 13 4
          whole year ending Mich. last...... ⎭

            My livery for the same year...............    1 13 4
            Pandoxator voted 1673.....................    5 0 0
            ½ dividend voted last audit 1674. ........   12 10 0

            In all................................ £21 16 8

                       By me, ISAAC NEWTON.

Nov. 20, 1675    R<sup>d</sup> the later moiety of Mr Newton's div.... £12 10 0
                       By me, JOHN BATTELY.

July 8, 1676    Wages for ¾ year ending Midsummer........ £2 0 0
            Livery for 1676................................    1 13 4
            Pandoxator's div. granted at audit 1674....    5 0 0
            ½ div....................last audit 1675....   12 10 0

                           21 3 4

Nov. 16, 1676    Rec<sup>d</sup> the later ½ of div. granted 1675........ £12 10 0
            Wages for quarter ending Mich<sup>e</sup>. last........    0 13 4

Dec. 13, 1677    Rec<sup>d</sup> wages for year ending Mich<sup>a</sup>............ £2 13 4
            Livery................................    1 13 4
            Pandox. Div. Audit 1675.....................    5 0 0
            Do..................... 1676.....................    5 0 0
            First ½ of div. Audit 1676....................   12 10 0

                           26 16 8

Nov. 22, 1678    Wages as fellow for year 1678................ £2 13 4
            Livery.............................................    1 13 4
            Later ½ of div. granted Audit 1676...........   12 10 0
            Pandox. div..................... 1677...........    5 0 0
            Later ½ of Mr Wickins's div. granted ⎫
               Audit 1676........................ ⎭   12 10 0

                           34 6 8

Dec. 30, 1678    Rec<sup>d</sup> div. granted Audit 1677.................. £12 10 0
            Also Mr Wickins's.............................   12 10 0

Besides the dividend Newton was in receipt of the following emoluments
    from the College:
    1.   Pandoxator's dividend (from the profits of the bakehouse and
        brewhouse) £2 10s. for year ending Mich<sup>t</sup>. 1668, and £5
        annually afterwards except when he did not reside the major
        part of the year as in 1689, 1697, &c.

2.  3s. 4d. weekly during residence "pro pane et potu." (This sum
    represents 10 penny loaves, 10 quarts of small beer at 1d. a
    quart and 10 quarts of ale at 2d. a quart.)

3.  13s. 4d. for livery for year ending Mich'. 1668 and £1 13s. 4d.
    annually afterwards until Mich'. 1701.  £2 3s. 4d. for stipend
    or wages for year ending Mich'. 1668, and 13s. 4d. a quarter
    until the quarter ending Dec. 21, 1701.

If to these sources of income it be added that, as fellow, he had nothing to pay for his dinners or room-rent, that his hereditary estate
brought him in £80 and his professorship £100 a year, every reader can
form his own opinion on the condition of Newton's worldly circumstances while he was a resident fellow of Trinity College.

## NEWTON'S EXITS AND REDITS.

[From the book in which the Fellows entered their names on going out of, or returning to, College. The entries are generally in Newton's own hand, but sometimes in that of North, the Master, or of Lynnet when Vice-Master, and occasionally they seem to have been written by a servant. There is a 4to. book in the Muniment Room containing the Exits and Redits of the Bachelor Fellows and Scholars, commencing with Octob. 1667. The first six names in it are those of the six fellows of Newton's year senior to him: the second leaf of the book, at the top of which Newton's name stood, with the dates of his Exits and Redits from Octob. 1667 until Midsummer of the following year, has been cut out, the lower portion of the D belonging to the "Ds" prefixed to his name being the only part of the entry relating to him that is left.]

| Year. | Exit. | Redit. | Year. | Exit. | Redit. |
|-------|-------|--------|-------|-------|--------|
| 1668 | | Sept. 29 | 1682 | Feb. 21 | Feb. 28 |
| 1669 | Nov. 26 | Dec. 8* | | Apr. 8 | Apr. 29 § |
| 1671 | Apr. 17 | May 11 | | May 10 | |
| 1672 | Jun. 18 | Jul. 19 | 1683 | March 27 | May 3 |
| 1673 | March 10 | Apr. 1 | | May 21 | |
| 1674 | Aug. 28 | Sept. 5 | 1685 | March 27 | Apr. 11 |
| 1675 | Feb. 9 | March 19 | | Jun. 11 | Jun. 20 |
| | Oct. 14 | Oct. 23 | 1687 | March 25 | |
| 1676 | May 27 | Jun. 1 | 1688 | March 30 | Apr. 25 |
| 1677 | Feb. 20 | March 3 | | Jun. 22 | Jul. 17 |
| | March 26 | | 1690 | | Feb. 4 |
| | Apr. 26 | May 22 + | | March 10 | Apr. 12 |
| | Jun. 8 | | | Jun. 22 | Jul. 2 |
| 1678 | May 6 | May 27 | 1691 | Sept. 12 | Sept. 19 |
| 1679 | May 15 | May 24 | | Dec. 31 ‖ | |
| | | Jul. 19 | 1692 | | Jan. 21 |
| | Jul. 28 | Nov. 27 ‡ | 1693 ¶ | May 30 | Jun. 8. |
| 1680 | March 11 | | 1695 | | Sept. 10 |
| | Apr. 28 | May 29 | | Sept. 14 | Sept. 28 |
| 1681 | March 15 | March 26 | 1696 | March 23 | |
| | May 23 | | | Apr. 20 | |

* Newton was making this entry under the Exits and had written more than half the first letter of his name when he found out his mistake.

† Newton has also entered Dr Lynnet's Redit, who returned to College the same day.

‡ Two entries, one in Newton's hand, the other by North. Newton had been down in Lincolnshire, and a friend of his availed himself of his return to Cambridge to employ him on a small commission, which it will be seen he lost no time in executing. "Nov. 28, 1679. Rec^d £11. 15s. 7d. by the hand of Mr Isaac Newton from Mr W. Walker, Rector of Grantham School." Dr Babington's Day-Book. Walker received the money from Mr Edw. Pawlet and he from James Thompson who owed it to Babington.

§ Newton had made nearly the whole of this entry under the year 1677, where on turning over the leaves of the book ample room offered itself, but when about to write the "8" he discovered the mistake.

‖ He was in London in Jan. 1692. On the 9th of that month we find Pepys inviting Evelyn to his usual Saturday evening party to meet Dr Gale and Mr Newton. (Memoirs, v. 181. 2nd ed.)

¶ Newton had entered an Exit for Apr. 15 of this year, but it was afterwards crossed out.

## NEWTON'S WEEKLY BUTTERY BILLS,

*From October* 1686, *to February* 1694, *and from June* 1698 *to March* 1702, *for Bread, Beer, &c.*

These relics of Newton's household expenditure are extracted from two mutilated Buttery Books in Trinity College Muniment Room. The Fellows' Buttery Books for the remainder of the period of his residence and all those of the Scholars during the time when he was undergraduate and bachelor, have, I fear, been destroyed by some person or persons, who, it is to be hoped, could not be supposed to know that books apparently so useless were indispensable for a correct history of the discovery of the new calculus and of the true theory of the world.

### EXPLANATION OF ABBREVIATIONS.

no co means *not in commons,* i. e. *not in residence,* or *out of college.*
di or dimi means *half the week.*

C. P. or Com. P. means *the fine for not delivering a Common Place in chapel after morning prayers.* This Fine continued to be levied until 1830, when the system of compounding was introduced.

M is supposed to stand for *man,* i. e. *servant.**

Ton. stands for *tonsor* (the College barber), ch. probably for *chapel,* Lett. for *letters.*

---

* Newton mentions his servant twice in his correspondence with Flamsteed. (Baily, pp. 139, 157). "As for the places calculated from the tables, I will give you no trouble about them : my servant has lately learnt arithmetic, and, if I go on with this business of the moon, he shall learn astronomical calculations and examine them, and I will send you his corrections." (Letter of Nov. 17, 1694). "I want not your calculations, but your observations only. For besides myself and my servant, Sr Collins { of Catharine Hall} (whom I can employ for a little money, which I value not) tells me that he can calculate an eclipse, and work truly." (Letter of June 29, 1695). This may have been the John Perkins "Astrologus Cantabrigiensis," to whom Vincent Bourne addressed a copy of elegiacs, beginning

> Lusit, amabiliter lusit Fortuna jocosa,
> Et tunc, siquando, tunc oculata fuit;
> Cum tibi, Joannes, Newtoni sternere lectum ;
> Cum tibi museum verrere diva dedit.

And ending

> Nec melior lex est, nec convenientior æquo,
> Quam siet astronomo servus ut astrologus.

| Week ending | s. | d. | Week ending | s. | d. | Week ending | s. | d. |
|---|---|---|---|---|---|---|---|---|
| 1686 Oct. 15 | 2 | 5½ | 1687 July 29 | 2 | 9½ | 1688 May 11 | 3 | 5½ |
| 22 | 7 | 2½ | Aug. 5 | 3 | 0½ | 18 | 4 | 2½ |
| 29 | 2 | 9½ | 12 | 3 | 2½ | 25 | 8 | 2 |
| Nov. 5 | 3 | 4½ | 19 | 3 | 1 | June 1 | 4 | 6 |
| 12 | 3 | 0½ | 26 | 4 | 2½ | 8 | 9 | 7½ |
| 19 | 3 | 10½ | Sept. 2 | 3 | 3½ | 15 | 12 | 7 |
| 26 | 2 | 9 | 9 | 16 | 10½ | 22 | 2 | 10½ |
| Dec. 3 | 5 | 0 | 16 | 13 | 6½ | no co. 8ᵈ. m. 29 | 0 | 11 |
| 10 | 10 | 0½ | 23 | 3 | 5½ | no co. July 6 | | ...... |
| 17 | 3 | 2 | 30 | 4 | 4½ | no co. 13 | | ...... |
| 24 | 2 | 11½ | Oct. 7 | 2 | 10½ | di. no co. 20 | 1 | 1 |
| 31 | 2 | 9½ | 14 | 4 | 3 | 27 | 3 | 2 |
| 1687 Jan. 7 | 2 | 11 | 21 | 3 | 5¾ | Aug. 3 | 4 | 3 |
| 14 | 7 | 0 | 28 | 3 | 4½ | 10 | 2 | 10¾ |
| 21 | 10 | 6 | Nov. 4 | 5 | 7½ | 17 | 3 | 8 |
| 28 | 3 | 11 | 11 | 6 | 1½ | 24 | 10 | 5¾ |
| Feb. 4 | 3 | 5½ | 18 | 3 | 0 | 31 | 3 | 2 |
| 11 | 2 | 5½ | 25 | 2 | 5½ | Sept. 7 | 18 | 8 |
| 18 | 3 | 5½ | Dec. 2 | 3 | 10¾ | 14 | 10 | 7½ |
| 25 | 3 | 0½ | 9 | 3 | 10¾ | 21 | 5 | 5¾ |
| March 4 | 4 | 9 | 16 | 2 | 11 | 28 | 4 | 0 |
| 11 | 14 | 5½ | 23 | 3 | 0½ | Oct. 5 | 6 | 1½ |
| 18 | 5 | 8½ | 30 | 3 | 1 | 12 | 4 | 11½ |
| 25 | 3 | 10½ | 1688 Jan. 6 | 3 | 5¾ | 19 | 2 | 8 |
| no co. Apr. 1 | 0 | 8 | 13 | 3 | 5½ | 26 | 3 | 2 |
| no co. 8 | | ...... | 20 | 16 | 1 | Nov. 2 | 3 | 1 |
| no co. 15 | | ...... | 27 | 3 | 0½ | 9 | 3 | 1 |
| di. no co. 22 | 2 | 4½ | Feb. 3 | 3 | 11 | 16 | 2 | 10½ |
| no co. 29 | | ...... | 10 | 3 | 5¾ | 23 | 4 | 10½ |
| no co. May 6 | | ...... | 17 | 3 | 0½ | 30 | 4 | 6 |
| no co. 13 | | ...... | 24 | 15 | 7½ | Dec. 7 | 2 | 8½ |
| d². ao co. 20 | 1 | 10 | March 2 | 8 | 0½ | 14 | 2 | 10½ |
| 27 | 2 | 7 | 9 | 10 | 7 | 21 | 3 | 3¼ |
| June 3 | 5 | 2½ | 16 | 3 | 1 | 28 | 6 | 8 |
| 10 | 10 | 0 | 23 | 3 | 4½ | 1689 Jan. 4 | 7 | 9 |
| 17 | 3 | 7½ | 6.8. C.P. 30 | 19 | 2 | 11 | 4 | 1¼ |
| 24 | 3 | 10½ | no co. Apr. 6 | 1 | 2½ | 18 | 8 | 5 |
| July 1 | 4 | 5½ | no co. 13 | 2 | 0 | no co. 25 | | ...... |
| 8 | 2 | 8½ | 20 | | ...... | no co. Feb. 1 | | ...... |
| 15 | 2 | 7½ | di. no co. 27 | 0 | 9½ | no co. 8 | | ...... |
| 22 | 2 | 11 | May 4 | 3 | 1 | no co. 15 | | ...... |

| | Week ending | *s.* | *d.* |
|---|---|---|---|
| 1689 | | | |
| no co. | Feb. 22 | ...... | |
| no co. | March 1 | 1 | 6 |
| no co. Ton. 7*s*. Ch. 3*s*. | 8 | 10 | 0 |
| no. co. | 15 | ...... | |
| no. co. | 22 | ...... | |
| no. co. | 29 | ...... | |
| no co. | Apr. 5 | ...... | |
| no co | 12 | ...... | |
| no co. | 19 | ...... | |
| no co. | 26 | ...... | |
| no co. | May 3 | ...... | |
| no co. | 10 | ...... | |
| no co. | 17 | ...... | |
| no co. | 24 | ...... | |
| no co. | 31 | 1 | 6 |
| no co. Ton. | June 7 | 5 | 0 |
| no co. | 14 | ...... | |
| no co. | 21 | ...... | |
| no co. | 28 | ...... | |
| no co. | July 5 | ...... | |
| no co. | 12 | ...... | |
| no co. | 19 | ...... | |
| no co. | 26 | ...... | |
| no co. | Aug. 2 | ...... | |
| no co. | 9 | ...... | |
| no co. | 16 | ...... | |
| no. co. | 23 | ...... | |
| no. co. | 30 | 6 | 8 |
| no co. | Sept. 6 | 1 | 6 |
| | 13 | 2 | 1 |
| | 20 | 2 | 10½ |
| | 27 | 3 | 6½ |
| | Oct. 4 | 3 | 1 |
| | 11 | 2 | 9½ |
| di. no co. | 18 | 5 | 3½ |
| no co. | 25 | ...... | |
| no co. | Nov. 1 | ...... | |
| no co. | 8 | ...... | |

| | Week ending | *s.* | *d.* |
|---|---|---|---|
| 1689 | | | |
| no co. | Nov. 15 | ...... | |
| no co. | 22 | ...... | |
| no co. | 29 | 1 | 6 |
| no co. | Dec. 6 | ...... | |
| no co. | 13 | ...... | |
| no co. | 20 | ...... | |
| no co. | 27 | ...... | |
| 1690 | | | |
| no co. | Jan. 3 | ...... | |
| no co. | 10 | ...... | |
| no. co. | 17 | ...... | |
| no co. | 24 | ...... | |
| no co. | 31 | ...... | |
| dimi. no co. | Feb. 7 | 1 | 9½ |
| | 14 | 3 | 11½ |
| | 21 | 6 | 2½ |
| | 28 | 5 | 8½ |
| | March 7 | 4 | 1 |
| dimi. no co. | 14 | 4 | 7½ |
| no co. | 21 | ...... | |
| no co. | 28 | ...... | |
| no co. | Apr. 4 | ...... | |
| no co. | 11 | ...... | |
| dimi. | 18 | 3 | 5½ |
| | 25 | 3 | 2 |
| | May 2 | 2 | 7½ |
| | 9 | 3 | 2½ |
| | 16 | 3 | 5½ |
| dimi no co. | 23 | 2 | 11½ |
| | 30 | 6 | 8 |
| | June 6 | 13 | 3 |
| | 13 | 10 | 10½ |
| | 20 | 8 | 6½ |
| | 27 | 2 | 3½ |
| dimi. no co. | July 4 | 2 | 3½ |
| | 11 | 19 | 6½ |
| | 18 | 15 | 6½ |
| | 25 | 7 | 5½ |

| | Week ending | *s.* | *d.* |
|---|---|---|---|
| 1690 | Aug. 1 | 6 | 5½ |
| | 8 | 11 | 11 |
| | 15 | 3 | 3½ |
| | 22 | 3 | 3½ |
| | 29 | 3 | 10½ |
| | Sept. 5 | 5 | 0 |
| Ton. 10*s*. | 12 | 14 | 2½ |
| no co. | 19 | 2 | 11½ |
| no co. | 26 | ...... | |
| | Oct. 3 | 4 | 1½ |
| | 10 | 4 | 10½ |
| | 17 | 4 | 3½ |
| | 24 | 4 | 2½ |
| 6*s*. 8*d*. Com. P. | 31 | 9 | 2½ |
| | Nov. 7 | 4 | 2½ |
| | 14 | 3 | 0½ |
| | 21 | 4 | 11½ |
| | 28 | 3 | 10½ |
| | Dec. 5 | 16 | 2½ |
| | 12 | 3 | 11 |
| | 19 | 3 | 8 |
| | 26 | 4 | 5½ |
| 1691 | | | |
| dimi. | Jan. 2 | 1 | 10½ |
| dimi. no co. | 9 | 4 | 0½ |
| | 16 | 4 | 4 |
| | 23 | 5 | 2½ |
| | 30 | 4 | 6 |
| | Feb. 6 | 4 | 10 |
| | 13 | 5 | 0 |
| | 20 | 3 | 9 |
| | 27 | 8 | 1½ |
| Ton. | March 6 | 13 | 3 |
| | 13 | 4 | 2 |
| | 20 | 3 | 4½ |
| | 27 | 5 | 7 |
| | Apr. 3 | 4 | 7½ |
| | 10 | 6 | 7½ |
| | 17 | 4 | 5½ |
| | 24 | 5 | 1½ |

| Week ending | s. | d. | Week ending | s. | d. | Week ending | s. | d. |
|---|---|---|---|---|---|---|---|---|
| 1691 May 1 | 5 | 0 | 1692 Feb. 5 | 7 | 0½ | 1692 Nov. 18 | 4 | 0 |
| 8 | 5 | 0 | 12 | 4 | 0 | 25 | 5 | 3½ |
| 15 | 4 | 6½ | 19 | 5 | 0 | Dec. 2 | 6 | 2 |
| 22 | 5 | 1½ | no co. 26 | 1 | 11 | 9 | 3 | 7½ |
| 29 | 6 | 0½ | March 4 | 7 | 6 | 16 | 7 | 9 |
| Ton. 10s. } June 5 | 13 | 6½ | 11 | 3 | 8 | 23 | 5 | 11½ |
| | | | 18 | 5 | 4 | 30 | 7 | 2½ |
| 12 | 3 | 11½ | 25 | 4 | 6½ | 1693 Jan. 6 | 4 | 10 |
| 19 | 3 | 9½ | Apr. 1 | 4 | 3½ | 13 | 8 | 4½ |
| 26 | 4 | 3½ | 8 | 10 | 7 | 20 | 6 | 1 |
| July 3 | 3 | 7½ | 15 | 4 | 10 | 27 | 13 | 6½ |
| 10 | 7 | 0 | 22 | 4 | 0 | Feb. 3 | 6 | 0 |
| 17 | 7 | 1 | 29 | 5 | 9 | 10 | 6 | 1 |
| no co. 24 | 0 | 6½ | May 6 | 4 | 8½ | 17 | 11 | 6½ |
| no co. 31 | ...... | | 13 | 8 | 0½ | 24 | 7 | 0 |
| no co. Aug. 7 | ...... | | 20 | 3 | 10 | March 3 | 8 | 7½ |
| dimi. no co. 14 | 1 | 6½ | 27 | 3 | 6 | 10 | 5 | 6½ |
| 21 | 4 | 0 | June 3 | 6 | 2 | 17 | 7 | 5½ |
| 28 | 3 | 0 | 10 | 5 | 1½ | 24 | 6 | 0½ |
| Sept. 4 | 4 | 2½ | 17 | 5 | 1 | 31 | 4 | 11 |
| 11 | 5 | 7½ | 24 | 3 | 7 | Apr. 7 | 5 | 6½ |
| no co. 18 | 1 | 4 | July 1 | 4 | 6 | 14 | 4 | 6 |
| 25 | 3 | 5½ | 8 | 6 | 6 | 21 | 8 | 0 |
| Oct. 2 | 4 | 5 | 15 | 4 | 8½ | 28 | 7 | 1½ |
| 9 | 3 | 3½ | 22 | 5 | 1½ | May 5 | 8 | 3 |
| 16 | 4 | 8½ | 29 | 4 | 6 | 12 | 4 | 8½ |
| 23 | 4 | 9½ | Aug. 5 | 5 | 0 | 19 | 5 | 0 |
| 30 | 4 | 4½ | 12 | 5 | 0½ | 26 | 7 | 2½ |
| Nov. 6 | 3 | 9½ | 19 | 4 | 4½ | no co. dimi. } June 2 | 7 | 0 |
| 13 | 3 | 0½ | 26 | 5 | 0 | | | |
| 20 | 2 | 7½ | Sept. 2 | 3 | 9½ | no co. dimi. 9 | 2 | 4 |
| 27 | 3 | 10½ | 9 | 5 | 8½ | 16 | 5 | 9½ |
| Dec. 4 | 6 | 7 | 16 | 4 | 0 | 23 | 5 | 5½ |
| 11 | 7 | 2½ | 23 | 4 | 8½ | no co. dimi. 30 | 3 | 3½ |
| 18 | 5 | 0 | 30 | 5 | 6 | no co. dimi. } July 7 | 3 | 7½ |
| 25 | 3 | 9 | Oct. 7 | 3 | 10 | | | |
| 1692 Jan. 1 | 3 | 9 | 14 | 6 | 2½ | 14 | 5 | 3 |
| no co. 8 | 1 | 0½ | 21 | 7 | 0 | 21 | 7 | 0 |
| no co. 15 | ...... | | 28 | 5 | 8½ | 28 | 5 | 0½ |
| 22 | ...... | | Nov. 4 | 5 | 6 | Aug. 4 | 11 | 2 |
| 29 | 4 | 0 | 11 | 5 | 0 | 11 | 4 | 5½ |

| Week ending | s. | d. |
|---|---|---|
| 1693 Aug. 18 | 5 | 3½ |
| 25 | 4 | 10 |
| Sept. 1 | 6 | 0 |
| 8 | 6 | 3 |
| no co. dimi. 15 | 4 | 0 |
| no co. M. 22 | 0 | 6½ |
| no co. dimi. 29 | 2 | 11½ |
| Oct. 6 | 5 | 0 |
| 13 | 5 | 4 |
| 20 | 5 | 3½ |
| 27 | 4 | 5½ |
| Nov. 3 | 4 | 3½ |
| 10 | 5 | 11½ |
| 17 | 8 | 1½ |
| 24 | 6 | 0 |
| Dec. 1 | 7 | 4½ |
| 8 | 3 | 7 |
| 15 | 5 | 7½ |
| 22 | 6 | 2 |
| 29 | 9 | 0 |
| 1694 Jan. 5 | 5 | 0 |
| 12 | 5 | 5½ |
| 19 | 5 | 6 |

| Week ending | s. | d. |
|---|---|---|
| 1694 Jan. 26 | 5 | 7½ |
| Feb. 2 | 6 | 0 |
| 9 | 6 | 2 |
| * * * * * | | |

1698 [In this and following years we have copied only those dates where a charge is put opposite his name. The sums consist principally of quarterly payments.]

| Week ending | s. | d. |
|---|---|---|
| no co. June 3 | 1 | 6 |
| no co. July 22 | ...... | |
| no co. dimi. 29 | 3 | 6 |
| no co. Aug. 5 | ...... | |
| no co. Sept. 9 | 1 | 6 |
| no co. Dec. 2 | 1 | 6 |
| 1699 | | |
| no co. Feb. 24 | 1 | 6 |
| no co. June 3 | 1 | 6 |
| no co. } June 23 | 6 | 8 |
| C. P. } | | |

| Week ending | s. | d. |
|---|---|---|
| 1699 | | |
| no co. Sept. 8 | 1 | 6 |
| no co. Dec. 1 | 1 | 6 |
| 1700 | | |
| no co. Mar. 1 | 1 | 6 |
| no co. May 31 | 1 | 6 |
| no co. Sept. 6 } | 1 | 6 |
| C. P. } | 6 | 8 |
| no co. Nov. 29 | 1 | 6 |
| 1701 | | |
| no co. Mar. 1 | 1 | 6 |
| no co. May 30 | 1 | 6 |
| no co. Sept. 12 | 1 | 6 |
| Nov. 21 | 6 | 4 |
| Nov. 28 | £1 0 | 10 |
| no co. Dec. 5 } | 1 | 6 |
| Lett. } | 0 | 5½ |
| no co. C. P. 26 | 6 | 8 |
| 1702 | | |
| no co. Mar. 6 | 1 | 6 |

May 8, name disappears from list of fellows.

# TABLE OF NEWTON'S LECTURES

## AS LUCASIAN PROFESSOR.

---

### NEWTON'S LECTURES ON OPTICS

#### (*MS. Univ. Libr. Dd. 9. 67.*)

[The numbers on the right designate the pages in the MS., those on tho
left the pages in the work as printed Lond. 1729.]

Jan. $16^{69}_{70}$.

| | | | |
|---|---|---|---|
| 1— 13 | Opticæ pars 1ª—varia. | Lect. | 1 ( 1— 6) |
| 13— 25 | Ex eodem—exigit. | Lect. | 2 ( 6—12) |
| 26— 34 | Jam liquet—determinentur. | Lect. | 3 (12—17) |
| 35— 41 | Sectio 2ᵈᵃ—reflexos. | Lect. | 4 (17—21) |
| 42— 52 | Cum eandem—attolluntur. | Lect. | 5 (21—28) |
| 53— 62 | Problematis—de aliis. | Lect. | 6 (28—33) |
| 62— 73 | Ad eundem—videar. | Lect. | 7 (33—39) |
| 74— 85 | Sectio 3ᵈᵃ—proxime. | Lect. | 8 (39—44) |
| | | | Octob. 1670. |
| 85— 95 | Prop. 12—æquales. | Lect. | 9 (45—49) |
| 95—105 | Lemma 5—$\mu x \nu$. Q . E. D. | Lect. | 10 (49—54) |
| 105—116 | Prop. 17—sufficiant. | Lect. | 11 (54—60) |
| 116—125 | De radiorum—*GXH*. | Lect. | 12 (60—64) |
| 126—136 | Sectio 4ᵗᵃ—possunt. | Lect. | 13 (64—69) |
| 137—146 | Prop. 32—definitur. | Lect. | 14 (69—73) |
| 146—152 | Prop. 36—censeam. | Lect. | 15 (74—77) |
| [145]-153 | Opticæ pars 2ᵈᵃ—disceptaturus. | Lect. | 1 ( 1— 5) |
| 153—164 | Prop. 1—nequeant. | Lect. | 2 ( 5—11) |
| 164—171 | Prop. 2—censeam. | Lect. | 3 (11—17) |
| | | | Octob. 1671. |
| 171—181 | Prop. 3—commisceantur sibi. | Lect. | 4 (17—23) |
| 182—189 | Adhæc—judicaveris | Lect. | 5 (23—29) |
| 189—197 | Verum—manifestum est. | Lect. | 6 (29—34) |
| 197—207 | Quinetiam—Prisma. | Lect. | 7 (34—41) |
| 207—215 | Ad hæc—cogantur. | Lect. | 8 (41—46) |
| 215—226 | Prop. 5—subjicient. | Lect. | 9 (46—54) |
| 227—239 | Sect. 2ᵈᵃ—emergentis. | Lect. | 10 (54—63) |
| 239—247 | Antequam—liceat. | Lect. | 11 (63—68) |
| 248—260 | 2 De Phænomenis—possint. | Lect. | 12 (69—77) |
| 261—269 | 3 De Phænomenis—dicere. | Lect. | 13 (78—84) |

Octob. 1672.

269—277  4 De Phænomenis—patebunt.          Lect. 14 (84—90)
278—285  Notissimum—inferioris.               Lect. 15 (90—96)
285—291  Superest—decrevi.                    Lect. 16 (96—101)

The MS. does not seem to be in Newton's hand, except some corrections here and there, almost all the marginal notes, the diagrams and
between 2 and 3 pages at the end. It was put into the hands of the
Vice-Chancellor and delivered by him to Robert Peachy to be placed
in the University Library, Octob. 21, 1674.

## LECTURES ON ARITHMETIC AND ALGEBRA.

### (MS. Univ. Libr. Dd. 9. 68.)

[The numbers on the left refer to the pages in the edition published
by Whiston, Cantab. 1707.]

Octob. 1673.

1—9  Computatio vel fit—in eadem ratione.      Lect. 1 ( 1— 5)

11—15  De Additione—·—$20a^3\sqrt{aa-xx}$.       Lect. 2 ( 5— 8)

15—17  De Subductione— $-\sqrt{3}+\dfrac{3}{5}$.      Lect. 3 ( 8—10)

18—21  De Multiplicatione— $-\dfrac{aab}{c}$.       Lect. 4 (10—12)

22—25  De Divisione—homogeneas.                Lect. 5 (13—15)
25—30  Quod si quantitas—sufficit.              Lect. 6 (15—18)
31—34  De extractione Radicum—279.              Lect. 7 (18—21)

Octob. 1674.

34—37  Extractionem radicis—observandum est.    Lect. 1 (21—22)
37—40  E simplicibus—radicibus.                Lect. 2 (22—24)
41, 42, 51, 52*  De Reductione— $-9bc$.           Lect. 3 (25—27)

53—55  Quod si divisor— $\dfrac{61}{21}$.           Lect. 4 (27—28)

55—57  De reductione Radicalium—et sic in aliis.   Lect. 5 (28—29)
62—66  De forma Æquationis— $= x^4$.             Lect. 6 (30—32)
66—68  Reg. 4—docere.                           Lect. 7 (32—33)
69—72  De duabus—linquo.                        Lect. 8 (33—35)
72—74  Exterminatio— $\times df = 0$.             Lect. 9 (35—37)
74—76  Reg. 3—asymmetria.                       Lect. 10 (37—38)

* The part De Inventione Divisorum—totam quantitatem, pp. 42—51, is taken from
the end of the MS.

---

* The matter in Lectures 1—4 is given in a modified form at the end of the MS., with a direction that it should be inserted at an earlier part of the volume. For these four lectures, therefore, it is impossible to give exact references to the pages of the printed book.

At the end of the Volume are corrections and additions by Newton, and "De Inventione divisorum—nihil relinquit" (pp. 42—51 of printed book.)

---

## LECTURES DE MOTU CORPORUM
### (MS. Univ. Libr. Dd. 9. 46.)

[The numbers on the left denote the pages in the 1st ed. of the *Principia*: those on the right the leaves in the MS.]

The title is " De motu corporum Liber primus." It forms the draught of the 1st book of the *Principia*, see p. 209, note.

Octob. 1684.

---

* In the corrections at the end of the MS. part of this Lecture is ordered to be transferred to an earlier place in the Volume, and accordingly it appears in pp. 58—61 of the printed book.

† The MS. in Lambeth Library, No. 592, (quoted by Rigaud, *Essay*, p. 97, note) entitled "Trigonometriæ Fundamenta a Viro Cl. Isaaco Newton, Matheseos Professore, anno 1683 data," contains merely rules for the solution of plane and spherical triangles given to Henry Wharton probably at one of those private lessons mentioned in p. xlv. It consists of two folio leaves (*i.e.* of two pages and seven lines on the last page, the second being blank), forming part of a volume entitled " Scripta Academica &c. annos inter 1682 et 1686, a me facta " &c. in Wharton's handwriting.

---

* "Demonstrationes hujus et præcedentis ut nimis obvias non adjungo." In the
*Principia* the demonstrations of these two propositions (17th and 18th MS., 18th and
19th *Princip.*) are given complete.

† This is Lemma XVI. of the *Princip.* p. 67. The Prop. which follows it in the MS.
is Prop. XIX. Prob. XI., being Prop. XXI. Prob. XIII. of the *Princip.* pp. 68, 69. The
reference to Lahire is not in the MS. having probably been suggested by Halley.
(Newton to Halley, Octob. 18, 1686. Rigaud's Appendix, p. 47).

Prop. XX. Prob. XII. in the MS. is Prop. XXX. Prob. XXII. in *Princip.* p. 104. The
difference in the numbers of the propositions arises from the circumstance of the 5th
section which contains eight propositions having been afterwards inserted.

After Prop. XX. comes a scholium containing the approximate solution of the same
problem for the ellipse and hyperbola. Then follows the clause "Hactenus...exponere"
as in *Princip.* p. 114.

137—144    Prop. xlv. Prob. xxxi....subinde determinamus.

Lect. 8 (90—95)

145—152    Artic. x. De Motibus Corporum...semper peragent. q. e. d.

Lect. 9 (95—99)

153—    Prop. lii. Prob. xxxiv.. ..                Lect. 10 (99—  )
breaks off in Prop. liv. with the words "quâvis altitudine CT per."
*Princip.* p. 159.

The MS., it will be seen, is imperfect, ending abruptly at the second page of fol. 102. Foll. 37—44 are repeated, one set being the first draught, the other as printed in the *Principia*, pp. 57—73. The nature of the former will be understood from the following outline. After Corol. 6. In Parabolâ, &c., and the other corollaries comes

Prop. xvi. Prob. viii. being Prob. xvii. Prob. ix. of *Princip.*
then, ...... xvii. ...... ix. ............. xviii. ....... x. ....... .........
without demonstration.
then Prop. xviii. ...... x. ............. xix. ....... xi. .............
without demonstration.
then Lem. xv. ...................,... Lem. xvi. of *Princip.*
.. ..... Prop. xix. ......................Prop. xxi. .............
............. xx. ............................. xxx. ...(see note † p. xcvi).
............. xxi. ........................... xxxii. .............
............. xxii. ............................. xxxiii. ............,....
.... ........... xxiii. ............................. xxxiv. ...... ........
............... xxiv. as far as "arcum Kk"... xxxv. .............

The latter set and foll. 55—58 as far as " absurdum est. q. e. d."
(*Princip.* p. 79) are not divided into Lectures. Fol. 45 is numbered 55 apparently by a clerical error, which is propagated through the remainder of the MS.

In binding the volume the sheets seem to have been taken at random. When the *disjecta membra* are brought together they form a whole, as follows:

1—  57    De motu corporum...ad tangentem.    ( 1—36)
57—  73    Corol. 6....in rectam quâ quævis    (37—44)
(The other 37—44 in the MS. is the rough draught of this.)
73—  88    quævis ex punctis...duæ evadent    (55—62)
88—118    parallelæ...arcum Kk                (63—78)
118—133    describere...moveri possunt        (79—86)
133—144    est in triplicata...usurpamus plana (87—94)
144—159    his parallela...altitudine *CT* per (95—102)

## LECTURES ON THE SYSTEM OF THE WORLD.

*From a Copy in Cotes's hand in Trin. Coll. Library,* (*R.* 16. 39).

### DE MOTU CORPORUM LIBER*.

| | | | |
|---|---|---|---|
| 1— 8† | Fixas in supremis—Astronomi. | Prælect. 1. | Sept. 29. 1687. |
| 8—16 | Martem—duplicatem. | Lect. 2. | |
| 16—22 | Stabilita—fuligine. | Lect. 3. | |
| 22—27 | Analogiæ—modum. | Lect. 4. | |
| 27—33 | Designet—intelligetur. | Lect. 5. | |

Here Cotes's copy ends. The remainder of the treatise, however, (not divided into Lectures) is bound up in the same volume, and was probably obtained by Professor Smith from Charles Morgan of Clare Hall, for in the Library of that College there is a MS. copy of the treatise which belonged to Morgan, who states in a note that the first 5 Lectures were communicated to him by Smith, and the remainder by Martin Folkes.

---

* This is the title in the MS., not "De Mundi Systemate" as in the printed book (Lond. 1731). This tract, drawn up "methodo populari ut a pluribus legeretur" was intended to form the 3d book of the *Principia*, but readers who have not mastered the principles, says the author, "vim consequentiarum minimè percipient, neque præjudicia deponent quibus à multis retro annis insueverunt," and therefore "ne res in disputationes trahatur, summam libri illius transtuli in Propositiones, more Mathematico, ut ab iis solis legantur qui principia prius evolverint." (Introduction to 3d book of *Princip.*)

† The numbers refer to the pages in the printed book.

# CORRESPONDENCE

OF

# SIR ISAAC NEWTON

AND

# PROFESSOR COTES,

*&c.*

# CORRESPONDENCE &c.

## LETTER I.

Dear Sir,

I waited to day on S$^r$ Isaac Newton, who will be glad to see you in town here, and then put into your hands one part of his Book corrected for y$^e$ press. I shall get of him a Character of M$^r$ Hussey; but we both apprehend y$^t$ Interest rather than Merit will prevail in y$^e$ Election, & y$^t$ one Coleson has y$^e$ best friends. D$^r$ Ayloff I suppose has given you a Bill * of 100$^{lb}$ payable here in London at 14$^{days}$ sight; I must desire you to transfer y$^r$ Bill to M$^r$ Smallwell in part of payment; for y$^e$ former bill I gave him upon y$^e$ Marquiss of Dorchester's Steward will not be p$^d$ yet. So y$^t$ if you send the Bill by Mascal y$^e$ Carrier to have it accepted, & from thence to bring it to me, I will take Smallwells receipt for so much money. Pray let me know, when you think of coming up hither.

I am,

Your affectionate friend & Serv$^t$

Cotton House. {May 21. 1709.}　　　　Ri : Bentley.

*For* Mr Cotes *Fellow of Trinity*
*College in Cambridg.*

The post mark of this letter, though at first sight scarcely legible, may I think be pronounced to be May 21, and the year is pretty clearly 1709. About the middle of July Cotes is in London (in his letter of Feb. 15, 1711, to Jones, he mentions his having been last

---

* I can discover no traces of this bill in any of the College Account Books. It may possibly have come into the Chapel Account, for which Cotes, as superintendent of the repairs of the Chapel, kept a separate book of receipts and disbursements. Whether this book is still in existence I am unable to say.

in town "about a year and a half ago") drawn thither, no doubt, by this note of Bentley's, and expecting to take with him down to Cambridge the first instalment of Newton's corrected copy of the Principia. Newton however is still reluctant to part with it, having probably some further improvements to make, but promises to send the copy down in about a fortnight. So Cotes returns to Cambridge without the "one part of the book corrected for the press," which Bentley's letter had informed him was ready to be put into his hands some eight weeks before. The copy does not arrive in that fortnight, nor in the next. The long vacation being nearly half over and no signs of the promised copy appearing, the young editor becomes impatient. Hence his letter of Aug. 18, which however produced no apparent effect, until his next-door neighbour Whiston, one evening probably in September, newly arrived from London, (he is known to have been in Cambridge on the 29th of that month) put into his hands "the greatest part of the copy of the Principia," ending at Prop. xxxiii Cor. 2 Lib. ii p. 320. That is followed some time afterwards by Newton's letter of Oct. 11, which apparently did not come through the post, being brought perhaps by some member of the University coming up on the beginning of term. Whiston, whose autobiography records so many other things certainly of not greater importance, makes no mention of his being employed as a messenger on this occasion: so absorbed was he in his Arian heresy and Apostolical Constitutions, with regard to which he tells us "his best friends began to be greatly affrighted this summer at what they had heard he was going about." It is not likely that he found his old patron wanting in the duties of friendship at this critical period of his life, and it is not impossible that Sir Isaac, in delivering to him a portion of a work containing so much close and profound reasoning, may have dropt a word of caution into his ear.

The "election" referred to in this letter is probably that of a Head Master of Sir Joseph Williamson's Free Mathematical School at Rochester, the electors to which post are some 17 in number, consisting of the Mayor, Recorder, eldest Resident Prebendary, &c. The Rev. John Colson was the first Head Master of this school, and was appointed June 1, 1709. He resigned the place March 1, $17\frac{39}{40}$, on being elected Lucasian Professor. He was entered at Emmanuel April 23, 1728, and was one of the 71 persons in the King's list (William Warburton was another of the number) on whom the degree of M.A. was conferred at George II's visit to the University, April 25, 1728. On coming to reside as Lucasian Professor at Cambridge, he was appointed Taylor Lecturer at Sidney College, where he was admitted "in convictum sociorum" 11 March $17\frac{39}{40}$, ætat. 60.

It was for the purpose of boarding with this same Colson, and being instructed by him "in Mathematics and Philosophy and humane learning," that Garrick set out from Lichfield on the morning of March 2, 173⅞ for London, accompanied by "one Mr Johnson," who was going "to try his fate with a tragedy, and to see to get himself employed in some translation, either from the Latin or French."

Christopher Hussey was a senior Bachelor of Arts of Trinity College, and was elected Fellow the following October. On Whiston's expulsion from the University (Oct. 30, 1710), he was appointed by him as his deputy in the Lucasian Chair, and "was ready to perform his duty, had not the heat of that time prevented him." Whiston's Memoirs I. 312. He was afterwards an unsuccessful candidate for the Professorship against "Blind" Saunderson (Nov. 20, 1711). See letter cviii, note.

A slightly different date is assigned to this letter in the Bentley Correspondence (p. 378), and a widely different one is mentioned as being suggested by Bishop Monk (ib. p. 787).

## LETTER II.

### COTES TO NEWTON.

S$^r$.                    Cambridge August 18$^{th}$, 1709.

The earnest desire I have to see a new Edition of Y$^r$ Princip. makes me somewhat impatient 'till we receive Y$^r$ Copy of it which You was pleased to promise me, about the middle of the last Month, You would send down in about a Fourtnights time. I hope You will pardon me for this uneasiness from which I cannot free my self & for giveing You this Trouble to let You know it. I have been so much obliged to You by Y$^r$ self & by Y$^r$ Book y$^t$ (I desire You to beleive me) I think my self bound in gratitude to take all the Care I possibly can that it shall be correct. Some days ago I was examining the 2$^d$ * Cor: of

---

* In this Corollary is determined the Attraction of a Spheroid on a point in its axis produced, the attractive force of each particle varying inversely as the square of the distance. A paper by Cotes containing the investigation is still preserved in the volume from which these letters are taken Nos. 24 and 25.

Prop 91 Lib ɪ and found it to be true by $y^e$ Quadratures of $y^e$ 1ˢᵗ & 2ᵈ Curves of $y^e$ 8ᵗʰ Form of $y^e$ second Table in Yʳ Treatise *De Quadrat.* At the same time I went over $y^e$ whole Seventh & Eighth Forms which agreed with my Computation excepting $y^e$ First of $y^e$ Seventh & Fourth of $y^e$ Eighth which were as follows

Form : 7. 1.
$$\frac{4de\,\dfrac{\Upsilon^3}{\xi} - 2df\,\dfrac{v^3}{x} - 8dee\sigma + 4dfgs}{4\eta eg - \eta ff} = t.$$

Form : 8. 4.
$$\frac{\begin{matrix} +36defg \\ -15df^3 \end{matrix}_s + \begin{matrix} +8degg \\ -2dffg \end{matrix}_{xxv} \begin{matrix} -28defg \\ +10df^3 \end{matrix}_{xv} \begin{matrix} -16deeg \\ +10deff \end{matrix}_v}{24\eta eg^3 - 6\eta ffgg} = t.$$

I take this Oportunity to return You my most hearty thanks for Yʳ many Favours & Civilitys to me who am

Yʳ most Obliged humble Servant

*For* Sʳ Isaac Newton *at his House in Jermin Street near St James's Church Westminster.*          Roger Cotes.

---

LETTER III.

NEWTON TO COTES.

Sʳ

I sent you by Mʳ Whiston the greatest part of $y^e$ copy of my Principia in order to a new edition. I then forgot to correct an error in the first sheet pag 3 lin 20, 21, & to write *plusquam duplo* for *quasi quadruplo* & *plusquam de- cuplo* for *quasi centuplo* *.

---

* These two corrections are not adopted literally in the 2d edition, the "quasi" for which Newton here substitutes "plusquam" being still retained in it. Perhaps Cotes had already altered the "quadruplo" and "centuplo" before receiving this letter, as so obvious an error could scarcely have escaped his attention. In the passage referred

I forgot also to add the following Note to the end of Corol. 1 pag. 55 lin 6. Nam datis umbilico et puncto contactus & positione tangentis, describi potest Sectio conica quæ curvaturam datam ad punctum illud habebit. Datur autem curvatura ex data vi centripeta: et Orbes duo se mutuo tangentes eadem vi describi non possunt.

I thank you for your Letter & the corrections of y$^e$ two Theorems in y$^e$ treatise de Quadratura. I would not have you be at the trouble of examining all the Demonstrations in the Principia. Its impossible to print the book w$^{th}$out some faults & if you print by the copy sent you, correcting only such faults as occurr in reading over the sheets to correct them as they are printed off, you will have labour more then it's fit to give you.

M$^r$. Livebody is a composer (I mean M$^r$ Livebody who made the wooden cutts) & he thinks that he can sett the cutts better for printing off then other composers can, and offers to come down to Cambridge & assist in composing if it be thought fit. When you have printed off one or two sheets, if you please to send me a copy of them I will send you a further supply of wooden cutts.

I am

Yo$^r$ most humble & faithful servant

London. Octob. 11. 1709.                                    Is. NEWTON.

For M$^r$ COTES *Professor of Astronomy*
*in the University of Cambridge at*
*his Chamber in Trinity College.*

Shortly after the date of the above letter, Newton changed his residence from Jermyn Street to Chelsea. Flamsteed, writing to Ab.

---

to, Newton, speaking of a ball shot horizontally with a given velocity from the top of a mountain to a distance of two miles before it reaches the ground, says, (as the words stand in the 2nd and 3rd editions) " dupla cum velocitate quasi duplo longius pergeret, et decupla cum velocitate quasi decuplo longius." When he wrote "quadruplo" and " centuplo," he was probably thinking of oblique projection. The passage in question occurs in some additional remarks in illustration of Def. v., which were not given in the 1st edition. The MS. of them, unfortunately, does not appear in the Newtonian Volume.

Sharp Oct. 25, (Baily, p. 272) says : " He |Sir Isaac| is now re-
moving to Chelsea, and has been lately much talked of; but not much
to his advantage.   Our Society |the Royal Society| is ruined by his
close, politic, and cunning forecast; I fear past retrieving, for our
Doctor's |Sloane| Transactions have been twice burlesqued publicly;
and now we have had none published I think this four months." This
burst of spleen would seem to be in anticipation of the resolution passed
by the Council of the Society Nov. 9, ordering Flamsteed's name "to
be left out of the list of the Society for next year for not having complied with the order of Council made 12 Jan. 170⅜" relative to the
payment of arrears.

_____

Here there is a break of 6 months in the correspondence until we
come to Cotes's letter of April 15, 1710, by which time nearly half
of the whole work was printed off, the part then finished ending at
p. 224, (2nd ed.) in the middle of the Lemma (II Lib. 2) in which
the principles of fluxions are explained.

A note by Mr Howkins states that there is wanting a letter of
Cotes to Newton, dated Apr. 9, 1710, " de Cor. 1 and 6 Prop. IX.
Lib. 2." No. 33 contains a draught in Cotes's writing of these two
Corollaries, and two additional steps in the proof of the Proposition,
but not (with the exception of the latter of the two steps) as they
stand in the second edition. On the same paper Cotes has also written
" dele Cor. 4 and 5, Prop. VIII." which are accordingly omitted in the
2nd ed. It is probable, therefore, that if this missing letter of Apr. 9
referred to the Corollaries mentioned by Mr. Howkins, the proposed
omission of Cor. 4 and 5, Prop. VIII. and the introduction of the two
steps into the reasoning of Prop. IX. in order to avoid a reference to the
latter of the cancelled corollaries would also form a part of its contents.

But besides this letter of Apr. 9 and Newton's answer to it, there
is good reason for supposing that at least one other pair of letters
passed between them during the interval from October to April. For
(1) it seems probable that Cotes would return some answer to Newton's
letter of Oct. 11, in explanation of his not adopting the precise language
of the emendations contained in it ; and at all events he would attend
to Newton's request to have one or two sheets sent to him, to say
nothing of the presumption that he would feel himself called upon to
take some notice of Mr Livebody's offer of his services. (2) The 2nd
method of finding the force to the centre of an ellipse given in p. 46
2nd Ed. is so much altered (in the opening part of it) from the form
in which it stands in Newton's MS. (No. 9), that Cotes would scarcely
have changed it without some communication from Newton on the
subject. At the head of this 2nd method Cotes has written " vid. fol.

sequ." but the leaf referred to is not to be found. (3) From a mark in No. 11 it appears that the first word in p. 49 in the proof sheet (H) was "corporis" which is now in the eighth and ninth lines lower down; so that some additional matter must have been introduced in a preceding page after the proof of H was printed*, and this almost necessarily implies the receipt of instructions from Newton to that effect, (perhaps at the end of November or beginning of December, if we may judge from the rate at which the press was working).

The loss of any letters in this interval is the more to be regretted, because if ever the celebrated Scholium to Lemma II. Book 2 was touched upon in the correspondence between Newton and his Editor, the place for doing so would lie within this period. The missing letter of April 9, as has been said, may have contained remarks connected with Prop. VIII. which immediately follows that Scholium. The only alteration in the Scholium† made in the 2nd Edition, consists in the addition of the words "et Idea generationis quantitatum" after "notarum formulis." The "annis abhinc decem" referring to his second letter through Oldenburg to Leibniz, in Oct. 1676, is still retained, though 26 years intervened between the publication of the 1st and 2nd editions.

In this interval, it may be remarked, the quarrel between Bentley and the Seniors broke out, and we read of Cotes being present at two conferences at the lodge between the conflicting parties, as a friend of the Master's. (See Monk's Bentley, p. 187.) On Jan. 18, 1710, Bentley cut Miller's name out of the boards. On Feb. 10, Miller presented the petition, signed by thirty of the fellows, to the Bishop of Ely. Great, however, as was the delay which retarded the second edition of the Principia in its passage through the press, Cotes had

---

* A comparison of Newton's MS. with sheet G of the 2nd Ed. shews that the addition must have been made in some sheet preceding that, but it is impossible to fix the exact place, as the part of the MS. which is preserved only begins with Prop. VI. Theor. v.

† This Scholium was completely remodelled in the 3rd Ed. and Leibniz's name suppressed. The reader of these pages will smile at the following piece of information with which Montucla favours us (III. 108): "On se demandera peut être pourquoi cette suppression ne fut pas faite lors de l'edition des *principes* de 1713, puisque alors la querelle étoit encore dans toute sa chaleur; en voici la raison, qui est une anecdote assez peu connue et que je tiens de la même main que ce que j' ai dit ci-dessus {the 'bonne main' that had informed him that the notes on the *Commercium Epistolicum* were written by Newton}. C'est que cette édition fut faite à Cambridge, loin de Neuton et presque en cachette, par les soins de Cotes et de Bentley, et que Neuton en fut très-mécontent. C'est, en effet, un procédé assez étrange de la part de ces deux hommes, d'ailleurs célèbres, que d'imprimer un ouvrage du vivant de son auteur sans prendre, pour ainsi dire, son attache sur les changemens ou additions à y faire."

brought his labours upon it to a conclusion nearly a year before Bentley's trial came on.

It may assist us still further to fill up this gap of six months, and to imagine the direction which the thoughts and conversation even of men engaged on a new edition of the Principia would occasionally take, if we remember that it was during this same period that the kingdom was plunged into the Sacheverell excitement, (the 2nd of the two obnoxious sermons was preached on Nov. 5, 1709, the trial began Febr. 27, 1710, and on March 21 the Doctor was suspended from preaching for three years); and that Marlborough, yielding to the solicitations of Godolphin, whose ministerial difficulties called for the support and authority of the Great Captain's presence, arrived from the Hague on Nov. 8, and, after experiencing in several mortifying instances the effects of Masham influence, against which even Malplaquet's recent laurels were powerless, was sent back to Holland towards the end of February, and that, on the failure of the negotiations with which Louis had been amusing the allies at Gertruydenberg, he and Eugene (*duo fulmina belli*) opened their magnificently planned campaign of 1710, by passing the French lines on the morning of Monday April 10, and proceeding to the investment of Douay.

---

## LETTER IV.

### COTES TO NEWTON.

S<sup>r</sup>.                                        |Saturday| Apr. 15. 1710.

We have printed so much of y<sup>e</sup> Copy You sent us y<sup>t</sup> I must now beg of You to think of finishing the remaining part assoon as You can with convenience. The last sheet y<sup>t</sup> is printed off ends at y<sup>e</sup> 251<sup>st</sup> page of y<sup>e</sup> old Edition & y<sup>e</sup> 224<sup>th</sup> page of y<sup>e</sup> new Edition. The whole y<sup>t</sup> is finished shall be sent You by the first oportunity. I have ventured to make some little alterations my self whilst I was correcting the Press such as I thought either Elegancy or Perspicuity or Truth sometimes required. I hope I shall have Y<sup>r</sup> pardon if I be found to have trusted perhaps too much to my own Judgment, it not being possible for me without great inconvenience to y<sup>e</sup> work & uneasiness to

$Y^r$ self to have $Y^r$ approbation in every particular. The Pages which are next to be printed being somewhat more $y^n$ usually intricate I have been looking over them before hand. Page 270* Reg. 1 I think should begin thus——Si servetur tum Medii densitas in $A$ tum velocitas quacum corpus projicitur & mutetur——.I must confess I cannot be certain $y^t$ I understand the design of Reg. 4 & $y^e$ last part of Reg. 7 and therefore dare not venture to make any alteration without acquainting You with it. I take it thus, $y^t$ in $y^e$ 4$^{th}$ Rule You are shewing how to find a mean among all $y^e$ Densitys through which $y^e$ Projectile passes, not an Arithmeticall mean between $y^e$ two extream Densitys $y^e$ greatest and least, but a mean of all $y^e$ Densitys considered together, which will be somewhat greater than $y^t$ Arithmeticall mean, $y^e$ number of Densitys which are greater $y^n$ it being greater $y^n$ $y^e$ Number of Densitys which are lesser $y^n$ $y^e$ same. If this be $Y^r$ design I would alter the 4$^{th}$ Rule thus, with $Y^r$ consent. Quoniam Densitas Medii prope verticem Hyperbolæ major est quam in loco $A$, ut habeatur Densitas mediocris debet ratio minimæ tangentium $GT$ ad tangentem $AH$ inveniri, & Densitas in $A$ augeri in ratione paulo majore quam semisummæ harum tangentium ad minimam tangentium $GT$. The latter part of $y^e$ 7$^{th}$ Rule I understand thus. Simili methodo ex assumptis pluribus longitudinibus $AH$ invenienda sunt plura puncta $N$ & per omnia agenda Curva linea regularis $NNXN$ secans rectam $SMMM$ in $X$. Assumatur demum $AH$ æqualis abscissæ $SX$ & inde denuo inveniatur longitudo $AK$; & longitudines quæ sint ad assumptam longitudinem $AI$ & hanc ultimam $AH$ ut longitudo $AK$ per experimentum cognita ad ultimo inventam longitudinem $AK$ erunt veræ illæ longitudines $AI$ & $AH$ quas invenire oportuit. Hisce vero datis, dabitur & resistentia Medii in loco $A$ quippe quæ sit ad vim gravitatis ut $AH$ ad $2AI$, augenda est

---

* pp. 270—274, (Schol. to Prop. x. Lib. 2.) contain Rules for the approximate determination of the motion of a projectile in the air.

autem densitas Medii per Reg. quartam & resistentia modo inventa, in eadem ratione aucta fiet accuratior. About y$^e$ end of the 8$^{th}$ Rule are these words—quorum minor eligendus est—which I would either leave out or print thus—quorum minor potius eligendus est. Page 274. l: 2 should be $\dfrac{2\,TGq}{nn - n \times GV}$; there are some others like this which I will not trouble You with. Prop. xiv Prob iv should be Prop xiv Theor xi. Two lines lower are these words—est ut summa vel differentia areæ per quam—I would leave {out} *summa vel.* Corol. page 281 I would print thus. Igitur si longitudo aliqua $V$ sumatur in ea ratione ad duplum longitudinis $M$, quæ oritur applicando aream $DET$ ad $BD$, quam habet linea $DA$ ad lineam $DE$; spatium quod corpus ascensu vel descensu toto in Medio resistente describit, erit ad spatium quod in Medio non resistente eodem tempore describere posset, ut arearum illarum differentia ad $\dfrac{BD \times VV}{4\,AB}$, ideoq: ex dato tempore datur. Nam spatium in Medio non resistente est in duplicata ratione temporis, sive ut $VV$, & ob datas $BD$ & $AB$, ut $\dfrac{BD \times VV}{4\,AB}$. Momentum hujus areæ, sive huic æqualis $\dfrac{DAq \times BD \times M^2}{DEq \times AB}$, est ad momentum differentiæ arearum $DET$ & $AbNK$ ut $\dfrac{DAq \times BD \times 2M \times m}{DEq \times AB}$ ad $\dfrac{AP \times BD \times m}{AB}$, hoc est, ut $\dfrac{DAq \times BD \times M}{DEq}$ ad $\frac{1}{2}$ $BD \times AP$ sive ut $\dfrac{DAq}{DEq}$ in $DET$ ad $DAP$, adeoq: ubi areæ $DET$ & $DAP$ quam minimæ sunt in ratione æqualitatis. Æqualis igitur— Page 286. l: 5* must be thus corrected

---

* Prop. xv. Lib. 2. On the motion of a body in a logarithmic spiral in a resisting medium, (force $\propto \dfrac{1}{(\text{dist.})^2}$, resist.$\propto(\text{vel.})^2$).

$Rr$ & $TQ$ seu ut $\dfrac{\frac{1}{2} VQ \times PQ}{SQ}$ & $\dfrac{\frac{1}{2} PQq}{SP}$ quas simul generant,

hoc est, ut $VQ$ & $PQ$ seu $OS$ & $OP$. This Corollary being thus corrected, the following must begin thus. Corol. 4. Corpus itaq: gyrari nequit in hac Spirali, nisi ubi vis resistentiæ minor est quam vis centripeta. Fiat resistentia æqualis vi centripetæ & Spiralis conveniet cum linea recta $PS$, inq: hac recta—&c. Tis evident by $y^e$ $1^{st}$ Corollary that $y^e$ descent along $y^e$ line $PS$ cannot be made $w^{th}$ an uniform velocity. Tis as evident I think $y^t$ it must be with an uniform velocity because $y^e$ resistance & force of gravity being equall, mutually destroy each other's effect and consequently no acceleration or retardation of motion can be produced. I cannot at present see how to account for this difficulty & I choose rather to own my ignorance to You $y^n$ to run $y^e$ hazard of leaving a blemish in a book I so much esteem*. Cor. 6. lin. ult. I would print thus—ut $\dfrac{OP}{OS}$,

id est, ut secans anguli ejusdem, vel etiam reciproce ut Medii densitas. If I mistake not $y^e$ design of $y^e$ $8^{th}$ Corollary, I would alter it thus—Centro $S$ intervallis continue proportionalibus $SA$, $SB$, $SC$, &c. describe circulos quotcunq: & statue tempus revolutionum omnium inter perimetros duorum quorumvis ex his circulis, in Medio de quo egimus, esse ad tempus revolutionum omnium inter eosdem in Medio proposito, ut Medii propositi densitas mediocris inter hos circulos ad Medii de quo egimus densitatem mediocrem inter eosdem quam proxime; sed & in eadem quoq: ratione esse secantem anguli quo Spiralis præfinita in Medio de quo egimus secat radium $AS$ ad secantem anguli quo Spiralis nova secat radium eundem in Medio proposito: Atq: etiam ut sunt eorundem angulorum tangentes ita esse numerum revolutionum inter circu-

---

* See the next and three following Letters.

los eosdem duos quam proxime. Si hæc passim— Prop.
16 must be altered for by my reckoning if $y^e$ centripetall
force be as $\dfrac{1}{SP^{n+1}}$ the Resistance will be as $\dfrac{1-\frac{1}{2}n,\ OS}{OP \times SP^{n+1}}$,
the Velocity as $\dfrac{1}{SP^{\frac{1}{2}n}}$, & therefore $y^e$ Density as $\dfrac{1-\frac{1}{2}n,\ OS}{OP \times SP}$.

With $Y^r$ consent I would add this Corollary. Si vis cen-
tripeta sit ut $\dfrac{1}{SP^{cub}}$, erit $1 - \frac{1}{2}n = 0$, adeoq : Resistentia &
Densitas Medii nulla erit ut in Prop ix Lib 1. Another
Corollary might be added to shew in what cases $y^e$ Resist-
ance is affirmative and in what cases negative. I beg of
You to pardon the freedom of this Letter.

$Y^r$ &c.

---

### LETTER V.

#### COTES TO NEWTON.

$S^r$                                                      Apr. 30$^{th}$ 1710

I suppose $M^r$ Crownfield our Printer has delivered to
You all $y^e$ Sheets that are already printed off. I desired
him to wait upon You before he return'd to Cambridge
$y^t$ I might have $Y^r$ answer to my former Letter or at least
to $y^e$ first part of it. The difficulty which I proposed to
You concerning $y^e$ 4$^{th}$ Corollary of Prop. xv I have since
removed. Upon examination of $y^t$ Proposition I think I
have observed another mistake in Cor. 3. which ballances
$y^t$ which I before mentioned * to You in $y^t$ Corollary. For
if I be not deceived $y^e$ force of Resistance is to $y^e$ Centripetall force as $\frac{1}{2}Rr$ to $TQ$ not as $Rr$ to $TQ$. You will
see my reasons in $y^e$ following alterations which I propose
to You. Page 284. 1: 6 Ponantur quæ in superiore Lem-

---

* viz. $TQ$ being erroneously put $= \dfrac{PQ^2}{SP}$ in the 1st ed. instead of $\frac{1}{2}\dfrac{PQ^2}{SP}$.

mate, & producatur $SQ$ ad $V$ ut sit $SV$ æqualis $SP$. Tempore quovis, in Medio resistente, describat corpus arcum quam minimum $PQ$, & tempore duplo arcum quam minimum $PR$; & decrementa horum arcuum ex resistentia oriunda, sive defectus ab arcubus qui in Medio non resistente iisdem temporibus describerentur, erunt ad invicem ut quadrata temporum in quibus generantur: est itaq: decrementum arcus $PQ$ pars quarta decrementi arcus $PR$. Postquam vero descriptus est arcus $PQ$ in Medio resistente, si areæ $PSQ$ æqualis capiatur area $QSr$, erit $Qr$ arcus quem tempore reliquo corpus describet absq: ulteriore resistentia, arcuumq: $QR$, $Qr$ differentia $Rr$ dupla erit decrementi arcus $PQ$; adeoq: vis resistentiæ & vis centripeta sunt ad invicem ut lineolæ $\frac{1}{2}Rr$ & $TQ$ quas simul generant. Quoniam vis centripeta, qua corpus urgetur in $P$ est—. Pag. 285. 1: 5 — $\frac{1}{2}VQ$ fit æqualitatis. Quoniam decrementum arcus $PQ$, ex resistentia oriundum, sive hujus duplum $Rr$ est ut resistentia & quadratum temporis conjunctim; erit Resistentia ut $\dfrac{Rr}{PQq \times SP}$. Erat autem $PQ$ ad—Page 286 1: 4.

Nam vires illæ sunt ut $\frac{1}{2}Rr$ & $TQ$ sive ut $\dfrac{\frac{1}{4}VQ \times PQ}{SQ}$ &

$\dfrac{\frac{1}{2}PQq}{SP}$, hoc est, ut $\frac{1}{2}VQ$ & $PQ$ seu $\frac{1}{2}OS$ & $OP$.—I satisfied my self more fully $y^t$ I am not mistaken in my reasoning after $y^s$ manner. If (as in Prop XVI) $y^e$ Centripetall force be as $\dfrac{1}{SP^{n+1}}$, the force of Resistance will be to $y^e$ Centripetall force as $\frac{1}{2}Rr$ to $TQ$ i é as $\overline{1 - \frac{1}{2}n}$, $OS$ to $OP$. Put $y^e$ Centripetall force as $\dfrac{1}{SP}$, & You will have $n = 0$, & consequently $\overline{1 - \frac{1}{2}n}$, $OS$ to $OP$ as $OS$ to $OP$. Therefore when $y^e$ Spiral coincides with $y^e$ line $PS$ $y^e$ Resistance will be equall to $y^e$ Centripetall force & $y^e$ Body will descend with an uniform Velocity as it ought to do, by Cor. 1 Prop xv, & Cor 5. Prop IV Lib I. compared together, and also upon $y^s$ consideration $y^t$ $y^e$ velocity in $y^e$ Spiral of Prop XVI is as $\frac{1}{SP^{\frac{1}{2}n}}$, i é, as $\frac{1}{SP^0}$. I have some things further to propose to You about $y^e$ remaining part of $Y^r$ copy, which I will not trouble You with till I have $Y^r$ answer to my former Letter

$Y^r$ &c.

## LETTER VI.

### NEWTON TO COTES.

$S^r$        Chelsea near London May 1<sup>st</sup> 1710.

I thank you for your letter with your remarks upon the papers now in the Press under your care. As soon as I could get some time to think on things of this kind, from $w^{ch}$ I have of late years disused myself, I examined them[*], & all your corrections may stand till you come at page 287. In page 286 lin 4 for $\frac{1}{2} OS$ read $OS$. In the same page let Corol. 4 stand thus. Corpus itaq : gyrari nequit in hac spirali nisi ubi vis resistentiæ minor est quam vis centripeta. Fiat resistentia æqualis vi centripetæ, et spiralis conveniet cum linea recta $PS$, et motus corporis cessabit. In page 287 & 288 the 8<sup>th</sup> Corollary may remain as in the Copy I sent you. In page 289 let the 16<sup>th</sup> Proposition end thus et resistentia in $P$ ut $\frac{Rr}{PQ^q \times SP^n}$, sive ut $\frac{1 - \frac{1}{2}n,\ VQ}{PQ \times SP^n \times SQ}$, adeoq : ut $\frac{1 - \frac{1}{2}n,\ OS}{OP \times SP^{n+1}}$, hoc est (ob datum $\frac{1 - \frac{1}{2}n,\ OS}{OP}$) reciproce ut $SP^{n+1}$. Et propterea densitas in $P$ est reciproce ut $SP^n$[†].

---

[*] Newton does not seem to have worked the problem out himself, but to have taken Cotes's results (in Letter IV.) for granted.

[†] The " $SP^n$ " is no doubt copied inadvertently from the 1st Ed. It should be $SP$.

Corol. 1. Si vis centripeta sit reciproce ut $SP^{cub}$, erit $1 - \frac{1}{2} n = 0$, adeoq: resistentia et densitas Medii nulla erit ut in Propositione nona Libri primi.

Corol. 2. Si vis centripeta sit reciproce ut radii $SP$ dignitas aliqua cujus index est major numero 3, resistentia affirmativa in negativam mutabitur.

When you sent me the sheets last printed off, I happened to be from home, but a{t} night found them left at my house, and thank you for them. I am going to finish the next part of the copy I am to send you, & I hope to have it ready in due time if some experiments* succeed. I thank you once more for your corrections & for your care of the edition.

I am

S$^r$ Your most humble & most obedient servant

Is. NEWTON.

After the writing of this Letter I received your second Letter dated Apr. 29. In the alterations you propose to be made in Prop xv you say. Postquam vero descriptus est arcus $PQ$ in Medio resistente, si areæ $PSQ$ æqualis capiatur area $QSr$, erit $Qr$ arcus quem tempore reliquo corpus describet absq: ulteriore resistentia. And this would be true if the velocity of the body at $Q$ were the same as when the arch $PQ$ is described in the same time in Medio non resistente. But the velocity at $Q$ being less in Medio resistente then in non resistente, the arch $Qr$ will be less in the same proportion & thereby reduce $Rr$ to half the bigness, & make the resistance to the centripetal force as $Rr$ to $TQ$. I hope therefore that what I have written on the other page of this Letter is right &

---

* Probably experiments with glass balls dropt from the dome of St Paul's with a view to test his theory of the resistance of fluids. See Letter XXV. *fin.* and note.

that yo$^r$ difficulty will be removed by the words & motus corporis cessabit.

I am Yo$^{rs}$

*May 2$^d$.                                I. N.

*For the* R$^{nd}$† M$^r$ ROGER COTES *Professor*
*of Mathematicks and Fellow of Trinity*
*College in Cambridge.*

---

## LETTER VII.

### COTES TO NEWTON.

S$^r$.                                              May 7. 1710

I received Y$^r$ Letter by y$^e$ last Post. I am not satisfied that Y$^r$ words [et motus corporis cessabit] will remove y$^e$ difficulty proposed. They cannot in my opinion be reconciled with Cor. 1. I acknowledge Y$^r$ objection to be just against those words of mine [erit Q$r$ arcus quem tempore reliquo corpus describet absq: ulteriore resistentia] I remember y$^t$ I inserted them into my Letter as I was hastily transcribing y$^t$ passage from another paper & was myself sensible of y$^e$ mistake soon after my Letter was gone from me. The alteration which I proposed, as it stood in y$^t$ Paper, was thus‡. [Ponantur quæ in superiore Lemmate et producatur *SQ* ad *V* ut sit *SV* æqualis *SP*. Tempore quovis in Medio resistente, describat corpus arcum quam minimum *PQ*, & tempore duplo arcum quam minimum *PR*; & decrementa horum arcuum ex resistentia oriunda, sive defectus ab arcubus qui in Medio non resistente iisdem temporibus describerentur, erunt ad invicem

---

* The post mark is May 4.

† Though addressed under this title by Newton here, and in the remainder of the correspondence, Cotes was not ordained until three years afterwards, (deacon, May 29, 1713, priest the following day).

‡ As may still be seen in the MS. of Letter V. (No. 41), the words " Unde etiam ......erit decrementum arcus *PQ* æquale dimidio lineolæ *Rr*," being crossed out and replaced by those which we have printed in p. 13, line 8, &c. " Postquam vero, &c."

ut quadrata temporum in quibus generantur. Est itaq:
decrementum arcus *PQ* pars quarta decrementi arcus *PR*.
Unde etiam si areæ *PSQ* æqualis capiatur area *QSr*, erit
decrementum arcus *PQ* æquale dimidio lineolæ *Rr*; adeoq:
vis resistentiæ & vis centripeta sunt ad invicem ut lineolæ
$\frac{1}{2}Rr$ & *TQ* quas simul generant.] I am yet of opinion y$^t$
this alteration is just & that the resistance is to y$^e$ centripetall force as $\frac{1}{2}Rr$ to *TQ*; Your own objection does I
think if You carefully consider it prove it to be so. To
avoid further misunderstanding I will put down my demonstration more at large thus ——————————————

P                    Q K              R  r  L

Tempore quovis in Medio resistente describat corpus arcum
quam minimum *PQ* & tempore duplo arcum quam minimum
*PR*; & decrementa horum arcuum ex resistentia oriunda
sive defectus [*QK, RL*] ab arcubus [*PK, PL*] qui in Medio
non resistente iisdem temporibus describerentur erunt ad
invicem ut quadrata temporum in quibus generantur; Est
itaq: decrementum [*QK*] arcus *PQ* pars quarta decrementi
*RL* arcus *PR*. Unde etiam si areæ *PSQ* æqualis capiatur
area *QSr* erit decrementum [*QK*] arcus *PQ* æquale dimidio
lineolæ *Rr*. [Nam ut *SQ* ad *SP* ita *PK* ad *KL* ita *PQ* ad
*Qr* ita dividendo *QK* ad *KL* − *Qr*; ergo componendo *PK*
ad *PL* ut *QK* ad (*QK* + *KL* − *Qr* sive)*rL*, unde *rL* = 2 *QK*:
sed erat *RL* = 4 *QK*, itaq: *Rr* = 2 *QK*] adeoq: vis resistentiæ
& vis centripeta sunt ad invicem ut lineolæ *QK* vel $\frac{1}{2}Rr$ &
*TQ* quas simul generant. This I take for a direct demonstration of the truth of what I proposed, & if You will be
pleased to consider what I offered at y$^e$ end of my second
Letter, You will {find} that also to amount to a demonstratio per absurdum. I did there assume y$^e$ proportion of
y$^e$ Resistance to y$^e$ Centripetall force to be as $\frac{1}{2}Rr$ to *TQ*
& from y$^t$ assumption I deduced a consequence whose truth
is very evident upon other considerations. But if You

take the proportion to be as $Rr$ to $TQ$ or any other way different from $y^t$ of $\frac{1}{2}Rr$ to $TQ$, the consequence will be as evidently false; Therefore the proportion can be no other than $y^t$ of $\frac{1}{2}Rr$ to $TQ$. You say in $Y^r$ Letter $y^t$ the 8$^{th}$ Corollary may remain as in $Y^r$ copy, but in $Y^r$ copy there are no alterations of $y^e$ first Edition. That You may see the reason I had for the alteration I proposed, I will put $N$ for the number of Revolutions, $T$ for $y^e$ Time of those Revolutions, $D$ for $y^e$ Density of the Medium, $t$ for $y^e$ tangent of $y^e$ Angle, $s$ for $y^e$ secant of $y^e$ same. Now in Cor. 6 You put $N$ as $t$, $T$ as $\frac{1}{D}$ or $s$, but in Cor 8 You put $N$ as $\frac{1}{D}$ or $t$, $T$ as $s$. The alteration which I proposed was to make $y^e$ 8$^{th}$ Corollary agree w$^{th}$ $y^e$ 6$^{th}$, for I am satisfied of $y^e$ truth of $y^e$ 6$^{th}$. In my first Letter I took notice of two mistakes in Prop xvi, You have consented $y^t$ one of 'em may be amended by putting $1 - \frac{1}{2}n$ for $\frac{1}{2}n$. The other You seem not to have observed which was $y^t$ $y^e$ Density is not reciprocally as $SP^{\iota}$ but reciprocally as $SP$: For the Resistance in $P$ being as $\dfrac{\overline{1 - \frac{1}{2}n},\ OS}{OP \times SP^{n+1}}$ and $y^e$ Velocity in $P$ as $\dfrac{1}{SP^{\frac{1}{2}n}}$, it follows $y^t$ $y^e$ Density in $P$ is as $\dfrac{\overline{1 - \frac{1}{2}n},\ OS}{OP \times SP}$ not as $\dfrac{\overline{1 - \frac{1}{2}n},\ OS}{OP \times SP^{n}}$, the Density being as $y^e$ Resistance directly & $y^e$ square of $y^e$ Velocity inversly. If You consent to this correction as I do not doubt You will, I desire You to send me the words of $y^e$ Proposition as You would have them altered. It seems to me not improper to add somewhere in this xvi Prop. or in a Corollary to it That $y^e$ force of resistance is to $y^e$ centripetall force as $\overline{1 - \frac{1}{2}n}$, $OS$ to $OP$

$Y^r$ &c.

LETTER VIII.

NEWTON TO COTES.

This letter is either misdated or was an unusually long time in arriving at its destination. It had not reached Cotes's hands when he penned his short note of May 17. It has no address, and was probably sent by a private hand, perhaps by Bentley.

M$^r$ Professor                    Chelsea. 13 May. 1710.

I have reconsidered the 15$^{th}$ Proposition with its Corollaries & they may stand as you have put them in yo$^r$ Letters. But in pag. 285 lin. 13 after the word *coincident* add the words, *et angulus PSV\* fit rectus.*

Let the 16$^{th}$ Proposition stand thus

Prop. xvi.     Theor. xii.

Si Medii densitas in locis singulis sit reciproce ut distantia locorum a centro immobili, sitq: vis centripeta reciproce ut dignitas quælibet ejusdem distantiæ : dico quod corpus gyrari potest in spirali quæ radios omnes a centro illo ductos intersecat in angulo dato.

Demonstratur eadem methodo cum Propositione superiore. Nam si vis centripeta in $P$ sit reciproce ut distantiæ $SP$ dignitas quælibet $SP^{n+1}$ cujus index est $n+1$ ; colligetur ut supra, quod tempus quo corpus describit arcum quemvis $PQ$ erit ut $PQ \times SP^n$†, et resistentia in $P$ ut

$$\frac{Rr}{PQ^q \times SP^n}, \text{ sive ut } \frac{1-\frac{1}{2}n, VQ}{PQ \times SP^n \times SQ}, \text{ adeoq: ut } \frac{1-\frac{1}{2}n, OS}{OP \times SP^{n+1}},$$

hoc est, ob datum $\dfrac{1-\frac{1}{2}n, OS}{OP}$, reciproce ut $SP^{n+1}$. Et propterea cum velocitas sit reciproce ut $SP^{\frac{1}{2}n}$, densitas in $P$ erit reciproce ut $SP$.

Corol. 1.   Resistentia est ad vim centripetam ut

$$1-\tfrac{1}{2}n \times OS \text{ ad } OP.$$

---

\* Cotes has written $PVQ$ in the margin.
† Cotes has written $SP^{\frac{1}{2}n}$ in the margin.

Corol. 2. Si vis centripeta sit reciproce ut $SP^{cub}$, erit $1 - \frac{1}{2} n = 0$, adeoq: resistentia et densitas Medii nulla erit, ut in Propositione nona Libri primi.

Corol. 3. Si vis centripeta sit reciproce ut dignitas aliqua radii $SP$ cujus index est major numero 3, resistentia affirmativa in negativam mutabitur.

Pag. 289, lin. 14. ffor *data lege*, read *data velocitatis lege*.

Your most humble servant

Is. NEWTON

## LETTER IX.

### COTES TO NEWTON.

S$^r$.                                          Cambridge May 17$^{th}$ 1710.

After I had received Y$^r$ Letter I wrote to You again about a week ago, about some difficultys which still remain with me. The Compositor is now at a stand, & I dare not let him go on till You shall be pleased to send me Y$^r$ answer.

Y$^r$ most Obedient and Faithfull Serv$^t$.

ROGER COTES.

## LETTER X.

### COTES TO NEWTON.

S$^r$.                                          May 20. 1710

I thank You for Y$^r$ Letter which came very seasonably. I now beg leave to propose to You some few alterations in the remaining part of Y$^r$ Copy. Page 293. 1: 1 — secunda *BFK* (per Prop xix) pro mensura sua æqualiter premuntur. 1: 4 Hac pressione, pro mensura sua,

& insuper — Page 303. 1: 6 — nisi forte per particulas intermedias virtute illa auctas — I think these words were better left out; for as I apprehend it, they alter $y^e$ case of $y^e$ Proposition. 1: 11 Ut si particula unaquæq:—quadrato-cubi Densitatis. I think also $y^t$ this whole Period ought to be omitted, the two propositions containd in it seeming to me to be erroneous, üless I mistake the sense of $Y^r$ words. Page 304. Coroll: 5 & 6 for [quadratum temporis directe] You have substituted in $Y^r$ copy [quadrato-quadratum temporis directe] I find written in $y^e$ margin of $Y^r$ book by a different hand [quadr. quadratum temporis (credo)] This marginal note, not $Y^r$ own judgment, was I beleive $y^e$ occasion of $Y^r$ making the alteration. Page 308 1: 10 I would omit $y^e$ words [si verbi gratia arcus alter sit altero duplo major]. With $Y^r$ leave I would begin the 311 page thus*. [Est itaq: incrementum velocitatis ut $V-R$ & particula illa temporis in qua factum est conjunctim: Sed & velocitas ipsa est ut incrementum contemporaneum spatii descripti directe & particula eadem temporis inverse. Unde cum resistentia (per Hypothesin) sit ut quadratum velocitatis, incrementum resistentiæ erit (per Lem: II) ut velocitas & incrementum velocitatis conjunctim, id est, ut momentum spatii & $V-R$ conjunctim; atq: adeo si momentum — In my Opinion this alteration is necessary to make the Demonstration accurate. When I first look'd over this passage upon account of it I thought the whole construction erroneous. I therefore set my self, after the following manner, to examine how it ought to be, which I here put down for a further use I have of it. Taking $x$, $\dot{x}$, $v$ for quantitys analogous to the Force arising from $y^e$ gravity of $y^e$ Pendulous body, the force of resistance, & $y^e$

---

* In Prop. xxix. Lib. 2. "Posito quod corpori in cycloide oscillanti resistitur in duplicata ratione velocitatis: invenire resistentiam in locis singulis." This Proposition contains the geometrical construction of the expression $\frac{1}{2kl}(2ks+1-\overline{2ka+1}\,e^{-2k\overline{a-s}})$, $a$ being the first arc of descent.

velocity in $D$, tis evident $y^t$ the arch $CD$ will also be as $x$, & the Fluxion of $y^e$ space $BD$ already described will be as $-\dot{x}$. If therefore $t$ be put for $y^e$ moment of time in which the fluxion of $y^e$ space $-\dot{x}$, the fluxion of $y^e$ velocity $\dot{v}$, the fluxion of $y^e$ resistance $\dot{z}$ are generated; You will have

$$v \parallel \frac{-\dot{x}}{t}, \quad \dot{v} \parallel \overline{x - z} \times t \quad \text{But } z \parallel vv \text{ & therefore } \dot{z} \parallel v\dot{v} \parallel$$

$-\dot{x} \times \overline{x - z} \parallel z\dot{x} - x\dot{x}$. Assuming therefore the determinate quantity $[a]$ of a just magnitude You will have this Æquation $a\dot{z} = z\dot{x} - x\dot{x}$. To construct this æquation I introduced another indeterminate quantity $[y]$ putting $z = p + qx + ry$ & $\dot{z} = q\dot{x} + r\dot{y}$; which values of $z$ & $\dot{z}$ being substituted in $y^e$ former æquation I obtained this other $aq\dot{x} + ar\dot{y} = p\dot{x} + qx\dot{x} + ry\dot{x} - x\dot{x}$. Then putting $q = 1$, $p = a$, I had the two following æquations $\dfrac{ay}{y} = \dot{x}$, $z = a + x + ry$ & $y^e$ construction of these two æquations agreed intirely with $Y^r$ own Solution of $y^e$ Problem*. Being satisfied by this Analysis of $y^e$ truth of $Y^r$ conclusion I easily saw $y^t$ my former difficulty lay in $y^e$ ambiguity of $y^e$ word [data] in line 1 & 5, & $y^e$ word [detur] in line 6. which I think may be remedied by the alteration which I propose. Page 312. 1: 21 I would leave out $y^e$ word [quamproxime]. Page 313. 1: 29† I would conclude the Demonstration thus — et ex æquo perturbate $Fh$ seu $MN$

---

* The analysis and construction of the problem will be found in Cotes's *Logometria*, (*Philos. Trans.* Jan.—March, 1714, pp. 40—42. *Harmonia Mensurarum*, pp. 36—38.)

† In Prop. xxx. Lib. 2. This Proposition contains the geometrical construction of the equation $\dfrac{g}{2l}(a^2 - b^2) = k \int_b^a v^n ds$ ($b$ being the first arc of ascent), which is obtained by one integration from the equation of motion $-\dfrac{v\,dv}{ds} = \dfrac{g}{l} \cdot s - kv^n$. Cotes's suggestion leads to further correspondence (see the next five letters). This and the preceding proposition may give us an idea of the trouble that Newton would take to exhibit his results in a synthetical form.

ad $Dd$ ut $DK$ ad $CF$ seu $CM$; Ideoq: summa omnium $MN \times CM$, id est, $\frac{1}{2}CAq - \frac{1}{2}Caq$ seu $Aa \times \frac{1}{2}aB$ æqualis erit summæ omnium $Dd \times DK$, id est, areæ $BKkVTa$, quam rectangula omnia $Dd \times DK$ seu $DKkd$ componunt. Q. E. D I was further satisfied y$^t$ there is no mistake in the Proposition or in this way of concluding it thus.

Taking $x$ for $CD$ & $z$ for $DK$ by y$^e$ abovementioned æquation $a\dot{z} = z\dot{x} - x\dot{z}$ it appeares y$^t$ $az + \frac{1}{2}xx$ is equall to the Fluent of $z\dot{x}$. Whence I conclude, if $CL$ be taken on y$^e$ other side of y$^e$ point $C$ equall to $Ca$ & y$^e$ ordinate $LQ$ be erected, y$^t$ the indeterminate area $DKVTa$ is equall to $\dfrac{DK}{LQ} \times LQTa + \frac{1}{2}CDq - \frac{1}{2}Caq$ & y$^e$ whole Area $BKVTa$ is equall to $\frac{1}{2}CBq - \frac{1}{2}Caq$ or $Aa \times \frac{1}{2}aB$. Page 315. 1 : 7 I would read thus—& Ellipsis $aBRVS$, centro $O$, semiaxibus $OB$, $OV$—1 : 22 Thus. Nam cûm Ellipsis vel Parabola $aBRVS$ congruat—1 : 24 thus alterutram $BRV$ vel $VSa$ excedit figuram—lin. penult. I would leave out [quamproxime]. pag. 319. 1 : 13 You say [cum distantiæ particularum Systematis unius sint ad distantias correspondentes particularum alterius, ut diameter particulæ vel partis in Systemate priore ad diametrum particulæ vel partis correspondentis in altero.] The same thing is implied in the Demonstration of Prop. 32. I think it ought also to be expressed in y$^e$ words of y$^e$ 32 Proposition.

Y$^r$ &c.

## LETTER XI.
### NEWTON TO COTES·

S<sup>r</sup>                                   Chelsea. May 30. 1710.

The corrections w<sup>ch</sup> you have sent me in your Letter of May 20 are right. But I fear least that w<sup>ch</sup> relates to Prop. xxx may render the Demonstration thereof too obscure. And therefore I think that the Proposition with its Demonstration may stand, & in the end of it, after the words et sic eidem æquabitur quam proxime, may be added these two sentences. Quinimo eidem æquabitur accurate, ideoq: conclusiones prædictæ sunt accuratæ. Nam si ad alteras partes puncti *C* capiatur *CL* æqualis ipsi *C a*, et erigatur normaliter *LQ* ad Curvam *a TVKB* terminata, et pro Curvæ hujus area indeterminata *a TVQL* ad ordinatam *LQ* applicata scribatur litera *M*; area indeterminata *a T VKD* æqualis invenietur quantitati *M, DK* + ½ *CD<sup>q</sup>* − ½ *Ca<sup>q</sup>*, et area tota *a T VKB* quantitati ½ *CB<sup>q</sup>* − ½ *Ca<sup>q</sup>*, seu *A a* × ½ *a B*.

The *Scholium Generale* w<sup>ch</sup> in the former edition was printed in the end of the seventh Section, I would have printed in the end of the sixt section next after Prop. xxxi. But it wants the following corrections

Pag. 339. lin 21, 22, 23 &c read

Scholium generale

Ex his Propositionibus per oscillationes Pendulorum in Mediis quibuscunq :, invenire licet resistentiam Mediorum. Aeris vero resistentiam investigavi per Experimenta sequentia. Globum ligneum pondere unciarum Romanarum 57 7/22, diametro digitorum Londinensium 6⅞ fab{r}icatum, filo tenui &c.

Pag. 340. lin 24, 25, blot out, omnino ut in Corollariis Propositionis xxxii demonstratum est.

Pag 341 lin. 18 for resistentia read resistentiæ.

Pag 342 lin 21 blot out, Unde cum corpus tempore, & what follows to the end of the words, longitudinem duplam 30,556 digitorum.

Pag. 343 lin 6 for pedum read digitorum. Ib lin 8 read vis resistentiæ eodem tempore uniformiter continuata. Ib lin 12 read posset.

Pag 344 lin 13, 14 for prima, secunda, tertia read tertia quinta septima & for $\dfrac{1}{193}$ read $\dfrac{\frac{1}{2}}{193}$.

Pag. 345 lin 7, 25 for dimidiata read subduplicata. Ib. lin. 8 read Nam ratio $7\frac{1}{2} - \frac{1}{3}$ ad $1 - \frac{1}{3}$ seu $10\frac{3}{4}$ ad 1, non longe *

Pag. 349 blot out the lines 17, 18, 19, 20, 21, 22, 23, 24, 25, 26, 27

Pag. 350 lin. 32 blot out Quare cum globus aqueus in aere movendo & what follows to the end of the words, probe tamen cum præcedentibus congruebat.

Pag 354 blot out the lines 11, 12, 13, 14, 15.

In the beginning of Sect vii pag. 317 lin. 5 after the words similes sint, insert the words & proportionales.

I am

Your most humble servant.

*For the* R<sup>nd</sup> M<sup>r</sup> Cotes, *Professor*          Is. Newton
*of Astronomy, & Fellow of Trinity*
*College in Cambridge.*

---

## LETTER XII.

### COTES TO NEWTON.

S<sup>r</sup>                                    June. 1<sup>st</sup> 1710

I received Your Letter last night, by which You give Y<sup>r</sup> consent to the other alterations which I proposed, but seem to fear least y<sup>t</sup> which relates to Prop xxx may render the Demonstration thereof too obscure & therefore at the end of y<sup>e</sup> Corollary after the words [et sic eidem æquabitur quamproxime] You add [Quinimo eidem æquabitur

---

* The words, "Ib. lin. 8......non longe", are crossed out, apparently by Cotes, in pursuance of Newton's orders in letters XIII. and XV.

accurate, &c] I beleive You designed those two sentences to be inserted pag. 314 lin 18 after the words [erit etiam æquale areæ $BKTa$ quamproxime, & $y^t$ by some inadvertency in $Y^r$ Letter You ordered them to be placed in page 315 l: 25 after $y^e$ words [eidem æquabitur quamproxime.] For though the Proposition it self & the first part of the Corollary ending $w^{th}$ the words [omnino ut in Propositione xxviii demonstratum est] be accurate, yet as I understand it the remaining part of the Corollary is still but an Approximation, the Ellipsis & Parabola mentioned in the latter part of $y^e$ Corollary not agreeing perfectly with the Figure $BKVTa$; but by placing those two sentences as in $Y^r$ Letter, even this latter part of the Corollary is declared to be accurate. I beg leave to express my self freely to You, I fear it will be look'd upon as a blemish in $Y^r$ book first to Demonstrate $y^t$ the Proposition is true & afterwards to assert it to be true $accuratè$. I am of opinion $y^t$ the alteration which I proposed pag. 313. l: 29 does make the Demonstration compleat to an intelligent Reader. If You think good it may be put down more at large some such way as this which follows — et ex æquo perturbate $(Fh$ seu$)$ $MN$ ad $Dd$ ut $DK$ ad $(CF$ seu$)$ $CM$; ideoq: summa omnium $MN \times CM$ æqualis erit summæ omnium $Dd \times DK$. Ad punctum mobile $M$ erigi semper intelligatur Ordinata rectangula æqualis indeterminatæ $CM$, quæ motu continuo ducatur in totam longitudinem $Aa$; & trapezium ex illo motu descriptum sive huic æquale rectangulum $Aa \times \frac{1}{2}aB$ æquabitur summæ omnium $MN \times CM$ adeoq: summæ omnium $Dd \times DK$, id est, areæ $BKkVTa$. q.e.d. Or if You think the Demonstration will even this way be too obscure, a new Scheme may be cut with $y^e$ addition of $y^e$ lines here drawn & the

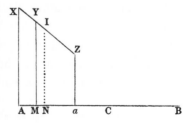

Demonstration may end thus — & ex æquo perturbate ($Fh$ seu) $MN$ ad $Dd$ ut $DK$ ad ($CF$ seu) $CM$: ideoq: $MN \times CM$ æquabitur $Dd \times DK$. Erigantur normales $AX$, $aZ$ æquales ipsis $AC$, $aC$ & jungatur $XZ$ occurrens normalibus $MY$, $NI$ in $Y$ & $I$; & erit $MY$ æqualis ipsi $CM$ atq: adeo $MN \times YM$ æquale $Dd \times DK$, & summa omnium $MN \times YM$, id est, trapezium $AaZX$ sive huic æquale rectangulum $Aa \times \frac{1}{2} aB$ æquabitur summæ omnium $Dd \times DK$, hoc est, areæ $BKkVTa$ Q E.D. I think the first of these two ways sufficiently clear ; but will wait for $Y^r$ resolution

$Y^r$ &c.

---

## LETTER XIII.
### NEWTON TO COTES.

$S^r$

I have reconsidered your emendation of the xxx[th] Proposition w[th] the Demonstration & approve it after the manner you propose in the first of the two ways set down in your Letter of June 1st. In my last letter, as I was sending it away, I crossed out four lines & should have struck out also these words relating to them [Ib. lin. 8, read, Nam ratio $7\frac{1}{2} - \frac{1}{3}$ ad $1 - \frac{1}{3}$ seu $10\frac{3}{4}$ ad 1, non longe]

I am

Yo[r] most humble Servant

Chelsea Jun 8. *                         Is. NEWTON.

1710.

I thank you for mending the Proposition

*For the* Rev[nd] M[r] COTES *Professor of Astronomy and fellow of Trinity College in Cambridge.*

This letter and the next must have crossed on the road.

---

* The post mark is Jun. 10.

## LETTER XIV.

### COTES TO NEWTON.

S$^r$                                  {Sunday} June 11 1710.

I received Y$^r$ Letter of May 30$^{th}$. In that which I wrote to You by y$^e$ next Post instead of y$^e$ alteration in page 316. l: 29 which You thought too obscure, I proposed the following—et ex æquo perturbate *Fh* seu *MN* ad *Dd* ut *Dk* ad *CF* seu *CM*; ideoq: summa omnium *MN* × *CM* æqualis erit summæ omnium *Dd* × *DK*. Ad punctum mobile *M* erigi semper intelligatur Ordinata rectangula æqualis indeterminatæ *CM*, quæ motu continuo ducatur in totam longitudinem *Aa*; & trapezium ex illo motu descriptum sive huic æquale rectangulum *Aa* × ½ *aB* æquabitur summæ omnium *MN* × *CM* adeoq: summæ omnium *Dd* × *DK*, id est, areæ *BKkVTa*. Q.E.D. We are now at a stand expecting Y$^r$ resolution. You gave me orders in Y$^r$ Letter to print the *Scholium Generale* after y$^e$ sixth section & sent me Y$^r$ corrections of it. I have not had leasure since I received Y$^r$ Letter to examine all the Calculations of y$^t$ Scholium, being at this time engaged in a Course of Experiments & having some other Buisness upon my Hands, but I have read it over & considered the reasoning of it. Page 345. l: 26 You say—{Si longitudo penduli.... augeretur.... arcuum descensu & subsequente ascensu descriptorum} differentia 0,4475 diminueretur in ratione velocitatis, adeoq: evaderet 0,4412. I do'nt see any reason for this diminution, but think it ought* to remain 0,4475 notwithstanding y$^t$ y$^e$ length of y$^e$ Pendulum is increased in the ratio of 126 to 122½, & thereby the time increased & y$^e$ velocity diminished in y$^e$ subduplicate ratio

---

\* This is also clear from the fact that the equation which connects the arcs of descent and ascent (the resistance varying as the square of the velocity) does not involve the length of the pendulum.

of 126 to 122$\frac{1}{2}$. You will see my reasons in what follows. Quæ tradita sunt in Prop xxxi & ejus Corollariis obtinent ubi Oscillationes sunt Isochronæ. At si oscillationum tempus quoq : mutetur, differentia inter arcum descensu descriptum & arcum subsequente ascensu descriptum erit ut resistentia & quadratum temporis conjunctim : Nam totius retardationis particulæ singulæ ex quibus differentia illa componitur sunt in hac ratione per Lem : x Libr. 1.

Unde si detur longitudo arcus descripti & resistentia sit ut quadratum velocitatis ; manebit differentia, utcunq : mutetur tempus atq : adeo velocitas corporis oscillantis. Nam ob datam longitudinem arcus descripti, tempus erit ut velocitas inverse ; adeoq : differentia illa, cum sit ut resistentia & quadratum temporis, erit ut quadratum velocitatis directe & quadratum velocitatis inverse, ac proinde magnitudinem datam habebit.

Idem aliter. (vide Fig Prop xxx) Manente longitudine arcus descripti $aB$ augeatur longitudo Penduli. Si mutata longitudine Penduli maneret Resistentia, maneret quoq : ratio resistentiæ ad vim gravitatis atq : huic æqualis ratio Ordinatæ $DK$ ad longitudinem Penduli ; adeoq : augenda esset Ordinata $DK$ in ratione longitudinis Penduli. Verum ob auctam Penduli longitudinem augetur quoq : tempus in ratione ejus subduplicata adeoq : diminuitur velocitas in eadem ratione subduplicata, & resistentia atq : huic proportionalis Ordinata $DK$ diminuitur in ratione integra. Itaq : Ordinata $DK$ diminuitur in eadem ratione qua prius augebatur ac proinde manet ejusdem longitudinis, manetq : adeo magnitudo areæ $BKVTa$ atq : huic æquale rectangulum $Aa \times \frac{1}{2}aB$ & differentia illa $Aa$. If You admit of this reasoning, it will not only affect this place in page 345 but also pag. 348 l: 1 and Pag. 353. l: 27 and page 341. l: 16. In Page 346: l: 23 You cite the Corollarys of Prop xl which are now to come after the Scholium ; there being no alteration of this place among the corrections You sent me, I do not know whether You took notice of it &

have therefore mentioned it to You. Page 348 l: 7 &c. You seem to confound the *Differentia arcuum* with $y^e$ *Resistentia Globi*; the former is represented by $AV + CV^2$ & $y^e$ latter ought I think to be represented by $\frac{7}{11}AV + \frac{3}{4}CV^2$. I desire $Y^r$ answer to this Letter, when I receive it I will examine & alter $y^e$ Calculation, if there be occasion, according to $Y^r$ direction

$$Y^r \ \&c.$$

---

## LETTER XV.

### NEWTON TO COTES.

$S^r$

I sent you a letter the last week in $w^{ch}$ I approved your correction of Prop xxx $w^{th}$ its demonstration according to the first of the two ways $w^{ch}$ you sent me in your Letter of June 1$^{st}$ & have now repeated in yours of June 11$^{th}$ $w^{ch}$ I received last tuesday morning {the 13th.} I thank you for that correction. In my last letter but one I crossed out four corrections $w^{ch}$ I had wrote down in it, & should have crossed out a fift $w^{ch}$ related to those four & was in these words. Pag. 345 lin. 8 lege, Nam ratio $7\frac{1}{2} - \frac{1}{3}$ ad $1 - \frac{1}{3}$ seu $10\frac{3}{4}$ ad 1.

The correction in the Scholium p. 345 lin 26, sent me in your last, is right, & I beg the favour that you would alter the calculations accordingly.

In pag. 346 lin 23 strike out the words et propterea (per corollaria Prop xl Libri hujus) resistentia quam Globi majores & velociores in aere movendo sentiunt & so on to the end of the sentence

In pag. 348 lin 7, 14, 15, 16 for $A$ & $C$ put other letters* suppose $F$ & $G$, writing, Designet jam $FV + GV^2$ resistentiam Globi &c because $AV + CV^2$ was used before for the differentia arcuum.

---

* Not adopted. Cotes altered this part of the Scholium in conformity with his remarks at the close of the preceding Letter.

You need not give your self the trouble of examining all the calculations of the Scholium. Such errors as do not depend upon wrong reasoning can be of no great consequence & may be corrected by the Reader.

I am w$^{th}$ many thanks

S$^r$ Your most humble servant

Chelsea June 15$^{th}$ 1710            Is. NEWTON.

*For the* R$^{nd}$ M$^r$ COTES *Professor of Astronomy*
*& Fellow of Trinity College in Cambridge*
*Cambridgeshire.*

## LETTER XVI.

### COTES TO NEWTON.

S$^r$                               June 30 1710

We have now finished all Y$^r$ Copy & y$^e$ Scholium Generale. I received Y$^r$ Letter of June 15$^{th}$ in which You consent to y$^e$ alterations y$^t$ I proposed in y$^t$ Scholium. I have examined the whole Calculation & done it anew where I thought it necessary. The discourse it self is also a little altered in those places which I mentioned in my last, as You will perceive by y$^e$ 2 inclosed sheets {Oo & Pp}. They are not yet printed off, but will stay for Your corrections if You shall think fit to make any. I could wish You would be pleased to look 'em over, for I fear I may possibly have injured You. The Press being now at a stand I will take this oportunity to visit my Relations in Lincolnshire & Leicestershire. I hope I shall come back again to College in 5 or 6 weeks. When I return I will write to You to desire y$^e$ remaining part of Y$^r$ Copy.

Y$^r$ &c.

## LETTER XVII.

### NEWTON TO COTES.

S<sup>r</sup>                                 Chelsea June 31*. 1710.

I received yours of June 30 this noon with the two inclosed proof sheets, & have perused them without observing any faults except in the last page of the second sheet lin 28 where vires autem motrices should be vires autem acceleratrices. And in the preceding page (pag. 295) upon reconsidering the words of Prop. xxxiii, I think the words will be better understood if they run as in the former edition, viz<sup>t</sup> Iisdem positis, dico quod Systematum partes majores resistuntur in ratione composita &c. The remaining part of the copy will be ready against your return from the visit you are going to make to your friends. I am w<sup>th</sup> my humble service to yo<sup>r</sup> Master & many thanks to yo<sup>r</sup> self for your trouble in correcting this edition,                        S<sup>r</sup>

Yo<sup>r</sup> most humble servant

*For the* R<sup>nd</sup> M<sup>r</sup> Cotes *Professor of*                    Is. Newton.
*Astronomy & Fellow of Trinity*
*College in Cambridge.*

---

" Wanting—a Letter from M<sup>r</sup>. Cotes to S<sup>r</sup>. Is. Newton—dated 11<sup>th</sup>. July 1710..." Note by Mr Howkins: who here and elsewhere informs us of the absence of letters, the dates of which we should otherwise (from any thing that can be gathered from the correspondence itself) have been in ignorance of. Smith had probably made a list of all the letters, and Howkins on collecting and arranging them when they came into his possession, noted such as were missing.

---

* This means July 1. Newton was not always exact in dating his letters. It may serve to make the *lapsus* in this case less incredible, though most persons will be able to supply instances for themselves, if I mention that the letters which were delivered by the morning post at Cambridge, on July 1, 1847, were stamped June 31, and that one of them, written the previous day by a distinguished prelate, was dated April 30.

## LETTER XVIII.
### COTES TO NEWTON.

S$^r$.                                    Monday Sept. 4$^{th}$ 1710.

I hope to be at Cambridge again on Wednesday next.
I have been somewhat longer in y$^e$ Country y$^n$ I at first intended, I hope You will excuse me : For the future I shall,
I hope, be ready without any further intermission to attend
upon y$^e$ Edition of Y$^r$ Principia. I desire You to send
the remaining part of Y$^r$ Copy assoon as You can.

Y$^r$ most Humble Servant

*For S$^r$ Isaac Newton at his House*          Roger Cotes.
*near the College in Chelsea near*
*London*

On his return to college Cotes would find that a slight change had
just been introduced into the daily habits of the place, which, for the
sake of those for whom the fact may possess an interest, may be
recorded here. "Sept. 4, at night Dr Smith the Senior Dean began
the custom of standing at grace, chiefly upon my sollicitation, and all
the Hall readily complied with the alteration." Rud's Diary.

## LETTER XIX.
### NEWTON TO COTES.

S$^r$

This Letter accompanies the next part* of the Principia. I am not certain that you have all y$^e$ cutts in wood,
but if any be wanting pray send me a draught in paper of
what is wanting & I'le get them cut {in} wood.

I am S$^r$

Yo$^r$ most humble Servant

Chelsea. Sept 13 1710.                    Is. Newton.

*For the R$^{nd}$ M$^r$ R. Cotes Professor of*
*Mathematicks & Fellow of Trinity Col-*
*lege in y$^e$ University of Cambridge.*

---

* Beginning at p. 321, with part of Cor. 2, Prop. xxxiii. Lib. 2, and ending at
p. 432, with Prop. xxiv. Lib. 3.

## LETTER XX.
### COTES TO NEWTON.

S$^r$.                                                    Sept. 21$^{st}$ 1710

I have received y$^e$ second part of Y$^r$ Copy, there are wanting only two wooden cutts which I can get done at Cambridge. I have read over what relates to y$^e$ resistance of Fluids, I thank You for the satisfaction I have received in seeing y$^t$ Theory so perfectly compleated. I confess I was not a little surprized upon y$^e$ first reading of Prop. 36 *; but I now begin to be better reconciled to it. One of my greatest difficulties was an Experiment of Mons$^r$. Marriotte which he says (page 245 Traite du Mouvment des Eaux †) he often repeated with great care. By his Experiment I concluded y$^t$ y$^e$ Velocity of y$^e$ effluent water was equall to y$^t$ gotten by an heavy body falling but from half y$^e$ Height of y$^e$ Vessel. He tells us y$^t$ 14 Paris Pints of water were evacuated in a Minute of time through a circular aperture of $\frac{1}{4}$ Inch diameter, the altitude of y$^e$ Vessel being 13 feet. He describes the Paris pint to be y$^e$ 35$^{th}$ part of y$^e$ Cube of y$^e$ Paris foot.

Therefore the water evacuated in a second was $\frac{14 \times 1728}{35 \times 60}$ or $\frac{2 \times 144}{25}$ Cubick inches. The Area of y$^e$ aperture was $\frac{11}{14 \times 16}$ inches. Hence y$^e$ length of a Cylinder equall in magnitude to y$^e$ evacuated water & having y$^e$ above mentioned Aperture for its Basis is $\frac{14 \times 16 \times 2 \times 144}{11 \times 25}$ Inches, and this length is y$^e$ space described in a second of time with y$^e$ uniform velocity of y$^e$ water as it passes

---

* Making the velocity of efflux of a fluid through an orifice in the base of a cylindrical vessel to be that due to the height of the surface of the fluid above the orifice, a result first stated by Torricelli, and adopted by him as a principle, ( *De motu Projectorum*, Florent. 1644. p. 191.) In the 1st Ed. (Prop. xxxvii.) the velocity had been made that due to half the height. The MS. of the Prop. which Cotes had before him when he wrote this Letter is wanting.

† New Edit. Paris 1700. The 1st Ed. is dated 1686.

through the aperture. The space described in a second of time with $y^e$ uniform Velocity acquired by any falling body in $y^e$ same time is (according to

\*     \*     \*     \*     \*     \*     \*     \*     \*

The remainder of this letter is wanting: at the point where it breaks off Cotes is saying that, according to Huygens's pendulum experiments, the velocity generated by gravity in $1''(g) = 30\frac{1}{6}$ Paris feet; and ∴ the height due to the velocity of efflux = $\left(\dfrac{14.16.2.12}{11.25}\right)^2 \cdot \dfrac{1}{2g}$ in feet, which lies between $6\frac{1}{2}$ and $6\frac{1}{3}$.

Some of the contents of the deficient part of the letter are mentioned in Newton's letter of March 24, $171\frac{0}{1}$. The letter which was actually sent will probably be found, with others that are wanting to complete this correspondence, in the Portsmouth Collection.

The above-mentioned result of Mariotte's experiment had been brought before the notice of the Royal Society by Halley at their meetings on March 18 and 25, 1691. On the latter day an experiment (inaccurately described in the Journal Book) was made, in which the jet was found to rise "far above the middle of the height of the liquor, whence it is to be noted that there is a mistake in the 37th Prop. of Mr Newton's 2nd Book, whereof it was ordered that Mr Newton should be certified." (It was probably in consequence of this suggestion that Newton revised the Proposition, and put it into the shape which surprised Cotes.) On Halley's recommendation, further experiments were made with a view to ascertain the cause of the discrepancy between the results derived from the observed height of the jet and the quantity of fluid discharged, but they only served to establish the fact, which remained unaccounted for until Newton (towards the end of 1710 and beginning of 1711), compelled by the statements of Cotes's letter of Oct. 5, 1710, to investigate the subject afresh, found the true explanation in the difference between the velocities at the orifice, and at that part of the vein of issuing fluid where it ceases to contract. See Newton's letter of March 24, $171\frac{0}{1}$. For an account of what has been done in this branch of Hydrodynamics, since Newton's time, see Rennie's Report to the British Association (meeting 1833) with the works there referred to, to which add Navier's Résumé des Leçons... sur l'Application de la Mécanique...Part. 2, 1838; and D'Aubuisson's Traité d'Hydraulique, 2nd Ed. 1840.

## LETTER XXI.

### NEWTON TO COTES.

S$^r$                                London. Sept. 30. 1710

Since the receipt of your Letter I have been removing from Chelsea to London, w$^{ch}$ has retarded my returning an answer to yo$^r$ last. I have not seen Mariots book concerning the motion of running water, but certainly there is something amiss in his experiment w$^{ch}$ you give me an account of. ffor I have seen this experiment tried & it has been tried also before the Royal Society *, that a vessel a foot & an half or two foot high & six or eight inches wide with a hollow place in the side next the bottom & a small hole in the upper side of the hollow, being filled with water; the water w$^{ch}$ spouted out of the small hole, rose right up in a small streame as high as the top of the water w$^{ch}$ stagnated in the vessel, abating only about half an inch by reason of the resistance of the air. The small hole was made in a thin plate of sheet tin and well polished, that the water might pass th{r}ough it with as little friction as possible. It was about the bigness of a hole made with an ordinary pin.

The corrections you have made are very well & I thank you for them, & am glad that the Theory of the resistance of fluids does not displease you provided the xxxvi$^{th}$ Proposition be true, as I think it is.

Direct your next Letters to me in S$^t$ Martins street neare Leicester fields.

I am Yo$^r$ most humble Servant

*For the* R$^{nd}$ M$^r$ COTES *Professor of Astro-*          Is. NEWTON
*nomy, & Fellow of Trinity College in*
*Cambridge in Cambridgeshire.*

---

* An experiment of this kind attended with the same result was tried by Hooke at a meeting of the Royal Society, April 1, 1691. The velocity of efflux was also the subject of experiment or discussion at several other meetings in that year. See the *Journal Book,* March 18, 25. April 8, 22.

" Wanting, two letters from Mr Cotes to Sir Isaac Newton, dated
5th and 26th Oct. 1710, concerning Prop. xxxvi. Lib. ii.

I remember to have seen the whole of this Prop. as it is now
printed in the 2nd Edition, fairly written in Mr Cotes's own hand ;
but I fear it is lost, or inadvertently destroyed; as I cannot find
it now.

E. Howkins, 1770."

## LETTER XXII.

### NEWTON TO COTES.

Sʳ

I received both your Letters & am sensible that I must
try three or four experiments before I can answer your
former*. My time has been taken up partly with removing to this house, partly with journeys about purchasing a
house† for the Royal society & partly wᵗʰ settling some
matters in the Mint in order to go on wᵗʰ yᵉ coynage‡ that
I have had no time to take these matters into consideration but hope wᵗʰin a fortnight to try the experiments &

---

* Of Oct. 5, containing probably, among other things, experiments Cotes had been
making on fluids issuing from an orifice in a vessel, and which went to confirm
Mariotte's. See letter of Newton in Macclesfield *Corresp.* ii. 437.

† In Crane Court. The Society met there for the first time on Nov. 8, having
previously held their meetings at Gresham College. The change, as is usual, was
opposed by some of the members. In 1782, Government assigned the Society apart-
ments in Somerset House. See Weld's *Hist. of Royal Soc.* i. 389, seqq. ; Ellis's *Letters of Eminent Literary Men*, 346, (where C. Wren's letter should evidently be dated,
1711.)

‡ The following table of gold and silver coined yearly from Christmas, 1708, to
Christmas, 1713, will shew approximately the times at which Newton's duties at the
Mint would experience a pressure during the years over which this correspondence on
the Principia extends.

|      | GOLD. lbs. | SILVER. lbs. |                          |
|------|-----------|--------------|--------------------------|
| 1709 | 2468      | 25423        | (in preceding year, 3751) |
| 1710 | 3716      | 817          |                          |
| 1711 | 9324      | 24768        |                          |
| 1712 | 2855      | 1784         |                          |
| 1713 | 13137     | 2333         |                          |

Macclesfield *Corresp.* ii. 434.

In the beginning of March 1711 the Royal Society changed their day of meeting to
Thursday at 4, the President " being obliged to attend the Mint on Wednesdays."

settle the matters in doubt & beg the favour that you will let the press stay till you hear from me again.

I am Yo<sup>r</sup> most faithfull friend

& humble Servant

London. Octob 27*. 1710.          Is. NEWTON

*For the* R<sup>nd</sup> M<sup>r</sup> COTES *Professor of Astronomy, at his chamber in Trinity College
in Cambridge.*

---

### LETTER XXIII.

#### NEWTON TO COTES.

S<sup>t</sup> Martins street by Leicester ffields. Mar. 24<sup>th</sup> 171⁹⁄₁₀.

S<sup>r</sup>

I send you at length the Paper for w<sup>ch</sup> I have made you stay this half year. I beg your pardon for so long a delay. I hope you will find the difficulty cleared, but I know not† whether I have been able to express my self clearly enough upon this difficult subject, & leave it to you to mend any thing either in the expression or in the sense of what I send you. And if you meet w<sup>th</sup> any thing w<sup>ch</sup> appears to you either erroneus or dubious, if you please to give me notice of it I will reconsider it. The emendations of Corol. 2 Prop 38 & Prop 40 are your own. You sent them to me in yours of Sept. 21, 1710, & I thank you for them. That you may have the clearer Idea of the experiments in the beginning of the inclosed paper, let *ABCD* represent a vessel full of water perforated in the side with a small hole *EF* made

---

* Post Mark 28.

† It is doubtful whether the "not" has not been added by another hand. If it be in Newton's handwriting, it is about the nearest approach to an instance of his crossing a 't', that I remember to have seen.

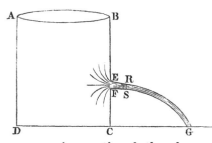

in a very thin plate of sheet tin. And conceive that the water converges towards the hole from all parts of the vessel & passes through the hole with a converging motion & thereby grows into a smaller stream after it is past the hole then it was in the hole. In my trial the hole $EF$ was $\frac{5}{8}$ths of an inch in diameter & about half an inch from the hole the diameter of the stream $RS$* was but $\frac{21}{40}$ of an inch. And therefore the streame had the same velocity as if it had flowed directly out of a hole but $\frac{21}{40}$ of an inch wide. And so in Marriotts experim$^t$ the stream had the same velocity as if it had flowed directly out of a hole but $\frac{21}{100}$ of an inch wide. In computing the velocity of the water w$^{ch}$ flows out we are not to take the diameter of the hole for the diameter of the streame, but to measure the diameter of the streame after it is come out of the hole & has formed itself into an eaven & uniform stream. And the velocity thus found will be what a body would get in falling from y$^e$ top of the water : as is manifest also by the distance $CG$ to which the stream will shoot it self, & also by the stream's ascending as high as the top of y$^e$ water stagnating in the vessel, if the motion be turned upwards.

I am

Your most humble & most obliged Servant

*For the R$^{nd}$ M$^r$ Roger Cotes Professor of Astronomy at his Chamber in Trinity College in the University of Cambridge.*

Is. NEWTON

---

* $RS$ is the diameter of the "sectio venæ contractæ," (a term first used by Jurin, *Philosoph. Transact.* Sept.—Oct. 1722, p. 185; and afterwards by Dan. Bernoulli, *Hydrodynam.* p. 65. Jurin also uses "vena contracta" to denote the same thing, and the expression is still retained in works on Hydrostatics, though differently defined by different writers, most of them describing it as that part of the issuing fluid between the orifice and the section whose diameter is $RS$.)

The " Paper" mentioned in the above letter seems to have consisted of four folio sheets, and to have included from Prop. xxxvi. to Prop. xxxix. with part of Prop. xl., and a page of corrections (No. 111.) to be made in the conclusions of "the Experiments set down in the Scholium to the 40th Proposition sent you formerly." The first three leaves are wanting, the portion which is preserved beginning with the latter part of the 37th Prop. (No. 72).

There were several things in this " Paper" which did not satisfy Cotes. (1) His " difficulty" about the 36th Prop. was not yet completely removed. This probably led to his (missing) letter of March 31, which, if no other letter passed between them in the interval, brought him a satisfactory answer from Newton. This answer, sent apparently in a parcel from Bentley, is also wanting. (2) Besides making other alterations of a minor character, Cotes has crossed out what is left of Prop. xxxvii., and written the Proposition out on another piece of paper (Nos. 70, 71) as it now stands in the 2nd Ed. with this note at the top : " Print this instead of what is blotted out in Prop. xxxvii." He has also modified part of the Scholium of this Prop. though not to the extent that Horsley (*Newtoni* Op. ii. 412) attributes to him. He has drawn his pen through almost the whole of Prop. xxxviii. and part of its 2nd Cor. and re-written the parts struck out as they now stand in the 2nd Ed. These were the materials of his letter of June 9. See introductory remarks to the fragment of that letter.

---

" Wanting, a letter from Mr Cotes to Sir Isaac Newton, dated 31st March 1711. Another dated 4th June 1711." Note by Mr Howkins.

---

### LETTER XXIV.
#### NEWTON TO COTES.

Sr

Yo^rs of June 4^th I received the next day & thank you for it. I am glad you received what D^r Bentley sent you & that you think the difficulty removed, except what you mention about the manner of delivering y^e 37^th Proposition. ffor clearing the sense of the first & second Paragraphs, these words may be added to the end of the second Paragraph after the word locatum. *Circellus autem*

*sustinendo vim aquœ defluentis minuet ejus velocitatem, idq :*
*in ratione qua minuit spatium per quod aqua jam transit.*
*Nam (per Cas. 5. Prop* xxxvi, *& ejus Corol.* 6) *aqua jam*
*transibit per spatium annulare inter circellum & latera canalis*
*eadem velocitate qua prius transibat per canalis cavitatem*
*totam* *.

And a little after where I have these words [augeatur
velocitas circelli in eadem ratione et resistentia ejus augebitur in ratione duplicata] may be written these [augeatur
velocitas circelli in eadem ratione & resistentia ejus augebitur in eadem ratione bis, nempe semel ob auctam quantitatem aquæ in quam circellus dato tempore agit & semel
ob auctum motum quem circellus in singulas aquæ partes
imprimit. Nam partes fluidi similibus motibus agitabuntur
atq : prius sed velocioribus et minore tempore *.]

But since you are considering how to set this xxxvii$^{th}$
Proposition in a cleare light I will suspend saying any
thing more about it till I see your thoughts. I am

Yo$^r$ humble servant

London 7$^{th}$ June 1711           Is. NEWTON.

*For the* R$^{nd}$ M$^r$ ROGER COTES *Professor of*
*Astronomy at his Chamber in Trinity*
*College in Cambridge*

---

## LETTER XXV.

### COTES TO NEWTON.

This is only the concluding part of a letter, which a note by Mr
Howkins states to have been dated June 9th, 1711. In the words with
which this fragment opens, Cotes is giving his emendation of Prop.
xxxviii. as it stood in the " Paper" which Newton sent him in his
letter of March 24. The former part of the letter must have contained
Prop. xxxvii. in the form in which Cotes had at last put it, and also

---

* Not adopted.

his modification of the construction in the Scholium, where Newton had made the latus rectum of the smaller parabolas 8 *AB* and that of the others 32 *CD*.

Horsley saw some of Cotes's actual letters in the Portsmouth Collection, and this of the 9th of June among others. In a note on the 37th Prop. he says (II. 404): Hæc demonstratio a Cotesio tota est.

Horsley also says that the 6th Cor. of Prop. XXXVII. is due to Cotes, and that in the letter in which it was sent (what the date of the letter was does not appear certain) after explaining this Corollary he adds: "Hoc Corollarium lucem aliquam tuo in Corollario decimo *quantum sentio* offundere possit." This 6th Corollary in Cotes's hand written on a slip of paper is still preserved (No. 67).

{June 9. 1711.}

\*        \*        \*        \*        \*        \*        \*        \*

{et propterea} Vis illa quæ tollere possit motum omnem Cylindri interea dum Cylindrus describat longitudinem quatuor diametrorum, Globi motum omnem tollet interea dum Globus describat duas tertias partes hujus longitudinis, id est, octo tertias partes diametri propriæ. Resistentia autem Cylindri est ad hanc vim quamproxime ut densitas Fluidi ad densitatem Cylindri vel Globi, per Prop XXXVII, & resistentia Globi æqualis· est resistentiæ Cylindri per Lemm: V. VI. VII. I will remember to alter the 2$^d$ Corollary of this Proposition which You had forgotten to do in Your last Copy. I have computed y$^e$ Table preceeding y$^e$ *Scholium* of Prop, XL & find some of the numbers to be amiss which I will take care to rectify; as over against 0,9$G$ the space should be 0,7196609$F$; over against 3$G$ the space should be 4,6186570$F$; over against 4$G$ should be 6,6143765$F$. I computed also all the Experiments & found my Calculations to agree nearly enough with Yours except in the 1$^{st}$ Experiment which I will alter throughout. Of the rest the greatest difference was in the 11$^{th}$, in which y$^e$ result was 46$\frac{5}{9}$ oscillations not 46 as You make it in Your corrections\*, I took care to make a right allowance

---

\* Sent March 24. See p. 40.

for $y^e$ narrowness of the Vessel. I desire You to send me the Altitude from which the Globes fell in the 9$^{th}$ Experiment. You had forgotten to mention it in Your Copy. The Six Experiments in $y^e$ Air *agree also very well with my computation, in the 5$^{th}$ the space should be 225$^p$. 5$^d$.

Your most humble Servant

R. C.

LETTER XXVI.

NEWTON TO COTES.

S$^r$

I have read over & considered your alterations, & like them very well & return you my thanks. In $y^e$ end of Exper. 9, add, *describentes altitudinem digitorum* 182. I thank you also for correcting the numbers. I hope there will be no more occasion of stopping the press. After you have read the objection of Muys† taken from

---

* These experiments were made by Hauksbee, June 9, 1710, with glass balls let fall from the top of the Cupola of St Paul's, (nearly 220 feet). See *Philosoph. Trans.* Oct. —Dec. 1710, p. 198. An account of them was read at a meeting of the Royal Society, June 14, at which Newton presided. At the previous meeting, June 7, (the President then also in the chair) Hauksbee read a paper on some experiments of the same kind, which are described in the article in the *Phil. Trans.* just referred to. Newton assisted at similar experiments, made by Desaguliers, April 27 and July 27, 1719, from the upper gallery in the lantern on the top of the Cupola, a height of 272 feet. He with some other persons was below, and noted the difference in the time of fall of the leaden and of the lighter balls. See *Phil. Trans.* Sept.—Oct. 1719, pp. 1071—1078. The experiments made on the latter day are introduced into the 3rd Ed. of the *Principia*, p. 353.

† In *Elementa Physices methodo Mathematica Demonstrata*, &c. Amstelod. 1711 : a heavy quarto, reviewed in the Leipsic *Acts* for Sept. 1711, and severely criticised by Leibniz and John Bernoulli in their Correspondence.

In the 1st Ed. of the *Principia* (p. 337), there is a Lemma which states that *if a spherical or other vessel, filled with fluid, move rectilinearly with an accelerated velocity, the molecules of the contained fluid participating equally in the motion of the vessel will remain at rest among themselves.* Muys (p. 355), in opposition to this quotes a passage from the 4th Dialogue of Galileo's *System. Cosmic.* (p. 315, Lyons, 1641,) where Salviati, in attempting to explain the tides, takes the case of a *vessel, which contains water,*

Galileo's experiment of the motion of a bucket full of water you will scarce expect very much from that author.

I am S$^r$

Yo$^r$ very humble servant

S$^t$ Martins street London.

June 18$^{th}$ 1711.  Is. NEWTON

*For the* R$^{nd}$ M$^r$ COTES *Professor of Astronomy, at his chamber in Trinity College in the University of Cambridge.*

---

## LETTER XXVII.

### COTES TO NEWTON.

S$^r$  June 23$^d$. 1711

I received Your Letter & have delivered Y$^r$ Papers to the Printer. I hope we shall now go on without any further intermission. As for Muys, I have look'd over what relates to the resistance of Fluids. He acknowledges that what he offers upon y$^t$ subject at present is but crude & indigested & I am very willing to agree with him. His Objections as far as I can understand 'em do not in any wise affect Your Book, much less the new Edition of it. One M$^r$ Green of Clare-Hall has now in the Press a book*

---

*moving horizontally,* and says that, *If a force be applied to retard the vessel, the molecules of the fluid will still retain their velocity, and the water will rise at the anterior part of the vessel. If, on the contrary, the velocity of the vessel be increased, the water will lag behind, and so will be higher at the hinder than at the fore part of the vessel.* This fact the speaker proceeds still further to illustrate by referring to the boats used to convey fresh water from Lizza Fusina to Venice.

* *The Principles of Natural Philosophy, in which is shewn the Insufficiency of the Present Systems,* &c. &c. Camb. 1712. With a Latin Tract at the end, entitled, *Geometria Solidorum,* &c. This eccentric writer also published *A Demonstration of the Truth and Divinity of the Christian Religion,* &c. Camb. 1711, and a large thick folio, (pp. 981) with the title of *The Principles of the Philosophy of the Expansive and Contractive Forces,* &c. Camb. 1727. In the Preface to this last work he says: "Our Philosophy, as it is now received and embraced, is the product of Popish countries,

of the like nature with Muys wherein I am inform'd he undertakes to overthrow the Principles of Your Philosophy. I do not expect very much from him, & I beleive You will not Your self when I have told You he is a Person who pretends to have solv'd $y^e$ grand Problem of $y^e$ Quadrature of the Circle. That the Press may not stop, I am now looking over Your Copy beforehand. I find nothing amiss till I come to Prop : 48. I will choose to make my Objection against the Corollary, wherein You have these words [Nam lineola Physica $\epsilon\gamma$, quamprimum ad locum suum primum $EG$ redierit, quiescet ;] This assertion cannot I think be reconciled with what You assert & prove in the Proposition [& propterea vis acceleratrix lineolæ Physicæ $\epsilon\gamma$ est ut ipsius distantia a medio vibrationis loco $\Omega$] I propose to alter the whole Proposition thus if You approve of it. [Propagentur pulsus in plagam $BC$ a $B$ versus $C$ & designet $BC$ intervallum eorundem ab invicem. Sint $E$, $F$, $G$ puncta tria Physica Medii quiescentis in recta $BC$ ad æquales distantias sita ; $ee$, $ff$, $gg$ spatia æqualia perbrevia per quæ puncta illa motu reciproco singulis vibrationibus eunt & redeunt ; $\epsilon$, $\phi$, $\gamma$ loca quævis intermedia eorundem punctorum ; & $EF$, $FG$ lineolæ Physicæ seu Medii partes lineares punctis illis interjectæ & successive

---

imported to us from Italy and France......All therefore which I design and intend, is to propose a Philosophy, which is truly *English*, a *Cantabrigian*, and a *Clarensian* one, as it was born, and educated, and studied in those places ;......And as my Name is not much worse in the Letters which belong to it, than those of Galileus or Des-Cartes,...... I shall venture to call the GREENIAN." Mr Green was not altogether a stranger to Newton when Cotes introduced a notice of him in this letter. On making the discovery that the area of a circle is equal to four-fifths of the square of its diameter, shortly after taking his B.A. degree (1700), "Dominum Newtonum accessi ut consulerem," says he, "orantem qui chartulas perlegeret, ipsis intactis, ne inspectis certe, rejecit, aggressus sum dein epistola, recusavit, (in the Preface to his *Geometria Solidorum*, his phrase is 'rescripsit nihil,') quid posthæc arbitremini me putasse? Saltem vel contemptum me vel Problema." (Ib. p. 940, 1st Lecture "ad Clarensem juventutem.") On the publication of Green's "Natural Philosophy" in 1712, where his quadrature of the circle was asserted, he tells us that Cotes was "so kind and obliging as to communicate to me with great candour and friendship a demonstration against it," which will be found Ib. pp. 924-5. Cf. Letter CVI.

translatæ in loca $\epsilon\phi$, $\phi\gamma$ & $ef$, $fg$. Rectæ
$ee$ æqualis ducatur recta $PS$, bisecetur eadem
in $O$, centro $O$ & intervallo $OP$ describatur circulus $SIPi$, & agatur diameter $QR$ ad diametrum $PS$ perpendicularis. Per circuli hujus
circumferentiam totam cum partibus suis expo-
natur tempus totum vibrationis unius cum ipsius partibus proportionalibus; sic ut completo
tempore quovis $QH$ vel $QHSh$,
si demittatur ad $PS$ perpendi-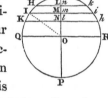
culum $HL$ vel $hl$, & capiatur
$E\epsilon$ æqualis $OL$ vel $Ol$, punctum Physicum $E$ reperiatur in
$\epsilon$. Hac lege punctum quodvis
$E$ eundo ab $E$ per $\epsilon$ ad $e$ atq : inde redeundo,
iisdem accelerationis ac retardationis gradib[9]
vibrationes singulas peraget cum oscillante Pen-
dulo. Probandum est quod singula Medii puncta
Physica tali motu agitari debeant. Fingamus
igitur Medium tali motu a causa quacunq : cieri,
& videamus quid inde sequatur.

In circumferentia $PQSR$ capiantur æquales
arcus $HI$, $IK$ vel $hi$, $ik$ eam habentes rationem
ad circumferentiam totam quam habent æquales
rectæ $EF$, $FG$ ad pulsuum intervallum totum
$BC$. Et demissis perpendiculis $IM$, $KN$ vel $im$,
$kn$; quoniam puncta $E$, $F$, $G$ motibus similibus
successive agitantur & vibrationes suas integras ex itu &
reditu compositas interea peragant dum pulsus transfertur
a $B$ ad $C$, si $QH$ vel $QHSh$ sit tempus ab initio motus
puncti $E$, erit $QI$ vel $QISi$ tempus ab initio motus puncti
$F$, & $QK$ vel $QKSk$ tempus ab initio motus puncti $G$; &
propterea $E\epsilon$, $F\phi$, $G\gamma$ erunt ipsis $OL$, $OM$, $ON$ in itu
punctorum, vel ipsis $Ol$, $Om$, $On$ in punctorum reditu
æquales respective. Unde $\epsilon\gamma$ seu $EG + G\gamma - E\epsilon$ in itu

punctorum æqualis erit $EG - LN$, in reditu autem æqualis $EG + ln$. Sed $\epsilon\gamma$ latitudo est seu expansio partis Medii $EG$ in loco $\epsilon\gamma$, & propterea expansio partis illius in itu, est ad ejus expansionem mediocrem ut $EG - LN$ ad $EG$; in reditu autem ut $EG + ln$ seu $EG + LN$ ad $EG$. Quare cum sit $LN$ ad $KH$ ut $IM$ ad radium $OI$, & $KH$ ad $EG$ ut circumferentia $PQSRP$ ad $BC$, id est, (si ponatur $V$ pro radio circuli peripheriam habentis æqualem intervallo pulsuum $BC$) ut $OI$ ad $V$, et ex æquo $LN$ ad $EG$ ut $IM$ ad $V$: erit expansio partis $EG$ punctive Physici $F$ in loco $\epsilon\gamma$ ad expansionem mediocrem quam habet in loco suo primo $EG$ ut $V - IM$ ad $V$ in itu, utq: $V + im$ ad $V$ in reditu. Unde vis Elastica puncti $F$ in loco $\epsilon\gamma$ est ad vim ejus Elasticam mediocrem in loco $EG$ ut $\dfrac{1}{V - IM}$ ad $\dfrac{1}{V}$ in itu, in reditu vero ut $\dfrac{1}{V + im}$ ad $\dfrac{1}{V}$. Et eodem argumento vires Elasticæ punctorum Physicorum $G$ & $E$ in itu sunt ad vires mediocres ut $\dfrac{1}{V - KN}$ & $\dfrac{1}{V - HL}$ ad $\dfrac{1}{V}$, & virium differentia sive excessus vis Elasticæ puncti $\gamma$ supra vim Elasticam puncti $\epsilon$ est ad Medii vim Elasticam mediocrem ut $\dfrac{KN - HL}{VV - V \times KN - V \times HL + KN \times HL}$ ad $\dfrac{1}{V}$, hoc est, ut $\dfrac{KN - HL}{VV}$ ad $\dfrac{1}{V}$ sive ut $KN - HL$ ad $V$, si modo (ob angustos limites vibrationum) supponamus $KN$ & $HL$ indefinite minores esse quantitate $V$. Quare cum quantitas $V$ detur, excessus ille est ut $KN - HL$, hoc est (ob proportionales $KN - HL$ ad $HK$ & $OM$ ad $OI$, datasq: $HK$ & $OI$) ut $OM$, id est, ut $F\phi$. Et eodem argumento excessus vis Elasticæ puncti $\gamma$ supra vim Elasticam puncti $\epsilon$ in reditu lineolæ Physicæ $\epsilon\gamma$ est ut $F\phi$. Sed excessus ille est vis qua hæc lineola acceleratur; & propterea vis acceleratrix lineolæ Physicæ $\epsilon\gamma$ est ut ipsius distantia a medio vibrationis loco $F$. Proinde tempus (per Prop xxxviii Libr. 1) recte exponitur per arcum $QI$; & Medii pars linearis $\epsilon\gamma$ perget lege præscripta moveri, id est, lege oscillantis Penduli : & par est ratio partium omnium linearium ex quibus Medium totū componitur. q.e.d.] I was going to propose an alteration of the Corollary but I choose rather to leave it to Your self. It must be made to correspond with what You have at the end of Page 372 where You cite it. I propose to alter Prop. 49 as follows. [p. 368. 1: 28 — ad lineolæ illius pondus ut $HK \times A$ ad $V \times EG$ sive ut $PO \times A$ ad $VV$, nam $HK$ erat ad $EG$ ut $PO$ ad $V$.] [1: 32 — urgente vi ponderis in subduplicata ratione $VV$ ad $PO \times A$ atq : adeo —] [1: ult — in subduplicata ratione $VV$ ad $PO \times A$ & subduplicata ratione $PO$ ad $A$ conjunctim, id est, in ratione integra $V$ ad $A$. Sed tempore vibrationis unius —.] [Ergo tempus — & reditu compositæ ut $V$ ad $A$, id est, ut $BC$ ad circumferentiam circuli &c.] I propose to add the 2 following Corollaries to Prop 49.

Cor. 1. Velocitas pulsuum ea est quam acquirunt Gravia æqualiter accelerato motu cadendo et casu suo describendo dimidium altitudinis $A$. Nam tempore casus hujus, cum velocitate cadendo acquisita, pulsus percurret spatium quod erit æquale toti altitudini $A$, adeoq : tempore oscillationis unius ex itu & reditu compositæ percurret spatium æquale circumferentiæ circuli radio $A$ descripti; est enim tempus casus ad tempus oscillationis ut radius circuli ad ejusdem circumferentiam.

Cor. 2. Unde cum altitudo illa $A$ sit ut Fluidi vis Elastica directe & densitas ejusdem inverse; velocitas pulsuum erit in ratione composita ex subduplicata ratione densitatis inverse & subduplicata ratione vis Elasticæ directe. I think the 47th Proposition is out of its place : for

the Demonstration of it proceeds upon the supposition of
the truth of the 48th, & therefore it ought to follow the
48th, & besides the 48th serves to form some Ideas which
are necessary to the understanding of the 47th*. If You
agree that these Propositions should change places I would
add the following words at y[e] end of the 47th which will
then be the 48th [Hæc Propositio ulterius patebit ex constructione sequentis]. I see nothing further in the 2[d] Book
which I could wish might be altered. In the 3[d] Book
under Phænom : 1, The Periodical times should be

$$1^d.18^h.27'.34'' \quad 3^d.13^h.13'.42''. \quad 7^d.3^h.42'.36''. \quad 16^d.16^h.32'.9''$$

and the Distantiæ ex temporibus periodicis may be

$$5,667 \qquad 9,017 \qquad 14,384 \qquad 25,299$$

I perceive You have made use of Cassini's Tables of Jupiter's Satellits printed in 1693 in the *Recueil d' Observations faites en plusieurs Voyages &c.* But Your numbers
give the times of the Revolutions to Jupiters shadow, not
to y[e] same point of y[e] Ecliptick. The Revolutions to the
same point of the Ecliptick are (by those Tables) as I have
set 'em down. Y[r] time of the Revolution of Saturns outer-
most Satellit differs from the time assigned by Hugenius
in his Cosmotheoros & by Cassini in the Philosophical
Transactions but I find it is y[e] time which was afterwards
determin'd by Cassini in y[e] *Memoires de l'Academ.* 1705.
You have made an addition to the 3[d] Proposition in which
are these words [Hæc ratio obtinet in Orbe Lunæ nostræ.

---

* The object of Prop. xlviii. is to shew that when pulses or undulations are propagated in a fluid, the particles vibrate according to the law of an oscillating pendulum.
Prop. xlvii. shews how the velocity of propagation varies, and Prop. xlix. determines
its quantity, the expression for which ($\sqrt{g}$ . height of homog. atmosph.) Laplace was
the first to prove, must (in the case of sound) be multiplied by

$$\sqrt{\frac{\text{spec. heat of air under a constant pressure}}{\text{volume}}}.$$

*Mécan. Célest.* v. 121, 129. Poisson, *Mécan.* ii. 716. Whewell's *Hist. Ind. Sci* ii.

In minore Orbe motus Aphelii minor esset in triplicata ratione minoris distantiæ Lunæ a Terra, & Fractio $\frac{4}{243}$ diminui deberet in eadem ratione. Et propter hanc diminutionem vis qua Luna retinetur in Orbe suo est ad vim eandem in superficie Terræ ut 1 ad $D^{2\frac{1}{243}}$ quamproxime, uti computum ineunti patebit] I should be glad to understand this place if it will not be too great a trouble to make it out to me. I do not at present so much as understand what it is that You assert.

I am S$^r$ Y$^r$ &c.

---

## LETTER XXVIII.
### COTES TO NEWTON.

S$^r$.                                                    July 19$^{th}$ 1711

I wrote to You about a Month ago concerning the 48$^{th}$ Proposition of Y$^r$ second Book, & the last week I ordered the Printer to send You all the sheets which were printed off. If You have received these sheets You will perceive by 'em that the Press is now at a stand. But having no Letter from You I fear the sheets have miscarried. The Compositor dunn's me every day, & I am forc'd to write to You again to beg Y$^r$ answer to my former Letter. I have received the last part* of Your Copy by D$^r$ Bently. I have now read over and examined all the calculations of the former part which ends in y$^e$ 432$^d$ page. I will write to You concerning it assoon as I receive Your answer to my last Letter.

I am S$^r$. Y$^r$ &c.

---

* Beginning at p. 433, with part of Prop. xxiv., Lib. 3, and terminating at p. 510 with Prop. xlii. (end of 1st Ed.) Bentley returned to College on the 7th, (Rud's Diary.)

## LETTER XXIX.

### NEWTON TO COTES.

S$^t$ Martins Street in Leicester ffields London July 28$^{th}$ 1711.

S$^r$

I received your Letters & the papers sent me by the
Printer But ever since I received yours of June 23 I have
been so taken up with other affairs that I have had no time
to think of Mathematicks. But now being obliged to keep
my chamber upon some indisposition w$^{ch}$ I hope will be
over in a day or two* I have taken your letter into consideration. You think that in the Corollary to the 48$^{th}$
Proposition these words [Nam lineola Physica $\epsilon\gamma$ quamprimum ad locum suum primum redierit, quiescet] consist
not w$^{th}$ what I assert & prove in the Proposition, viz$^t$ [&
propterea vis acceleratrix lineolæ Physicæ $\epsilon\gamma$ est ipsius
distantia a medio vibrationis loco $\Omega$] But I suspect that
you take the words [ad locum suum primum] in another
sence then I might intend them. ffor when all the lineolæ
physicæ $\epsilon\gamma$ are returned to their first places or places in
w$^{ch}$ they were before the vibrations began, the medium will
be uniform as before & the vis acceleratrix of the lineola
physica $\epsilon\gamma$ will cease, whether that lineola arrived to its
first place in the beginning middle or end of the vibrations.
For making the Corollary more intelligible, these words
may be added to the end of it. Partes fluidi non quies-
cent nisi in locis suis primis. Quamprimum in loca illa
motu retardato redierint, component Medium uniforme
quietum quale erat ante vibrationes excitatas.

In altering the 48$^{th}$ Proposition you have shortned
the Demonstration. If you had proposed your alteration
of the Corollary I should have been better able to compare
the whole w$^{th}$ mine.

---

* He was sufficiently recovered by the following Thursday, (Aug. 2,) to preside at
a meeting of the Council of the Royal Society on that day.

Your emendations of Prop 49 are very well & the two Corollarys you propose may be added to it. And the 47<sup>th</sup> & 48<sup>th</sup> Propositions may change places, & at the end of the 47<sup>th</sup> these words may be added [Hæc Propositio ulterius patebit ex constructione sequentis.

I will write to you about {the} third book in my next.

I am S<sup>r</sup> Your very humble servant

*For the* Rever<sup>nd</sup> M<sup>r</sup> ROGER COTES *Professor*     Is. NEWTON.
*of Astronomy, at his Chamber in Trinity*
*College in Cambridge.*

---

## LETTER XXX.

### COTES TO NEWTON.

S<sup>r</sup>.                                            July 30<sup>th</sup> 1711.

I have read Y<sup>r</sup> Letter & find my self obliged to trouble You once more. I must beg leave to tell You I am not as yet satisfied as to the Inconsistency which I mention'd in my former Letter. You seem to say that when the Lineola Physica $\epsilon\gamma$ is return'd to its first place, which You take to be the beginning of the Vibration, the Medium will be uniform as at first & consequently its Vis acceleratrix will cease. If upon the return of the Lineola to its first place it be granted that the Medium will be uniform I confess it must also be granted that the Vis Acceleratrix will cease: but then if the Vis acceleratrix does cease in this place it must likewise be granted that its quantity is less than in places nearer the middle of y<sup>e</sup> Vibration where it does not cease, & of consequence its quantity will not be proportionable to the distance of the Lineola from the middle of the Vibration, for to be proportionable it ought not to cease in the beginning of the Vibration, but on the contrary it should be greater there than in any other place, & if it be greater there than in any other place the Medium will not then be uniform.

This consideration was to me the occasion of altering the
Proposition. By making the middle of the Vibration the
*locus primus* I saw this inconsistency might be avoided.
But besides this, it appeares altogether reasonable upon
other accounts that the *locus primus* should be the middle
of the Vibration. Suppose a Musical Chord to be put into
motion; tis certain its *locus primus* is the middle of its
Vibration & consequently also $y^e$ *locus primus* of any *lineola
Physica* of Air which is contiguous to the Chord is in the
middle of its own space of Vibration; for the motion of
this *Lineola Physica* follows & depends upon the motion
of the contiguous Chord. And for the same reason, a
second *Lineola Physica* not contiguous to the Chord but to
the first *Lineola* will have its *locus primus* in the middle of
its own Vibration, since its motion depends upon the first
as the first did upon the Chord it self; & the same may be
said of other *Lineolæ* which are yet more remote from the
Chord. Now assoon as the motion of the Chord ceases in
its *locus primus* iè, in the middle of its Vibration, though it
should perhaps be said $y^t$ the motion of the first Lineola
would not cease of it self at the same time with it, yet tis
evident it will be made to cease by the resistance of the
Chord, for being contiguous to the Chord when it is
arriv'd at its *locus primus* or the middle of its Vibration it
can proceed no further towards the Chord whilst $y^e$ Chord
maintains its rest, & it cannot return back again from the
Chord as having no *Vis Acceleratrix* or acquired Impetus
that way. And as this first *Lineola* ceases by $y^e$ resistance
of $y^e$ Chord, so $y^e$ second ceases by $y^e$ resistance of $y^e$ first,
& so on. By this You will understand how I would alter
the Corollary; but I chose rather to refer it to Your self,
as fearing I could not express my thoughts with sufficient
clearness & brevity & exactness at $y^e$ same time. What I
have represented above is not so exact as it should be, for
$y^e$ motions of the *Lineolæ* must be suppos'd gradually to

cease with the motion of the Chord; but I chose to express my self as I have done that You might the more clearly understand me. In altering the Proposition I altered the 4$^{th}$ line of Page 366 by putting *Pl, Pm, Pn* instead of *Pn, Pm, Pl*; & in the 2$^d$ line of Page 367 instead of [ob brevitatem pulsuum] I have put it [ob angustos limites vibrationum] for it would be truer & more to the purpose to say *ob magnam pulsuum distantia{m}* than to say *ob brevitatem pulsuum.* In Your Example taken from M$^r$ Sauveur the latitude of the Pulse is about 10 foot, when perhaps y$^e$ space of Vibration is not above y$^e$ 10$^{th}$ of an Inch at y$^e$ utmost. If You consent to my Alteration of the Proposition the Figure must be altered. I propose to have it cut like y$^e$ Figure I sent You, which does better express the disproportion of y$^e$ breadths of y$^e$ Pulses & Vibrations than the former Figure.

I am S$^r$. Y$^r$ &c.

---

## LETTER XXXI.
### COTES TO NEWTON.

S$^r$   Sept. 4$^{th}$ 1711

I received a Letter from you about a Month ago, & sent You an Answer to it the next day by y$^e$ Carrier, in which I gave You my reasons why I was not yet satisfied as to y$^e$ Inconsistency in the 48$^{th}$ Proposition & its Corollary which I formerly mention'd to You. I have not heard from You since y$^t$ time, & therefore I fear that either my Letter or Your Answer to it has miscarried. I shall be glad to know Your resolutions concerning this 48$^{th}$ Proposition assoon as You have leasure that the Press may go on. There were some things relating to the 3$^d$ Book in my former Letter, I hope You will not forget to let me know Your mind concerning them also.

I am S$^r$ Y$^r$. &c.

Newton's occupations at the Mint (see note ‡ p. 37) coupled with his duties as President of the Royal Society will probably be sufficient to account for his not having had leisure to attend to the two preceding letters until after the lapse of 5 or 6 months (Feb. 2, 1712, the date of the next letter). The following dates will give us a glimpse or two of him during some of these months.

October 16, 1711 : "The President |of the Royal Society| appointed a Council {a meeting of the Committee of Visitors of Greenwich Observatory} to be called on Friday come sevennight (the 26th) when Mr Hunt is ordered to desire Mr Flamsteed to meet the Council on that day at 11 o'clock, at their house in Crane Court in Fleet Street ; to know of him if his instruments be in order, and fit to carry on the necessary celestial observations." (Baily's Flamsteed, p. 96, 97 note). Three accounts of this meeting from Flamsteed's pen are extant, (Baily, p. 96, 228, 294), which bear painful marks of his unhappy temper soured by the mortification he felt at having a board of Visitors " set over him."

Jan. 31, 171½. Leibniz's 2nd letter to Dr Sloane (dated 29 Dec. 1711), complaining of " Keill's unfair dealing with him in his last letter, relating to the dispute between him and Sir Isaac Newton, was read : the letter was delivered to the President to consider the contents thereof." (Journal Book of Royal Soc.) This letter, in which Leibniz, speaking of the obnoxious passage in the Leipsic Acts for Jan. 1705, in the review of Newton's tract *De Quadratura,* says " in illis circa hanc rem quicquam cuiquam detractum non reperio, sed potius passim suum cuique tributum," led to the appointment of a Committee (March 6, 171½), to inspect the letters and papers relating to the subject, who delivered in their Report, Apr. 24.—

A great part of Cotes's correspondence with Jones falls within this interval (letters ciii—cx) and may be conveniently read here as contributing towards filling up the blank.—

With the next letter the correspondence begins to be carried on with briskness. In a letter of Saunderson to Jones, March 16, 1712, (Macclesfield Corresp. i. 264, where it is printed out of its chronological place,) a postscript adds that " Sir Is. Newton is much more intent upon his Principia than formerly, and writes almost every post about it, so that we are in great hopes to have it out in a very little time."

## LETTER XXXII.

### NEWTON TO COTES.

S$^r$                                    London 2$^d$ Feb. 171$\frac{1}{2}$.

I have at length got some leasure to remove the difficulties w$^{ch}$ have stopt the press for some time, & I hope
it will stop no more. ffor I think I shall now have time to
remove the rest of your doubts concerning the third book
if you please to send them.

In reveiwing yo$^r$ letters I do not see but that y$^e$ XLVIII$^{th}$
Proposition of the second Book with its Corollary may
stand. ffor the particles of air go from their loca prima
with a motion accelerated till they come to the middle of
the pulses where the motion is swiftest. Then the motion
retards till the particles come to the further end of the
pulses. And therefore the loca prima are in the beginning
of the pulses. There the force is greatest for putting y$^e$
particle into motion if any new pulses follow. But if no
new pulse follows the force ceases & the particle continues
in rest. In this Proposition pag. 366. lin. 12, this emendation may be made. Quare cum sit *LN* ad *KH* ut *IM* ad
radium *OP*, et *KH* ad *EG* ut circumferentia *PHShP* ad

*BC*; id est (si circumferentia dicatur $\varkappa$ et $\dfrac{OP \times BC}{\varkappa}$ dica-

tur *V\**,) ut *OP* ad $\dfrac{OP \times BC}{\varkappa}$ seu *OP* ad *V*. Et ex æquo

*LN* ad *EG* ut *IM* ad *V*: erit expansio partis *EG*, punctive physici *F*, in loco $\epsilon\gamma$, ad expansionem mediocrem
quam pars illa habet in loco suo primo *EG* ut *V – IM* ad
*V* in itu, utq: *V + im* ad *V* in reditu. Vnde vis elastica
puncti *F* in loco $\epsilon\gamma$ est ad vim ejus elasticam mediocrem

---

\* Cotes did not adopt the part where $\varkappa$ is brought in, but printed it as he proposed
in his Letter of June 23, "(si ponatur *V* pro radio circuli circumferentiam habentis
æqualem intervallo pulsuum *BC*), &c." His suggestion of " ob angustos limites vibrationum," (Letters June 23, July 30,) of which Newton takes no notice, is also
introduced into the 2nd Ed.

in loco *EG* ut $\dfrac{1}{V-IM}$ ad $\dfrac{1}{V}$ in itu, in reditu vero ut

$\dfrac{1}{V+im}$ ad $\dfrac{1}{V}$. Et eodem argumento vires elasticæ &c

See lin 27.

You stuck at a difficulty in the third Proposition of the third Book. I have revised it & the next Proposition & sent you them inclosed* as I think they may stand. What further Observations you have made upon the third Book or so many of them as you think fit if you please to send in yo$^r$ next Letters, I will dispatch them out of hand. I shall be glad to have them all because I would have {the} third Book correct.

I am Yo$^r$ most humble Servant

*For the* R$^{nd}$ M$^r$ Cotes, *Professor of Astro-*       Is. Newton
*nomy, at his chamber in Trinity College*
*in Cambridge.*

---

### LETTER XXXIII.

#### COTES TO NEWTON.

S$^r$.                                  Febr. 7$^{th}$ 17$\frac{11}{12}$

I have received Your Letter & as to the buisness of sounds, I do intirely agree with You upon considering that matter over again. By Your alteration of y$^e$ 3$^d$ Prop: of y$^e$ 3$^d$ Book, it is now very intelligible. What I have observed concerning the remaining part of Your Copy I will send You in the most convenient order I can. I begin with the 37$^{th}$ Proposition, in the 3$^d$ section of which You have these words [Eo autem tempore Luna distat a Sole 15 $\frac{1}{2}$†gr. circiter. Et Sol in hac distantia minus

---

* A folio sheet, Nos. 127—129. To the 4th Proposition, a Scholium beginning "*Picartus* mensurando arcum, &c." is subjoined, which is a modification of what he had previously sent down in the second instalment of his copy of the *Principia*, Sept. 13, 1710. He afterwards, (Letter XLI.), determined on omitting this Scholium, and placing it after Prop. xxxvii. Eventually, however, part of it was transferred to Prop. xix., and a smaller part to Cor. 7 of Prop. xxxvii.

† It should be 15¼, as it stands in Newton's MS. No. 193. See p. 78.

auget ac minuit motum maris a vi Lunæ oriundum quam
in ipsis Syzygiis & quadraturis, in ratione Radii ad cosinum distantiæ hujus duplicatæ seu anguli $30\frac{1}{2}$ gr.
hoc est, in ratione 7 ad 6 circiter ideoq: in superiore Analogia
pro $S$ scribi debet $\frac{6}{7}$ $S$. I suppose You intended to have
said [in ratione duplicata Radii ad cosinum distantiæ
hujus] or [in ratione diametri ad sinum versum duplicati
complementi hujus distantiæ]. After the same manner
in $y^e$ foregoing proposition, at $y^e$ bottom of $y^e$ 463 page,
You have added these words *. [In aliis solis positionibus
vis ad mare attollendum est ut cosinus duplæ altitudinis
Solis supra horizontem loci directe & cubus distantiæ Solis
a Terra inverse] I suppose You intended to have said
[ut sinus versus duplæ altitudinis]. This alteration being
made in Prop 37, You will have $\frac{13}{14}$ $S$ instead of $\frac{6}{7}$ $S$,
whence $S$ will be to $L$ as 1 to $5\frac{3}{25}$, & in $y^e$ $4^{th}$ Corollary
You will have a different proportion from $y^t$ of 1 to 38.
In $y^e$ $3^d$ Corollary You make use of $31'.27''$ & $32'.12''$
for $y^e$ apparent diameters of $y^e$ Sun & Moon: I query
whether it would not be more adviseable to use $y^e$ numbers
of Your new Theory † $32'.15''$ for $y^e$ Sun, $31'.16''\frac{1}{2}$ for $y^e$
Moon. Making use of these numbers, & of $57'.5''$ for
$y^e$ Moons Horizontal Parallax, & taking $y^e$ density of $y^e$
Sun to be to $y^e$ density of $y^e$ earth as 100 to 398 $\frac{1}{17}$ as my
computation gives it; the quantity of matter in $y^e$ Moon
will be to $y^e$ quantity of matter in $y^e$ Earth as 1 to
$176\frac{2}{5} \times \dfrac{S}{L}$, or as 1 to $34\frac{2}{3}$. This alteration will very
much disturb Your Scholium of $y^e$ $4^{th}$ Proposition as it
now stands; neither will it well agree with Proposition
$39^{th}$, in which I further observe that You take $y^e$ proportion of $y^e$ semidiameters of $y^e$ earth to be as 689 to
692; But if their difference be 32 Miles, there will be

---

* No. 191.
† "Lunæ Theoria Newtoniana," printed in David Gregory's *Astronomiæ Elementa*,
(Oxford, 1702), p. 332.

another proportion, & I query whether here ought not to be some allowance made upon that score.

I have not examin'd all the calculations of $y^e$ Scholium to $y^e$ $iv^{th}$ Proposition but I formerly observ'd a small difference from Your Numbers as to $y^e$ descent of heavy bodies. If $y^e$ length of a Pendulum which vibrates seconds be 3 feet & $8\frac{5}{9}$ lines, the descent in that time will be 15 feet 1 inch $2\frac{1}{18}$ lines: You have it $2\frac{1}{4}$ lines. And when I examin'd $y^e$ $xix^{th}$ Proposition I found the vis centrifuga to be in proportion to the vis gravitatis as 1 to $288\frac{7}{9}$, You have it as 1 to $290\frac{4}{5}$. In this computation I took $y^e$ measure of a degree to be 57200 Toises as You had formerly stated it, the descent of heavy bodies in a second to be 15,0976 feet, the time of $y^e$ earths revolution to be $23^h.56'.4''$. If this Vis centrifuga be increased in $y^e$ proportion of 57230 to 57200, it will be to $y^e$ vis gravitatis as 1 to $288\frac{5}{8}$. I will send You some things further as I can recollect them from my loose papers of $y^e$ computations which I made about $\frac{1}{2}$ an Year ago; In Your next You may be pleasd to send me Your Answer to what I formerly proposed concerning $y^e$ periodical times of $y^e$ Satellits, for I do not yet know Your resolution as to that part of my Letter.

---

### LETTER XXXIV.

#### NEWTON TO COTES.

$S^r$            London Feb. 12. $17\frac{11}{12}$.

In the third Book under Phænom. I, the periodical times may be

$1^d.18^h\ 27'\ 34''$.    $3^d\ 13^h\ 13'\ 42''$.    $7^d\ 3^h\ 42'\ 36''$.    $16^d\ 16^h\ 32'\ 9''$ & the distances, ex temporibus periodicis 5,667    9,017 14,384   25,299 as you have put them in $yo^{rs}$ of June 23 last. But the numbers in the Corollaries of Prop. viii must be altered accordingly. And so must one or two of $y^e$ numbers in Prop. XII & XIII.

In $y^e$ $3^d$ section of $y^e$ XXXVII$^{th}$ Proposition, I think my proportion is right. ffor the force of the Sun increases the force of the Moon in the Syzygies, diminishes it in the Quadratures & neither increases nor decreases it in the Octants: & therefore the distance of the Moon from the Sun must be doubled that the cosine thereof may vanish in the Octants.

In the $3^d$ Corollary of that Proposition lin 5, 6, the words should run thus [et cubus diametri Lunæ ad cubum diametri Solis inverse, id est, (cum diametri mediocres apparentes Lunæ et Solis sint 31′ 27″ & 32′ 12″) ut &c.] But instead of the Moons mean diameter 31′ 27″ may be written 31′ 16½, & the Suns mean diameter 32′ 12″ may be every where retained, even in the Moons Theory. For 32′. 15″ is too bigg.

In the Scholium to the IV$^{th}$ Proposition, if the length of a Pendulum w$^{ch}$ vibrates seconds in vacuo be put 3 feet & 8⅖, the descent in that time will be 15 feet 1 inch & 2¼ lines.

And in the XIX$^{th}$ Proposition the vis centrifuga may be put in proportion to the vis gravitatis as 1 to 289, & then these corrections must be made. Neare the end of the Scholium of Prop IV. for the numbers 290⅘, 669 & $\frac{1}{669}$ write 289, 665, & $\frac{1}{665}$. Also pag 422 lin 9 write, ut 1 ad 289. lin. 13, ut 289 ad 288. lin 15, 289. lin 16, 288. Pag 423 lin 27, ut 1 ad 288. lin 28, pars $\frac{1}{288}$. lin 31, vis centrifuga $\frac{1}{288}$. lin ult. pars tantum $\frac{1}{288}$. Pag. 424 lin 1, ut 229 ad 228. lin 3, 19674224, seu millia{r}ium 3935. lin 5, pedum 86101 seu milliarium 17. lin 16, ut $\dfrac{29 \times 1 \times 5}{5 \times 228}$ ad 1, seu 1 ad 8. lin 29, ut 229 ad 228.

The XXXIX$^{th}$ Proposition must be corrected by putting the semidiameters of the earth as 228 to 229 instead of 689 to 692, or perhaps as 3919 to 3951 the difference being 32 miles. I think [228 to 229] should be put for [689 to 692] & the difference of 32 miles may be allowed for in the latter part of the Proposition. But I have lost my copy of the emendation I made to that Proposition & the Lemmas preceding, & so know not how to make this correction. If you can mend the numbers so as to make $y^e$ precession of the Equinox about 50″ or 51″, it is sufficient. I am

Yo<sup>r</sup> most humble Servant

*For the* R<sup>nd</sup> M<sup>r</sup> COTES *Professor of*                         Is. NEWTON
*Astronomy, at his chamber in*
*Trinity College in Cambridge.*

---

LETTER XXXV.

COTES TO NEWTON.

S<sup>r</sup>                                        Cambridge Feb 16<sup>th</sup> 17$\frac{11}{12}$

I received Your last of $y^e$ 12<sup>th</sup> of this Month. Tis very evident that $y^e$ 3<sup>d</sup> section of Proposition xxxvii<sup>th</sup> ought not to be altered. I had observ'd, that in an addition which You have made at $y^e$ bottom of page 463, *cosinus* ought to be chang'd into *sinus versus;* & thereupon, (without any further consideration), I had applied the same change to $y^e$ 3<sup>d</sup> section of $y^e$ following Proposition. I will observe Your directions as to $y^e$ Diameters of $y^e$ Sun & Moon in Corol. 3; retaining in all other places 32′. 12″ for $y^e$ Sun. In $y^e$ Scholium of iv<sup>th</sup> Proposition I think the length of $y^e$ Pendulum should not be put 3 feet & 8$\frac{2}{5}$ lines; for the descent would then be 15 feet 1 inch 1$\frac{1}{3}$ line. I have considered how to make y<sup>t</sup> Scholium appear to the best advantage as to $y^e$ numbers, & I propose to alter it thus. To take 57220 Toises for $y^e$ measure of a degree, instead of 57230; for 57220 is $y^e$ nearest round number to a mean amongst 57060, 57292, 57303. To take 3 feet $8\frac{10}{19}$ lines for $y^e$ length of $y^e$ Pendulum; for $y^e$ French sometimes make it $8\frac{1}{2}$ sometimes $8\frac{5}{9}$, & $8\frac{10}{19}$ is a mean betwixt these measures. To take $48^{gr}$. $50'$ for $y^e$ Latitude of Paris instead of $48^0$. $45'$ as You had put it. From these principles the following alterations may be made. [semidiameter Terræ 19670787 ped] [distantia mediocris Lunæ a Terra 1190082614 ped] [distantia)$^{æ}$ a communi centro gravitatis 1159567675 ped] [Sinus Versus ped. 14, dig. 9, lin. $5\frac{5}{14}$] — id est, in ratione 1 ad 3680,84502 ; ideoq: corpus ad superficiem Terræ vi illa cadendo describet pedes Parisienses 15, dig. 1, lin. $5\frac{1}{6}$.

Observatum est longitudinem Penduli ad minuta secunda oscillantis in vacuo, esse pedum trium Parisiensium & linearum $8\frac{1}{2}$ seu linearum $8\frac{5}{9}$. Sumatur longitudo mediocris pedum trium & linearum $8\frac{10}{19}$: & altitudo quam grave in vacuo cadendo tempore minuti unius secundi describit, (cum sit ad dimidiam longitudinem Penduli hujus in duplicata ratione circumferentiæ ad diametrum circuli, ut indicavit Hugenius,) erit pedum Parisiensium 15, dig. 1, lin. $1\frac{9}{10}$. Hic est descensus gravium in Latitudine Lutetiæ Parisiorū seu $48^{gr}$. $50'$.

Ad Æquatorem vis centrifuga corporum a diurna rotatione Terræ oriunda est ad vim gravitatis ut 1 ad 289 circiter; & in Latitudine Lutetiæ minor est, idq: in duplicata ratione sinus complementi Latitudinis $48^0$. $50'$ ad Radium adeoq: est ad vim gravitatis ut 1 ad 667. Et hac vi descensus gravium in latitudine Lutetiæ diminuitur. Descensus igitur pedum 15, dig. 1, lin. $1\frac{9}{10}$ augeatur parte $\frac{1}{666}$ seu lineis $3\frac{4}{15}$, & habebitur totus gravium descensus pedum 15, dig. 1, lin. $5\frac{1}{6}$ quem gravitas sola, tempore minuti unius secundi in Latitudine $48^{gr}$. $50'$ efficere posset, si modo Terra quiesceret.

I have gone over the computation of $y^e$ $\text{viii}^{th}$ Proposition again taking $32'$. $12''$ for $y^e$ Suns diameter, for I had

formerly made use of 32′. 15″.  I propose these alterations.
[Satellitis extimi Jovialis tempus periodicum dierum 16 &
horarum $16\frac{8}{15}$] Pondera ad æquales distantias a centris

Solis, Jovis, Saturni ac Terræ 1. $\dfrac{1}{1033}$. $\dfrac{1}{2411}$. $\dfrac{1}{227512}$.

Semidiametri Solis, Jovis, Saturni ac Terræ 10000. 1077.
889. 104.  Pondera ad superficies Solis, Jovis, Saturni ac
Terræ 10000. 835. 525. 410.  Densitates Solis, Jovis, Saturni
ac Terræ 100. 78. 59. 396*.

The xii$^{th}$ Proposition may be altered thus [Nam cum,
per Corol. 2. Prop viii. materia in Sole sit ad materiam in
Jove ut 1033 ad 1, & distantia Jovis a Sole sit ad semidiametrum Solis in ratione paulo majore; incidet commune
centrum gravitatis Jovis & Solis in punctum paulo supra
superficiem Solis.  Eodem argumento cum materia in Sole
sit ad materiam in Saturno ut 2411 ad 1, & distantia Saturni a Sole sit ad semidiametrum Solis in ratione paulo
minore : incidet &c.]  The xiii$^{th}$ Proposition may be altered
thus, pag. 419. 1 : 18 [ut 1 ad 1033]. lin : 21. [ut 81 ad 16×1033

seu 1 ad 204 circiter] lin : antepenult. $\left[\&\ \dfrac{16\times81\times2411}{25}\right.$

---

* All the figures which Cotes proposes in this paragraph, duly appear in their places
in the first three Corollaries of Prop. viii., in the 2nd Edit., though Newton in his
answer to this Letter takes no notice of his suggestions with respect to them.

Cotes has made about half a dozen other alterations (adopted in the 2nd Ed.) in
the MS. of the four Corollaries of this 8th Prop., which are not noticed in this rough
draught, though some of them would probably be mentioned in the letter actually sent.
The most important of them are the following, (Nos. 133, 134):

In Cor. 1.  The last sentence is, "Pondera corporū in superficie Lunæ fere duplo
minora esse quam pondera corporum in superficie Terræ dicemus in sequentibus," as it
stands in the 1st Ed.  Cotes has altered it to " Quanta sint pondera corporum in superficie Lunæ dicemus in sequentibus."

In Cor. 3, the words " Densitas Terræ *hic posita* non pendet a parallaxi Solis, &c."
are altered to " Densitas Terræ *quæ prodit ex hoc computo* non pendet, &c."

In Cor. 4, Newton had written " Sed et densiores sunt Planetæ, cæteris paribus,
qui sunt Soli propiores; ut Jupiter Saturno, et Terra Jove.  Oritur utiq: densitas materiæ ex calore solis eam decoquentis.  Et collocandi erant Planetæ in diversis a Sole
distantiis ut quilibet pro gradu densitatis calore solis majore vel minore frueretur."
Cotes has drawn his pen through the words " Oritur......Planetæ," and has altered the
last clause to " In diversis utiq : a Sole distantiis collocandi erant Planetæ, ut quilibet,
&c."  In the 1st Ed. the last clause runs thus : " Collocavit igitur Deus Planetas in
diversis distantiis a Sole......vel minore fruatur."

seu 124986] lin : ult. [ut 65 ad 124986 seu 1 ad 1923]. I
observe that You have added* to the xiv$^{th}$ Proposition a
Scholium concerning y$^e$ motion of y$^e$ Aphelia of y$^e$ Planets,
in which by supposing y$^t$ of Mars to go forward 35′ in 100
Yeares You deduce the motion of y$^e$ Earths Aphelium to
be 18′, 36″. I should be glad to know whether You have
found these motions to be nearly so by Observations or
whether these numbers are propos'd barely as an Example ;
for in Your new Theory published by D$^r$ Gregory You
make y$^e$ motion of y$^e$ Earths Aphelium to be 21′. 40″ in an
100 Yeares   The Rule delivered in this Scholium puts me
in mind of a mistake in the New Edition of Your book
which I did not observe till it was too late.   In y$^e$ 16$^{th}$
Corollary of y$^e$ lxvi$^{th}$ Prop : of Lib : 1, or in page 166, line
9$^{th}$ of y$^e$ New Edition You will find *ut quadratum temporis
periodici corporis P directe* &c.   So You had altered it in
Your Copy, but I think it should be as in y$^e$ former Edition
*ut tempus periodicū.*   Over against Your alteration there is
written in y$^e$ margin with a black lead pencil by another
hand *quadr. temporis period.* which I suppose You depended
upon without considering the thing Your self.   I will write
to You concerning the xix$^{th}$ & xx$^{th}$ propositions in my
next.   I come now to y$^e$ xxxix$^{th}$ Proposition, it stands thus
in Your Copy. pag : 470. lin : 10 dele *reciproce.* lin : 26.
*ut 474721 ad 4143 seu 114584 ad 1000* pag : 471. lin : 20
[evaderet minor quam prius in ratione 2 ad 5.   Ideoq :
annuus æquinoctiorū regressus jam esset ad 20$^{gr}$. 11′. 46″. ut
1 ad 7330, ac proinde fieret 9″, 55‴. 8′′.   Cæterum hic motus,
ob inclinationem plani Æquatoris ad planum Eclipticæ,
minuendus est, idq : in ratione &c.]   You have left out all
from pag : 471. 1 : 22 to pag. 473. lin : 13.   Then in pag.
473. lin : 27 You have [diminuendus est motus 9″. 55‴. 8′′
in ratione sinus 91706 (qui sinus est complementi graduum

---

$23\frac{1}{2}$) ad radium 100000. Qua ratione motus iste jam fiet $9''.5'''.46''$. Hæc est annua Præcessio Æquinoctiorum a vi Solis oriunda.

Vis autem Lunæ ad mare movendum &c.*

I should be glad to have this Proposition settled before we print any thing which may in any wise relate to it

Y<sup>r</sup> humble Servant

ROGER COTES.

Before I conclude this Letter, I will take notice of an objection which may seem to be against y<sup>e</sup> 3<sup>d</sup> Corol: of Prop: vi. Lib: iii *Itaq: Vacuum necessario datur* &c. Let us suppose two Globes *A* & *B* of equall magnitudes to be perfectly fill'd with matter without any interstices of void Space; I would ask the question whether it be impossible that God should give different vires inertiæ to these Globes. I think it cannot be said that they must necessarily have y<sup>e</sup> same or an equal Vis inertiæ. Now You do all along in Your Philosophy & I think very rightly estimate the quantity of Matter by the vis inertiæ & particularly in this vi<sup>th</sup> Proposition, in which no more is strictly proved than y<sup>t</sup> y<sup>e</sup> Gravitys of all Bodies are proportionable to their Vires inertiæ. Tis possible then that the equal spaces possessed by the Globes *A* & *B* may be both perfectly filld with Matter so as no void interstices may remain & yet that

---

* This being merely the draught of a letter, Cotes has not taken the trouble of transcribing the whole of the passage, though of course in the letter which was actually sent, he would copy it entire. It stands as follows in Newton's MS. No. 204.

Vis autem Lunæ ad mare movendum erat ad vim Solis ut $4\frac{2}{7}$ ad 1 circiter. Et in eadem proportione est vis Lunæ ad vim Solis ad Æquinoxia movenda. Indeq: prodit annua Æquinoctiorum Præcessio a vi Lunæ oriunda $42''.52'''.54''$, ac tota Præcessio annua a vi utraq: oriunda $51''.58'''.40''$.

Si vis Lunæ ad mare movendum esset ad vim Solis ut $4\frac{4}{7}$ ad 1 (nam proportionem harum virium nondum satis accurate ex phænomenis definire licuit) prodiret annua Æquinoctiorum præcessio $50''.40'''.43''$. Quod cum phænomenis congruit. Nam præcessio illa ex observationibus Astronomicis est $50''$ vel $51''$ circiter.

Descripsimus jam systema Solis Terræ & Planetarum; superest ut de Cometis nonnulla adjiciantur.

the quantity of matter in each space shall not be the same Therefore when You define or assume $y^e$ quantity of matter to be proportionable to its vis inertiæ You must not at the same time define or assume it to be proportionable to $y^e$ space which it may perfectly fill without any void interstices unless You hold it impossible for the two Globes $A$ & $B$ to have different Vires Inertiæ. Now in $y^e$ $3^d$ Corollary I think You do in effect assume both these things at once

## LETTER XXXVI.

### NEWTON TO COTES.

$S^r$

In the scholium to $y^e$ $iv^{th}$ Proposition I should have put the length of $y^e$ Pendulum in vacuo 3 feet & $8\frac{3}{5}$ lines. It was by an accidental error that I wrote $8\frac{2}{5}$ lines. The Pendulum must be something longer in Vacuo then in Aere to vibrate seconds. You may put it either $8\frac{3}{5}$ or $8\frac{10}{19}$ as you shall think fit, the difference being inconsiderable. If you chuse $8\frac{10}{19}$, the numbers computed from thence may stand.

In the new Scholium to the $xiv^{th}$ Proposition, I took the motion of the Aphelium of Mars to be what $D^r$ Halley had computed it & thence deduced the motion of the Earth's Aphelium to be $18'.36''$ in an 100 years. $D^r$ Halley had formerly given me the motion of $y^e$ Aphelium of $\mars$ $40'$ in 100 years & thence I computed the motion of the Earths Aphelium $21'.40''$: but I account the latter recconing to be more confided in, & therefore in the Theory of $y^e$ Moon you may put the motion of $y^e$ earths Aphelium $18'.36''$ in 100 years.

In $y^e$ $16^{th}$ Corollary of Prop. LXVI Lib. 1 (or in pag 166 lin 9 of $y^e$ new Edition) it should be ,[ut tempus periodicum corporis *P* directe &c] as you well observe, & not [ut quadratum temporis periodici] as it is now printed. In the xxxix<sup>th</sup> Proposition these emendations may be made. Pag. 470 lin 26 [ad diametrum majorem *AC* ut 228 ad 229) ut 51984 ad 457 seu 11375 ad 100.] Pag 471 lin 1 [ut 100 ad 11375 et 1000000 ad 925275 conjunctim, hoc est, ut 1000 ad 105042, ideoq: motus annuli esset ad summam motuum annuli et globi ut 1000 ad 106042.] Ib. lin 7 [ut 1000 ad 106042;] Ib. lin 10 [ut 1436 ad 39343 et 1000 ad 106042 conjunctim, id est, ut 1 ad 2919. Ib. lin. 20 [evaderet minor quam prius in ratione 2 ad 5. Ideoq: annuus æquinoxiorum regressus jam esset ad 20° 11′ 46″ ut 1 ad 7298, ac proinde fieret 9″ 57‴ 42 .] Pag 473 lin 27 [Cum igitur inclinatio illa sit 23½ graduum, diminuendus est motus 9″ 57‴ 42<sup>IV</sup> in ratione sinus 91706 (qui sinus est complementi graduum 23½) ad radium 100000. Qua ratione motus ille jam fiet 9″ 8‴ 8<sup>IV</sup>. And a little after. Præcessio a vi Lunæ oriunda 43″. 4‴. 4<sup>IV</sup>½, ac tota Præcessio annua a vi utraq: oriunda 52″ 12‴. 13<sup>IV</sup>.

Si vis Lunæ ad Mare movendum esset ad vim Solis ut 4½ ad 1 (nam proportio harum virium nondum satis accurate ex phænomenis definire licuit) prodiret annua æquinoxiorum præcessio 50″ 14‴. 45<sup>IV</sup>. Quæ cum phænomenis congruit. Nam præcessio illa ex observationibus Astronomicis est vel 50″ vel 51″ circiter.

Si altitudo Terræ ad Æquatorem superet altitudinem ejus ad polos milliaribus plusquam 17, materia ejus rarior erit ad circumferentiam quam ad centrum, et præcessio æquinoxiorum ob altitudinem illam augebitur & vicissim ob raritatem diminuetur.

Descripsimus jam systema Solis Terræ et Planetarum: superest ut de Cometis nonnulla adjiciantur.

ffor obviating the objection you make against the 3<sup>d</sup> Corollary of Prop. vi Lib. iii, you may add to the end of that Corollary these words. Hoc ita se habebit si modo materia sit gravitati suæ proportionalis & insuper impenetrabilis adeoq: ejusdem semper densitatis in spatiis plenis.

I am $Y^r$ most humble Servant

London Feb. 19 $17\frac{11}{12}$.　　　　　　　　　　Is. Newton.

*For the $R^{nd}$ $M^r$ Cotes Professor of Astronomy, at his Chamber in Trinity College in Cambridge.*

---

## LETTER XXXVII.

### COTES TO NEWTON.

$S^r$.　　　　　　　　　　Febr. $23^d$ $17\frac{11}{12}$ Cambridge

I have received Your last. As I reviewd the $xx^{th}$ Proposition I perceiv'd it was by a slip of the Pen that You had put $8\frac{2}{5}$ instead of $8\frac{2}{3}$ lines in Your former Letter. I choose this number rather than $8\frac{10}{19}$ for the reason which You gave & because the fraction is more simple & already in use amongst the French. I am satisfied that these exactnesses, as well here as in other places, are inconsiderable to those who can judge rightly of Your book: but $y^e$ generality of Your Readers must be gratified $w^{th}$ such trifles, upon which they commonly lay $y^e$ greatest stress. I thank You for the information You have given me concerning the new Scholium to the $xiv^{th}$ Proposition. You have very easily dispatch'd the 32 Miles in Prop. $xxxix^{th}$, I think You have put that matter in the best method which the nature of the thing will bear.

Your addition to $y^e$ $3^d$ Corollary of Prop. $vi^{th}$ does not seem to come fully up to $y^e$ Objection. Your words are [Hoc ita se habebit si modo materia sit gravitati suæ proportionalis & insuper impenetrabilis adeoq: ejusdem semper densitatis in spatiis plenis]. Now by *materia* You mean the quantity of Matter & this You had always estimated by its Vis inertiæ, & therefore it will be supposed that You do in this place so estimate it: but if *materia* be here taken in this sense the Objection will not be obviated. Perhaps with some alteration of my words, which You may be pleased to make, the addition may stand thus [Hoc ita se habebit si modo magnitudo vel extensio materiæ in spatiis plenis, sit semper proportionalis materiæ quantitati & vi Inertiæ atq: adeo vi gravitatis: nam per hanc Propositionem constitit quod vis inertiæ & quantitas materiæ sit ut ejusdem gravitas]

In the xix$^{th}$ Proposition pag. 422. lin 9 I will put [1 ad 289] & in line 13$^{th}$ [ut 289 ad 288] in line 15$^{th}$ [289], in line 16$^{th}$ [288] according to Your former directions*. In the 25$^{th}$ & 28$^{th}$ lines I would omit y$^e$ fractions $\frac{2}{15}$ & write [ut 126 ad 125] & [ut 125 ad 126]: for my computation makes the former proportion to be 126,44024 ad 125,44024 & the latter to be 124,80397 ad 125,80397. In Page 423 line 11$^{th}$ I would put [hæ tres rationes 126 ad 125, 126 ad 125$\frac{1}{2}$, & 100 ad 101]. Ib. lin 27$^{th}$ [ut 1 ad 289]. lin 28$^{th}$ [est tantum pars $\frac{1}{289}$] line 31$^{st}$ [vis centrifuga $\frac{1}{289}$] in y$^e$ last line [pars tantum $\frac{1}{229}$]. Page 224, line 1$^{st}$ I would put [per polos 230 ad 229] & y$^e$ rest accordingly taking the measure of a mean degree to be 57230 Toises.

In the xx$^{th}$ Proposition, page 425, line 8$^{th}$, You have altered thus [Unde tale confit Theorema—vel, quod perinde est, ut quadratum sinus recti Latitudinis. Et in eadem circiter ratione triplicata† augentur arcus graduum Latitudinis in Meridiano. Ideoq: cum Latitudo Lutetiæ &c.] I think the word *triplicata* ought to be omitted: it should be [Et in eadem circiter ratione augentur arcus graduum &c]. I suppose by some inadvertency the mistake arose from this, That the degree under y$^e$ Æquator is

---

* In letter of Feb. 12.
† In Newton's MS. the word is *triplice* (No. 138.)

to y$^e$ degree under the Pole as $CP$ cub to $CA$ cub (fig:
page 422). This proportion is no where mentioned in
Your additional papers, but I guess You designed to have
added it or something to y$^e$ same effect to make Your
Rule compleat for finding the measure of a degree under
any Latitude.

When I was formerly upon this place I made the following alteration in order to examine the numbers of Your
Table. [Unde tale confit Theorema, quod incrementum
ponderis ut et mensuræ gradus unius in Meridiano pergendo ab Æquatore ad Polos sit quam proxime ut sinus
versus latitudinis duplicatæ, vel, quod perinde est, ut quadratum sinus recti latitudinis. Nam si $M$ ponatur pro
$$\frac{AB \times PQ \, cub - PQqq}{ABqq}, \quad N \text{ pro } \frac{ABqq - PQqq}{ABqq}, \quad \& \; O \text{ pro}$$
$$\frac{ABq - PQq}{ABq} \text{ (vid: fig: p. 422) erit gravitas sub Æquatore}$$
ad excessum gravitatis in alio quovis loco cujus sinus rec-
tus latitudinis est $S$ existente $R$ radio, ut 1 ad $\dfrac{M}{RR} SS$
$$+ \frac{MN}{R^4} S^4 + \frac{MNN}{R^6} S^6 + \&c.$$ Mensura vero gradus unius in
Meridiano ad Æquatorem, erit ad excessum ejus in alio
loco ut 1 ad $\dfrac{3O}{2RR} SS + \dfrac{3 \times 5OO}{2 \times 4R^4} S^4 + \dfrac{3 \times 5 \times 7O^3}{2 \times 4 \times 6R^6} S^6 + \&c.$
Itaq: cum sit $AB$ ad $PQ$ ut 230 ad 229, & Lutetiæ Parisi-
orum in latitudine 48$^{gr}$. 50′ longitudo penduli singulis minutis secundis oscillantis sit pedum trium Parisiensium &
linearum 8$\frac{5}{9}$; longitudines vero pendulorum æqualibus
temporibus in locis diversis oscillantium sint ut gravitates :
longitudo penduli sub Æquatore erit pedum trium &
linearum 7,48, sub Polo erit pedum trium & linearum 9,39 :
mensura vero gradus unius ad Æquatorem erit Hexapedarum 56783, ad Polum erit Hexapedarum 57530, si modo
inter gradus latitudinis 48 & 49 ponatur esse Hexapedarum
57200. Et simili computo confit Tabula sequens.]

In making these rules I take the measure of a degree at any point of the Meridian to be proportionable to y$^e$ Radius of the curvature of y$^e$ Ellipsis at that point, or which is y$^e$ same thing to be proportionable to y$^e$ Cube of y$^t$ part of the Radius of y$^e$ curvature which is intercepted between y$^e$ point proposed in y$^e$ Ellipsis & the point where the Radius intersects y$^e$ greater Axis; and y$^e$ angle made by that intersection I take for the measure of the Latitude. Thus I had then altered y$^e$ place, but I think this exactness is not necessary; for y$^e$ following terms of these series are inconsiderable in respect of the first, & the figure of the Earth is not exactly Elliptical & the solution of the Problem will be more simple without it, by taking y$^e$ length of y$^e$ Pendulum under the Æquator to y$^e$ length under the Poles in the proportion of 229 to 230, & the Measure of a degree at the Æquator to y$^e$ measure at y$^e$ Poles in the triplicate proportion of 229 to 230 or as 228 to 231 or 76 to 77, & in both cases by making the increment from the Æquator to be as the square of y$^e$ sine of y$^e$ Latitude or as the versed sine of the doubled Latitude.

As to the Table of the lengths of Pendulums & the measures of Degrees I beleive Your Readers would rather desire it were computed to y$^e$ difference of 32 Miles than to that of 17 Miles, & I do not see any use of it as it now stands for which the Table made to the difference of 32 Miles may not serve. If You agree to this Proposal, I will compute it as you shall direct either by the Series or the other way. It must be placed after Your account of the Observations & thereby some small changes will be made in the context which You may be pleased to send me.

What I have further observed as to this Proposition is as follows. You have put down Goreæ Latitudo 14°. 15′. by y$^e$ observations of Des Hayes tis 14°. 40′. In Your account of Picard's experiment of an heated wire You say [in igne posita] De la Hire says only [car M: Picard ayant exposé

les corps a gelée, les mettoit ensuite aupres du feu] or near
the fire.   By my computation the observation at Guada-
loupe reduced to the Æquator gives the difference of 2,29
lines, that at Martinique 2,31 lines, exceeding Your limit of
$2\frac{1}{4}$ lines; the rest fall within Your limits.   After [auctus
in ratione differentiarum fiet milliarium 32] I would add
[& diameter secundum æquatorem erit ad diametrum per
polos ut 123 ad 122] for as 1,07 to 2 so is $\frac{1}{230}$ to $\frac{1}{123}$.
Speaking of the Shadow of the Earth in Lunar Eclipses
You say [diameter ejus ab Oriente in Occidentem ducta,
major erit quam diameter ejus ab Austro in Boream ducta
excessu 56″ fere] I think it should be 41″; for the mean
Horizontal Parallax of $y^e$ Moon in Syzygiis being 57′. 30″,
the Parallax of $y^e$ Sun 10″, & the Suns mean diameter
32′. 12″; the diameter of $y^e$ Shade will be 4988″, add 70″
upon account of the Atmosphære & the diameter will be
5058″, which divided by 123 gives 41″.   At the end of this
Paragraph You have [Et distantia mediocris centrorum
Terræ & Lunæ erit $60\frac{1}{5}$ semidiametrorum Terræ] which I
do'nt well understand.   In $y^e$ last Paragraph You have [et
Pendula isochrona longiora forent in Observatorio Regio
Parisiensi quam ad Æquatorem excessu semissis digiti cir-
citer] I suppose it should be [longiora forent ad Æquatorem
quam in Observatorio] And a little lower You have [Sed
& diameter umbræ Terræ — major foret—excessu 2′. 45″
seu parte duodecima diametri Lunæ] I think it should be
[excessu 2′, seu parte decima sexta diametri Lunæ]
    In the Memoires of the Royale Academie for the Year
1708 there are one or two observations of the lengths of
Pendulums, besides those which You have related in Your
History from other Memoires & from the Observations
faites en plusieurs voyages.
    Taking $y^e$ semidiameters of the Earth to be as 229 &
230 instead of 228 & 229, I have made a small alteration in
Proposition xxxix$^{th}$ which I will not trouble You with since

I think I do understand Your thoughts as to that Proposition. The conclusion of it puts me in mind of an allowance which ought to be made in Prop. xxxvii$^{th}$ on account of the Moons coming nearer to y$^e$ Earth in Syzygiis & going further from it in Quadraturis than in her mean distance at the Octants. But this allowance would increase the number $4\frac{5}{7}$ so much as to give some disturbance to the xxxix$^{th}$ Proposition & the Scholium of the iv$^{th}$ as they now stand, unless You think fit to ballance it some other way, for there is a latitude in that xxxvii$^{th}$ Proposition.

I am, S$^r$, Your most Humble Servant

*For* S$^r$ ISAAC NEWTON *at his House*       ROGER COTES.
*in S$^t$ Martin's Street in Leicester*
*Fields London*

---

LETTER XXXVIII.

NEWTON TO COTES.

S$^r$

I have reconsidered the third Corollary of the vi$^{th}$ Proposition. And for preventing the cavils of those who are ready to put two or more sorts of matter you may add these word{s} to the end of the Corollary. Vim inertiæ proportionalem esse gravitati corporis constitit per experimenta pendulorum. Vis inertiæ oritur a quantitate materiæ in corpore ideoq: est ut ejus massa. Corpus condensatur per contractionem pororum, & poris destitutum (ob impenitrabilitatem materiæ) non amplius condensari potest; ideoq: in spatiis plenis est ut magnitudo spatii. Et concessis hisce tribus Principiis Corollarium valet.

Your emendations of the xix$^{th}$ Proposition may all of them stand.

In the emendation of the xx$^{th}$ Proposition pag 425 lin. 8 the word triplicata should be struck out as you observe. The rest may stand unto the words [Et simili computo fit Tabula sequens] correcting only the numbers as you propose & putting the numbers 229 & 230 instead of 689 & 692. The Table is computed to $y^e$ excess of 17 miles rather then to that of 32 miles, because that of 17 is the least that can be & is certain upon a supposition that the earth is uniform, that of 32 is not yet sufficiently ascertained, & I suspect that it is too big.

After the last observations of Des Hayes ending $w^{th}$ these words [et quod in insula S. Dominici eadem esset ped. 3, lin. 7] add this Paragraph.

Deniq: anno 1704, P. Fuelleus invenit in Porto-belo in America longitudinem Penduli ad minuta secunda oscillantis esse pedum trium Parisiensium et linearum $5\frac{7}{12}$, id est tribus circiter lineis breviorem quam in Latitudine Lutetiæ Parisiorum; & subinde ad insulam Martinicam navigans invenit longitudinem Penduli isochroni esse pedum trium Parisiensium et linearum $5\frac{5}{6}$.

Latitudo autem Paraibæ est $6^{gr}$ 38' in austrum et ea Porto-beli $9^{gr}$ 33' in boream, et Latitudines insularum &c. You may here put the Latitude of Goree $14^{gr}$ 40'. I have not books by me to examin it.

Let the next Paragraph run thus. Observavit utiq: ......... ad ignem calefacta evasit pedis unius cum quarta parte lineæ......... In priore casu calor major fuit quam in posteriore, in hoc vero major fuit quam calor externarum partium corporis humani. Nam metalla ad solem æstivum valde incalescunt......... sed excessu quartam partem lineæ unius vix superante......... differentia prodiit non minor quam $1\frac{19}{20}$ lineæ non multo major quam linearum $2\frac{2}{3}$. Et inter hos limites quantitas mediocris est $2\frac{3}{10}$. Propter calores locorum in Zona torrida negligamus tres decimas partes lineæ et manebit differentia duarum linearum circiter......... jam autus in ratione differentiarum fiet milliarium plus minus 32. Est igitur excessus ille non minor quam milliarium 17, non multo major quam milliarium 32.

I think the words [excessu 56″ fere] are right. ffor the Moons parallax 57′ 30″ must have an increase in the proportion of 32 miles to the earths semidiameter, that is an increase of 28″, w$^{ch}$ doubled give 56″ to be added to y$^e$ diameter of the earths shadow. ffor the Suns diameter & parallax remain without sensible alteration. And for y$^e$ same reason I take [excessu 2′ 45″] to be right.

In the calculation of the Moons force (Prop. xxxvii) your scruple may be eased (I think) by relying more upon the observation of the tyde at Chepstow then on that at Plymouth, but I have mislaid my copy of the calculation. If the nearer access of the Moon to the earth in the Syzygies then in the Quadratures create any difficulty be pleased to send me a copy of the calculation & I will reconsider it. The Latitude of Paris should be 48$^{gr}$ 50′.

I am S$^r$

Yo$^r$ most humble Servant

London Feb. 26 17$\frac{11}{12}$.

Is. NEWTON

*For the* R$^{nd}$ M$^r$ ROGER COTES *Professor*
*of Astronomy at his chamber in*
*Trinity College in Cambridge.*

---

## LETTER XXXIX.

### COTES TO NEWTON.

S$^r$

Febr. 28$^{th}$ 17$\frac{11}{12}$

I have look'd over Your new addition to y$^e$ 3$^d$ Corollary of y$^e$ vi$^{th}$ Proposition, but I am not yet satisfied as to the difficulty, unless You will be pleased to add, that it is true upon this concession that the Primigenial particles out of which the world may be supposed to have been fram'd (concerning which You discourse at large in y$^e$ additions to Your Opticks pag. 343 & seqq.) were all of them created equally dense, that is, (as I would rather speak,) have all the same vis Inertiæ in respect of their real magnitude or extension in spatio pleno. I call this a concession, because I cannot see how it may be certainly proved either a priori by bare reasoning from the nature of the thing, or be inferrd from Experiments. I am not certain whether You do not Your self allow the contrary to be possible. Your words seem to mean so in pag: 347. lin: 5 Optic: [forte etiam & diversis densitatibus diversisq: viribus]

I do not clearly understand how You would have y$^e$ alteration settled in Prop: xx$^{th}$, I mean that which begins with [Unde tale confit Theorema] & ends with [et simili computo confit Tabula sequens]. You may be pleased to send me a transcript of y$^e$ Context leaving void spaces for the Numbers. You may let me know at y$^e$ same time time whether You choose 57200 or 57230 Toises for the Measure of a degree between the Latitudes 48°. 49°. I suppose You retain 8$\frac{5}{9}$ lines for y$^e$ length of y$^e$ Pendulum.

I am satisfied that 56″ is the right increase of y$^e$ shadow of y$^e$ Earth, 'twas my oversight in making the figure of y$^e$ shadow to be similar to that of y$^e$ Earth.

As to the xxxvii$^{th}$ Proposition, I take it that the Moons force must be augmented in her Syzygies & diminished in her Quadratures in the proportion of 47 to 46 nearly. Whence by my computation, if nothing else be altered in the Proposition, $S$ will be to $L$ nearly as 1 to 5$\frac{2}{7}$. To make $S$ to $L$ as 1 to 4$\frac{7}{10}$ or 4$\frac{5}{7}$; instead of putting $L + \frac{6}{7} S$ to $\frac{6}{7} L - \frac{6}{7} S$ as 7 to 4, it may be put $\frac{47}{46} L + \frac{6}{7} S$ to $\frac{46,6}{47,7} L - \frac{6}{7} S$ as 11 to 6. But this proportion of 11 to 6 falls without y$^e$ Limits at Bristol & Plymouth. I shall therefore leave it to Your self to settle y$^e$ whole Proposition as You shall judge it may best be done. In y$^e$ xxviii$^{th}$ Proposition I shall hereafter take notice, that I find the proportion to be as $69\frac{1}{24}$ to $70\frac{1}{24}$ instead of $68\frac{10}{11}$ to $69\frac{10}{11}$. I think 69 to 70 may every where be used. Your Copy of y$^e$ xxxvii$^{th}$ Proposition is as follows *. {Vis Lunæ ad mare movendum colligenda est ex ejus proportione ad vim Solis, et hæc proportio colligenda est ex proportione motuum maris qui ab his viribus oriuntur. Ante ostium fluvii *Avonæ* ad lapidem tertium infra *Bristoliam*, tempore verno et autumnali totus aquæ ascensus in conjunctione et oppositione Luminarium, observante *Samuele Sturmio*, est pedum plus minus 45, in Quadraturis autem est pedum tantum 25. Altitudo prior ex summa virium posterior ex eorundem† differentia oritur. Solis igitur et Lunæ in Æquatore versantium et mediocriter a Terra distantium sunto vires $S$ et $L$, et erit $L + S$ ad $L - S$ ut 45 ad 25 seu 9 ad 5.

In portu *Plymuthi* æstus maris (ex observatione *Samuelis Colepressi*) ad pedes plus minus sexdecim altitudine mediocri attollitur, ac tempore verno et autumnali altitudo æstus in syzygiis superare potest altitudinem ejus in quadraturis pedibus plus septem vel octo. Si maxima harum altitudinum differentia sit pedum octo, erit $L + S$ ad $L - S$ ut 20 ad 12 seu 5 ad 3. Donec aliquid certius ex phænomenis constiterit, assumamus $L + S$ esse ad $L - S$ (proportione mediocri) ut 7 ad 4.

Cæterum ob aquarum reciprocos motus æstus maximi non incidunt in ipsas Luminarium syzygias sed sunt tertii a syzygiis ut dictum fuit, et incidunt in horam Lunarem plus minus tricesimam sextam a syzygiis, id est, in horam solarem tricesimam septimam circiter. Oritur hic æstus ab actione Lunæ in ejus præcedente appulsu ad meridianum

---

* I have transcribed the Proposition from Newton's MS. Nos. 193, 194, Cotes not having copied it into this draught of his letter. The heading is " Invenire vim Lunæ ad Mare movendum."

† sic.

loci et hic appulsus præcedit æstum in portu *Bristoliæ* horis plus minus septem, ideoq: incidit in horam solarem post syzygias et quadraturas tricesimam circiter. Eo autem tempore Luna distat a Sole $15\frac{1}{4}$ gr. circiter. Et Sol in hac distantia minus auget ac minuit motum maris a vi Lunæ oriundum quam in ipsis syzygiis et quadraturis, in ratione Radii ad cosinum distantiæ hujus duplicatæ seu anguli $30\frac{1}{2}$ gr. hoc est, in ratione 7 ad 6 circiter; ideoq: in superiore analogia pro $S$ scribi debet $\frac{6}{7}$ $S$

Sed et vis $L$ in Quadraturis ob declinationem Lunæ diminui debet. Nam Luna in Quadraturis tempore verno et autumnali extra æquatorem in declinatione graduum plus minus $23\frac{1}{2}$ versatur, et Luminaris ab Æquatore declinantis vis ad mare movendum diminuitur in duplicata ratione sinus complementi declinationis quamproxime, & propterea vis Lunæ in his Quadraturis est tantum $\frac{6}{7}$ $L$. Est igitur $L + \frac{6}{7} S$ ad $\frac{6}{7} L - \frac{6}{7} S$ ut 7 ad 4. Et inde fit $S$ ad $L$ ut 7 ad 33 vel 1 ad $4\frac{5}{7}$.

Est igitur vis Solis ad vim Lunæ ut 1 ad $4\frac{5}{7}$ quam proxime. Et hanc proportionem donec aliquid certius ex observationibus accuratius institutis constiterit, usurpare licebit. Unde cum vis Solis sit ad vim gravitatis in superficie Terræ ut 1 ad 12868162, vis Lunæ erit ad vim gravitatis ut 1 ad 2729610 circiter.

Corol. 1. Cum aqua maris vi Solis agitata ascendat ad altitudinem pedis unius & undecim digitorum cum quadrante, eadem vi Lunæ ascendet ad altitudinem pedum novem, & vi utraq: ad altitudinem pedum undecim circiter, et ubi Luminaria sunt in perigæis, ad altitudinem pedum duodecim & ultra, præsertim ubi æstus ventis spirantibus adjuvantur. Tanta autem vis ad omnes maris motus excitandos abunde sufficit, et quantitati motuum probe respondet. Nam in maribus....}

## LETTER XL.

### COTES TO NEWTON.

S$^r$                                              March 13$^{th}$ 17$\frac{11}{12}$

I received Your last of the 26$^{th}$ of February in due time & by the next post I sent You w$^{th}$ one or two other things a Transcript of the xxxvii$^{th}$ Proposition as it now stands in Your Copy. Having received no Letter from You since that time I fear there has been some miscarriage. About two sheets of the iii$^d$ Book are composed, but expecting Your answer I have not yet given leave to print them off.               Your most humble.

## LETTER XLI.

### NEWTON TO COTES.

S$^r$

I have not yet been able fully to settle the Theory of the xix$^{th}$, xx$^{th}$, xxxvi$^{th}$ xxxvii$^{th}$ & xxxix$^{th}$ Propositions & that of the Scholium to the iv$^{th}$. But I think to let the Scholium of iv$^{th}$ Proposition be set at the end of the xxxvii$^{th}$ because it depends on a Corollary of that Proposition. And therefore you may let the Press go on at present without it & set it aside till you come to the xxxvii$^{th}$ Proposition. But let the new Corollary* to y$^e$ iii$^d$ Proposition be printed at the end of that Proposition. And in the third Corollary to y$^e$ v$^{th}$ Proposition strike out the word [novissimam,] & let the words in the latter part of y$^e$ Corollary run thus [Et hinc Jupiter & Saturnus prope conjunctionem se invicem attrahendo sensibiliter perturbant motus mutuos, Sol perturbat &c]. In my copy it is prope

---

* Sent Feb. 2. See p. 57, note *.

conjunctionem novissimam. If it be so in yours, the word novissimam is better omitted.

I thank you for explaining yo$^r$ objection against y$^e$ third Corollary of the sixt Proposition. That Corollary & the next may be put in this manner. Corol. 3. Spatia omnia non sunt æqualiter plena. Nam si spatia omnia æqualiter plena essent, gravitas specifica fluidi quo regio aeris impleretur, ob summam densitatem materiæ, nil cederet gravitati specificæ argenti vivi vel auri vel corporis cujuscunq : densissimi, et propterea nec aurum neq : aliud quodcunq : corpus in aere descendere posset. Nam corpora in fluidis, nisi specifice graviora sint, minime descendunt. Quod si quantitas materiæ in spatio dato per rarefactionem quamcunq : diminui possit, quidni diminui possit in infinitum? Corol. 4. Si omnes omnium corporum particulæ solidæ sint ejusdem densitatis neq : absq : poris rarefieri possint, Vacuum datur. Ejusdem densitatis esse dico quarum vires inertiæ sunt ut magnitudines. Corol. 5. Vis gravitatis diversi est generis a vi magnetica. Nam attractio magnetica non est ut materia attracta. Corpora aliqua magis trahuntur, alia minus, plurima non trahuntur; Et vis magnetica in uno et eodem corpore intendi potest & remitti, estq : nonnunquam longe major pro quantitate materiæ quam vis gravitatis, et in recessu a magnete decrescit in ratione distantiæ non duplicata sed fere triplicata quantum ex crassis quibusdam observationibus animadvertere potui*.

In the tenth Proposition pag. 417 lin 11 for [viginti et unius] read [triginta.] & lin. 12 for [320] read [459] & lin 17 for [800] read [850].

---

* At the meeting of the Royal Society two days afterwards, Newton proposed that Halley and Hauksbee should make experiments with "the great loadstone," in order to find the true law of the decrease, "which he believed would be nearer the cubes than the squares." See also *Journal Book*, March 27, Apr. 3, May 15, Jun. 12, 26. *Phil. Trans.* Jul.—Sept. 1712. June—Aug. 1715. Coulomb's experiments with the Torsion Balance first established the law to be as the squares.

I hope to send you the xix & xx$^{th}$ Propositions emended within a Post or two.   I am S$^r$

Yo$^r$ most humble Servant

\* Mar. 18$^{th}$ 17$\frac{11}{12}$$^{th}$        Is. NEWTON.

For the R$^{nd}$ M$^r$ COTES Professor of
Astronomy in the University of
Cambridge
To be left at Trinity College.

---

## LETTER XLII.

### NEWTON TO COTES.

S$^r$                                      London Apr. 3 1712.

I have been diverted a few days w$^{th}$ some other intervening business, but now send you the emendations of y$^e$ xix$^{th}$ xx$^{th}$ & xxv$^{th}$† Propositions, as follows.

Prop. xix. Prob. ii.

*Invenire proportionem axis Planetæ ad diametros eidem perpendiculares.*

*Picartus* mensurando arcum gradus unius et 22′. 55″ inter *Ambianum* & *Malvoisinam,* invenit arcum gradus unius esse hexapedarum Parisiensium 57060.   Unde ambitus Terræ est pedum Parisiensium 123249600, ut supra.   Sed cum error quadringentesimæ partis digiti tam in fabrica instrumentorum quam in applicatione eorum ad observationes capiendas sit insensibilis, et in Sectore decempedali quo *Galli* observarunt Latitudines locorum respondeat minutis quatuor secundis, et in singulis observationibus incidere possit tam ad centrum Sectoris quam ad ejus circumferentiam, et errores in minoribus arcubus sint majoris

---

\* The date is in Cotes's hand.

† This is an oversight, as this letter does not contain any emendations of Prop. xxv. and in his next letter he speaks of his having sent his corrections of the 19th and 20th Propositions, making no mention of the 25th.

momenti: *ideo *Cassinus* jussu Regio    * *Vide Historiam Aca-*
mensuram Terræ per majora locorum    *demiæ Regiæ Scientiarū*
intervalla aggressus est, et subinde    *anno* 1700.
per distantiam inter Observatorium Regium *Parisiense* et
villam *Colioure* in *Roussillon* & latitudinum differentiā
6$^{gr}$. 18′, supponendo quod figura Terræ sit sphærica, invenit
gradum unum esse hexapedarum 57292, prope ut *Norwoodus*
noster antea invenerat. Hic enim circa annum 1635 mensurando distantiam pedum Londinensium 905751 inter
*Londinum* et *Eboracum* & observando differentiam Latitudinum 2$^{gr}$. 28′ collegit mensuram gradus unius esse pedum
Londinensium 367196, id est, hexapedarum Parisiensium
57300. Ob magnitudinem intervalli a *Cassino* mensurati,
pro mensura gradus unius in medio intervalli illius id est
inter Latitudines 45$^{gr}$ & 46$^{gr}$ usurpabo hexapedas 57292.
Unde, si Terra sit sphærica, semidiameter ejus erit pedum
Parisiensium 19695539.

    Penduli in Latitudine *Lutetiæ Parisiorum* ad minuta
secunda oscillantis longitudo est pedum trium Parisiensium &
linearum 8$\frac{5}{9}$. Et longitudo quod {*sic*} grave tempore minuti
unius secundi cadendo describit est ad dimidiam longitudinem penduli hujus in duplicata ratione circumferentiæ
circuli ad diametrum ejus (ut indicavit *Hugenius*) ideoq : est
pedum Parisiensiū 15, dig. 1, lin. 2$\frac{1}{4}$†, seu linearum 2174$\frac{1}{4}$†.

    Corpus in circulo, ad distantiam pedum 19695539 a
centro, singulis diebus sidereis horarum 23.56′. 4″ uniformiter revolvens, tempore minuti unius secundi describit
arcum pedum 143,6223‡, cujus sinus versus est pedum
0,05236558, seu linearum 7,54064. Ideoq : vis qua gravia
descendunt in Latitudine *Lutetiæ* est ad vim centripetam
corporum in Æquatore a Terræ motu diurno oriundam ut
2174$\frac{1}{4}$† ad 7,54064.

---

† $\frac{1}{4}$ is altered by Cotes in the MS. to $\frac{1}{13}$.
‡ Altered by Cotes to 1436,223.

Vis centrifuga corporum in Æquatore est ad vim
centrifugam qua corpora directe tendunt a Terra in
Latitudine *Lutetiæ* in duplicata ratione Radii ad sinum
complementi Latitudinis illius, id est, ut 7,54064 ad 3,27*.
Addatur hæc vis ad vim qua gravia descendunt in Latitudine *Lutetiæ*, et corpus in Latitudine *Lutetiæ* vi tota
gravitatis cadendo, tempore minuti unius secundi describet
lineas 2177,52† seu pedes Parisienses 15, dig. 1, & lin 5,52†.
Et vis tota gravitatis in Latitudine illa erit ad vim centripetam corporum in Æquatore Terræ ut 2177,52† ad 7,54064,
seu 289 ad 1.

Unde si *APBQ* figuram Terræ designet jam non amplius sphæricā sed revolutione Ellipseos circum axem
minorem *PQ* genitam, sitq : *ACQqca* canalis aquæ plena,
a polo *Qq* ad centrum *Cc*, & inde ad Æquatorem *Aa* per-
gens : debebit pondus aquæ in canalis crure *ACca* esse ad
pondus aquæ in crure altero *QCcq* ut 289 ad 288, eò quod
vis centrifuga ex circulari motu orta partem unam e ponderis partibus 289 sustinebit ac detrahet, et pondus 288 in
altero crure sustinebit reliquas.  [In the rest of the xix^th
Proposition proceed according to the former corrections
untill you come at page 484‡, where read] ad ipsius
diametrum per polos ut 230 ad 229.   Ideoq: cum Terræ
semidiameter mediocris juxta mensuram *Cassini* sit pedum
Parisiensium 19695539, seu milliarium 3939 (posito quod
milliare sit mensura pedum 5000) Terra altior erit ad
Æquatorem quam ad Polos excessu pedum 85820, seu mil-
lia{ri}um 17⅛.

Si Planeta major sit vel minor quam Terra manente
ejus densitate ac tempore periodico revolutionis diurnæ,
manebit proportio vis centrifugæ ad gravitatem, & propterea manebit etiam proportio diametri inter polos ad

---

* Altered by Cotes to 3,267.        † A ˙red by Cotes to 32.
‡ This should be 424.

diametrum secundum æquatorem. At si motus diurnus in
ratione quacunq : acceleretur vel retardetur, augebitur
vel minuetur vis centrifuga in duplicata illa ratione, et
propterea differentia diametrorum augebitur vel minuetur
in eadem duplicata ratione quamproxime. Et si densitas
Planetæ augeatur vel minuatur in ratione quavis, gravitas
etiam in ipsum tendens augebitur vel minuetur in eadem
ratione, et differentia diametrorum vicissim minuetur in
ratione gravitatis auctæ vel augebitur in ratione gravitatis
diminutæ. Unde cum Terra respectu fixarum revolvatur
horis 23.56' Jupiter autem horis 9.56', sintq : temporum
quadrata ut 29 ad 5 et densitates ut 5 ad 1 : differentia
diametrorum Jovis erit ad ipsius diametrum minorem ut
$\dfrac{29}{5} \times \dfrac{5}{1} \times \dfrac{1}{229}$ ad 1, seu 1 ad 8 quamproxime. Est igitur
diameter Jovis ab oriente in occidentem ducta ad ejus
diametrum inter polos ut 9 ad 8 quamproxime, et propterea
diameter inter polos est 35"½. Hæc ita se habent ex hypothesi quod uniformis sit Planetarū materia. Nam si
materia densior sit ad centrum quam ad circumferentiam,
diameter quæ ab oriente in occidentem ducitur erit adhuc
major.

Jovis vero diametrum quæ polis ejus interjacet minorem
esse diametro altera *Cassinus* dudum observavit, et Terræ
diametrum inter polos minorem esse diametro altera patebit per ea quæ dicentur in Propositione sequente.

In the xx<sup>th</sup> Proposition page 425 lin. 8, read. Unde
tale confit Theorema, quod incrementum ponderis pergendo
ab Æquatore ad Polos, sit quam proxime ut sinus versus
Latitudinis duplicatæ, vel, quod perinde est, ut quadratum
sinus recti Latitudinis. Et in eadem circiter ratione augentur arcus graduum Latitudinis in Meridiano. Ideoq :
cum Latitudo *Lutetiæ Parisiorum* sit 48<sup>gr</sup>. 50', ea locorum
sub Æquatore 00<sup>gr</sup>. 00', et ea locorum ad Polos 90<sup>gr</sup> &

duplorum sinus versi sint 11334, 00000 et 20000, existente Radio 10000, et gravitas ad Polum sit ad gravitatem ejus sub Æquatore ut 229 ad 228, & excessus gravitatis ad polum ad gravitatem sub Æquatore ut 1 ad 228 : erit excessus gravitatis in Latitudine *Lutetiæ* ad gravitatem sub Æquatore, ut $1 \times \dfrac{11334}{20000}$ ad 228 seu 5667 ad 2280000. Et propterea gravitates totæ in his locis erunt ad invicem ut 2285667 ad 2280000. Quare cum longitudines pendulorum æqualibus temporibus oscillantium sint ut gravitates, et in Latitudine *Lutetiæ Parisiorum* longitudo penduli singulis minutis secundis oscillantis sit pedum trium Parisiensium & $8\frac{4}{9}$ linearum, longitudo penduli sub Æquatore superabitur a longitudine synchroni penduli *Parisiensis*, excessu lineæ unius et 92 partium millesimarū lineæ. Et simili computo confit Tabula sequens.

| Latitudo Loci. | Longitudo Penduli. | | Mensura gradus unius in Meridiano |
|---|---|---|---|
| Grad. | Ped. | Lin. | Hexaped. |
| 0 | 3 . | 7,463 | 56907 |
| 5 | 3 . | 7,478 | 56913 |
| 10 | 3 . | 7,521 | 56930 |
| 15 | 3 . | 7,592 | 56957 |
| 20 | 3 . | 7,689 | 56995 |
| 25 | 3 . | 7,808 | 57041 |
| 30 | 3 . | 7,945 | 57095 |
| 35 | 3 . | 8,098 | 57154 |
| 40 | 3 . | 8,260 | 57218 |
| 45 | 3 . | 8,427 | 57283 |
| 46 | 3 . | 8,461 | 57296 |
| 47 | 3 . | 8,494 | 57309 |
| 48 | 3 . | 8,528 | 57322 |
| 49 | 3 . | 8,561 | 57335 |
| 50 | 3 . | 8,594 | 57348 |
| 55 | 3 . | 8,756 | 57412 |
| 60 | 3 . | 8,909 | 57471 |
| 65 | 3 . | 9,046 | 57525 |
| 70 | 3 . | 9,165 | 57571 |
| 75 | 3 · | 9,262 | 57602 |
| 80 | 3 . | 9,383 | 57626 |
| 85 | 3 . | 9,376 | 57653 |
| 90 | 3 . | 9,391 | 57659 |

Constat autem per hanc Tabulam &c
Hæc ita se habent ex hypothesi quod Terra &c
Jam vero Astronomi aliqui in longinquas regiones &c.
Deinde anno 1682 D. Varini &c.
Posthac D. Couplet filius anno 1697
Annis proximis (1699 & 1700) D. Des Hayes &c
Annoq : 1704 P. Feuelleus invenit in Po{r}tobelo in
America Longitudinem Penduli ad minuta secunda oscillantis esse pedum trium Parisiensium et linearum tantum
$5\frac{7}{12}$, id est tribus fere lineis breviorem quam Lutetiæ
Parisiorum, sed errante Observatione. Nam deinde ad
insulam Martinicam navigans invenit longitudinem Penduli
isochroni esse pedum tantum trium Parisiensium et linearum $5\frac{10}{12}$.

Latitudo autem Paraibæ est $6^{gr}$ $38'$ ad austrum et ea
Portobeli $9^{gr}$ $33'$ ad boream, et Latitudines insularum
Cayennæ, Goreæ, Guadaloupæ, Martanicæ, Granadæ, $S^{ti}$
Christophori & $S^{ti}$ Dominici sunt respective $4^{gr}$ $55'$, $14^{gr}$ $40'$,
$14^{gr}$ $00'$, $14^{gr}$ $44'$, $12^{gr}$ $6'$, $17^{gr}$ $19'$ & $19^{gr}$ $48'$ ad boream. Et
excessus longitudinis Penduli ...... auxerint.

Observavit utiq : D. Picartus quod virga ferrea, quæ
tempore hyberno ubi gelabant frigora erat pedis unius
longitudine, ad ignem calefacta evasit pedis unius cum
quarta parte lineæ. Deinde D. de la Hire ...... cum duabus tertiis partibus lineæ. In priore casu calor major fuit
quam in posteriore, in hoc vero major fuit quam calor
externarum partium corporis humani. Nam metalla ad
solem æstivum valde incalescunt. At virga penduli ......
quam hyberno, sed excessu quartam partem lineæ unius
vix superante. Proinde ...... differentia illa prodiit haud
minor quam $1\frac{18}{20}$ lineæ, haud major quam $2\frac{1}{2}$ linearum. Et
inter hos limites quantitas mediocris est $2\frac{9}{40}$ linearum.
Propter calores locorum in Zona torrida negligamus $\frac{9}{40}$
partes lineæ et manebit differentia duarum linearum.

Quare cum differentia illa per Tabulam præcedentem

ex hypothesi quod Terra ex materia uniformiter densa constat, sit tantum 1 $\frac{87}{1000}$* lineæ: excessus altitudinis Terræ ad æquatorem supra altitudinem ejus ad polos, qui erat milliarium 17$\frac{1}{6}$, jam auctus in ratione differentiarum, fiet milliarium 31$\frac{1}{3}$†. Nam tarditas Penduli sub Æquatore defectum gravitatis arguit; et quo levior est materia eo major esse debet altitudo ejus ut pondere suo materiam sub Polis in æquilibrio sustineat.

Hinc figura umbræ Terræ per eclipses Lunæ determinanda, non erit omnino circularis sed diameter ejus ab oriente in occidentem ducta, major erit quam diameter ejus ab austro in boream ducta, excessu 55″ circiter. Et parallaxis maxima Lunæ in Longitudinem paulo major erit quam ejus parallaxis maxima in Latitudinem. Ac Terræ semidiameter maxima erit pedum Parisiensium 19764030, minima pedum 19609860 & mediocris pedum 19686945 quam proxime.

Cum gradus unus mensurante *Picarto* sit hexapedarum 57060, mensurante vero *Cassino* sit hexapedarum 57292 : suspicantur aliqui......seu parte duodecima diametri Lunæ. Quibus omnibus experientia contrariatur. Certe Cassinus, definiendo gradum unum esse hexapedarum 57292, medium inter mensuras suas omnes, ex hypothesi de æqualitate graduum assumpsit. Et quamvis *Picartus* in *Galliæ* limite boreali invenit gradum paulo minorem esse, tamen Norwoodus noster in regionibus magis mensurando majus intervallum, invenit gradum paulo majorem esse quam *Cassinus* invenerat. Et *Cassinus* ipse mensuram *Picarti* ob parvitatem intervalli mensurati non satis certam & exactam esse judicavit ubi mensuram gradus unius per intervallum longe majus definire aggressus. Differentiæ vero inter mensuras *Cassini, Picarti* & *Norwoodi* sunt prope

---

* Newton had written 92, but Cotes has altered it to 87. See Cotes's next letter.
† Cotes has drawn a line round the $\frac{1}{3}$ and written $\frac{7}{12}$ by the side of it.

insensibiles & ab insensibilibus observationum erroribus facile oriri potuere, ut nutationem axis Terræ præteream.

Pag. 424 lin penult. read 229 ad 228.

The rest of the Propositions to Prop. xxxvi, may continue as they are, w$^{th}$ y$^e$ corrections already sent you. I will speedily send you the corrections of y$^e$ xxxvi, xxxvii, & xxxix Propositions.

I am

Yo$^r$ very humble Servant

ISAAC NEWTON.

The following is in Cotes's hand.

| | | |
|---|---|---|
| " Maxima | 19767630 | 19688725 |
| Minima | 19609820 | 19714886 |
| Mediocris | 19688725 | 39403611 |
| Sem$^d$. Sph : Æqu : | 19714886 | 19701805 Media Mediarum." |

---

LETTER XLIII.

NEWTON TO COTES.

S$^r$                                      London Apr 8$^{th}$ 1712.

I sent you by D$^r$ Bently my emendations of the 19$^{th}$ & 20$^{th}$ Propositions, & now send you those of the 36$^{th}$ & 37$^{th}$. When you have perused them I should be glad to have your thoughts upon them, & if any thing else want to be corrected before you come at y$^e$ 39$^{th}$ Proposition. In my next I intend to send you my emendations of that Proposition.

I am

Yo$^r$ most humble Servant

*For the* R$^{nd}$ M$^r$ COTES *Professor of Astro-*                Is. NEWTON.
*nomy, at his chamber in Trinity College*
*in Cambridge.*

All that is preserved of the emendations of Prop. xxxvi. is contained in a small slip of paper (No. 192); it relates to the Corollary and is as follows:

"In Prop. xxxvi. pag. 464 lin. 3, read 85820; & lin. 9 read, et digitorum undecim cum triente, Est enim hæc mensura ad mensuram pedum 85820 ut 1 ad 44038."

Cotes, however, afterwards (letter of Apr. 26) altered the numbers in the Corollary otherwise, and the changes together with his other suggestions were approved of by Newton (letter of May 10). The emended form of Prop. xxxvii. coincides with Newton's previous copy (a transcript of which Cotes sent him Feb. 28), as far as the middle of the 2nd paragraph except that "earundem" appears in the right gender. It is not necessary therefore to print that common part again, but it will be sufficient to begin our transcript at the point where the first correction shews itself. (Nos. 195-198) ... "Si maxima harum altitudinum differentia sit pedum novem, erit $L + S$ ad $L - S$ ut $20\frac{1}{2}$ ad $11\frac{1}{2}$ seu 41 ad 23. Quæ proportio satis congruit cum priore. Ob magnitudinem æstus in Portu *Bistoliæ**, observationibus *Sturmii* magis fidendum esse videtur, ideoq: donec aliquid certius constiterit, proportionem 9 ad 5 usurpabimus.

Cæterum ob aquarum reciprocos motus, æstus maximi non incidunt in ipsas Luminarium syzygias, sed sunt tertii a syzygiis ut dictum fuit, seu proxime tertium Lunæ post syzygias appulsum ad meridianum loci, vel potius tertium post tertiam circiter vel quartam a syzygiis horam appulsum ad meridianum loci. Æstas et hyems maxime vigent, non in ipsis solstitiis, sed ubi sol distat a novissimis solstitiis decima circiter vel undecima parte totius circuitus, seu gradibus plus minus 35. Et similiter maximus æstus maris oritur ab appulsu Lunæ ad meridianum loci ubi Luna distat a Sole decima vel undecima parte motus totius ab æstu ad æstum, seu gradibus plus minus septendecim cum dimidio. Et Sol in hac distantia minus auget vel diminuit motum maris a vi Lunæ oriundum quam in ipsis syzygiis et quadraturis in ratione Radii ad sinum complementi distantia* hujus duplicatæ seu anguli graduum 35, hoc est, in ratione 1000000 ad 819152 ; ideoq : in analogia superiore pro $S$ scribi debet $0,819152 S$.

Sed et vis Lunæ in Quadraturis, ob Declinationem Lunæ ab Æquatore, diminui debet. Nam Luna in Quadraturis vel potius in gradu $17\frac{1}{2}$ post Quadraturas, tempore Æquinoctiorum, in Declinatione graduum plus minus 22 & 21′ versatur. Et Luminaris ab Æquatore Declinantis vis ad mare movendum diminuitur in duplicata ratione sinus complementi Declinationis quamproxime. Et propterea vis Lunæ in his Quadraturis est tantum $0,85539968 L$. Est igitur $L + 0,81952 * S$ ad $0,85539968 L - 0,81952 * S$ ut 9 ad 5.

Præterea diametri Orbis in quo Luna absq: excentricitate moveri

---

* sic.

deberet sunt ad invicem ut 69 ad 70 (per Prop. xxviii,) ideoq : distantia Lunæ a Terra in Syzygiis est ad distantiam ejus in Quadraturis ut 69 ad 70 cæteris paribus. Et distantia ejus in gradu $17\frac{1}{2}$ a syzygiis ubi æstus maximus generatur est ad distantium ejus in gradu $17\frac{1}{2}$ a Quadraturis ubi æstus minimus generatur ut 83,8317 ad 84,8317, id est, ut 1 ad 1,0119286 vel 0,9882125 ad 1. Unde fit $1,0119286 L + 0,819152 S$ ad $0,9882125 \times 0,85539968 L - 0,819152 S$ ut 9 ad 5. Et $S$ ad $L$ ut 1 ad $4\frac{1}{2}$.

Corol. 1. Cum igitur aqua vi Solis agitata ascendat ad altitudinem pedis unius et digitorum undecim cum triente, eadem vi Lunæ ascendet ad altitudinem pedum octo et digitorum novem. Tanta autem vis &c.

Corol. 2. Cum vis Lunæ ad mare movendum &c.

Corol. 3. Quoniam vis Lunæ ad mare movendum est ad Solis vim consimilem ut $4\frac{1}{2}$ ad 1, et vires illæ (per Corol. 14 Prop. lxvi Libr. i) sunt ut densitates corporum Lunæ & Solis & cubi diametrorum apparentum conjunctim : erit densitas Lunæ ad densitatem Solis ut $4\frac{1}{2}$ ad 1 directe et cubus diametri Lunæ ad cubum diametri Solis inverse, id est, (cum diametri mediocres apparentes Lunæ et Solis sint $31'.16''$ et $32' 12''$) ut 49112* ad 10000. Densitas autem Solis erat ad densitatem Terræ ut 100 ad 396 et propterea densitas Lunæ est ad densitatem Terræ ut 49112* ad 39600 seu 31 ad 25. Est igitur corpus Lunæ densius et magis terrestre quam Terra nostra.

Corol. 4. Et cum vera diameter Lunæ (ex observationibus Astronomicis) sit ad veram diametrum Terræ ut 100 ad 365, erit massa Lunæ ad massam Terræ ut 1 ad $39\frac{1}{5}$.

Corol. 5. Et gravitas acceleratrix in superficie Lunæ erit triplo minor quam gravitas acceleratrix in superficie Terræ.

Corol. 6. Et distantia centri Lunæ a centro Terræ erit ad distantiam centri Lunæ a communi gravitatis centro Lunæ ac Terræ ut $40\frac{1}{5}$ ad $39\frac{1}{5}$.

Corol. 7. Et distantia mediocris centrorum Lunæ ac Terræ æqualis erit maximis Terræ semidiametris $60\frac{1}{4}$ quam proxime. Nam Terræ semidiameter maxima fuit pedum Parisiensium 19764030. Et hujusmodi semidiametri $60\frac{1}{4}$ æquantur pedibus 1190782815. Et si hæc sit distantia centrorum Solis et Lunæ, eadem (per Corollarium novissimū) erit ad distantiam centri Lunæ a communi gravitatis centro Lunæ ac Terræ ut $40\frac{1}{5}$ ad $39\frac{1}{5}$, quæ proinde est pedum 1161161352. Et cum Luna revolvatur respectu fixarum diebus 27 horis 7 & minutis primis $43\frac{1}{5}$, sinus versus anguli quem Luna tempore minuti unius primi motu suo medio circa commune gravitatis centrum Lunæ ac Terræ describit

---

* The last two figures are altered by Cotes to 51. The " n " in " sint " (lin. 18), the " 2 " in lin. 37 & the " 5 " in lin. 6 (p. 91) seem also due to him.

est 1,275235, existente Radio 100,000000,000000. Et ut Radius est ad hunc sinum versum ita sunt pedes 116116135 ad pedes 14,807536. Luna igitur vi illa qua retinetur in orbe, tempore minuti unius primi cadendo describeret pedes 14,807536. Et hæc vis (per Corol. Prop. III est ad vim gravitatis nostræ in orbe Lunæ ut 177$\frac{29}{40}$ ad 178$\frac{29}{40}$; proindeq: corpus grave in orbe Lunæ ad distantiam pedum 1190782815 a centro Terræ, vi gravitatis nostræ in Terram cadendo, tempore minuti unius primi describeret pedes 14,8908, & ad sexagesimam partem distantiæ illius, id est ad distantiam pedum 1984638 a centro Terræ, vi gravitatis in Terram cadendo tempore minuti unius secundi describeret etiam pedes 14,8908, et ad distantiam pedum 19694278 a centro Terræ cadendo eodem tempore minuti unius secundi describeret pedes 15,1217 seu pedes 15, dig. 1, et lin. 5$\frac{1}{2}$. Et hac vi gravia cadunt in superficie Terræ in Latitudine urbis Lutetiæ Parisiorum, ut supra ostensum est. Et distantia pedum 19694278 paulo major est quam Terræ semidiameter mediocris, et paulo minor quam semidiameter globi cui Terra æqualis est, suntq: differentiæ insensibiles; ac proinde vis qua Luna retinetur in Orbe suo ad distantiam prædictam semidiametrorum 60$\frac{1}{4}$, si descendatur in Terram, congruit cum vi gravitatis quam experimur in superficie Terræ.

Corol. 8. Distantia mediocris centrorum Lunæ ac Terræ æqualis est mediocribus Terræ semidiametris 60$\frac{1}{2}$ quamproxime. Nam tot semidiametri mediocres sunt pedum 1191060172.

Siquando mensuræ graduum in meridiano, longitudes* pendulorum isochronorum in diversis parallelis Terræ, leges fluxus & refluxus maris, diametri apparentes Solis et Lunæ, & Lunæ parallaxis horizontalis ex phænomenis accuratius determinatæ fuerint: licebit calculum hunc omnem accuratius repetere."

---

## LETTER XLIV.

### COTES TO NEWTON.

S$^r$.    Cambridge Aprill y$^e$ 14$^{th}$ 1712

I have received Your Letter by D$^r$ Bentley & the other which You wrote since. I have sent You two Proof Sheets† for Your revisal, having made some alterations in them different from Your Copy.

In Page 379 line 6 I have put [lin. 2 $\frac{1}{18}$] instead of

---

* sic.    † Ccc, Ddd, pp. 377—392.

[lin. 2 $\frac{1}{4}$]. In line 10$^{th}$ 1436,223 instead of 143,6223. In line 21$^{st}$, 2177,32 instead of 2177,52

In Page 382 I have put the proportion of 230 to 229 instead of 229 to 288* and altered the latter part of y$^e$ Page accordingly & computed the Table anew in the next Page. The Latter Column supposes the measure of a degree at y$^e$ Latitude of 45°. 41' to be 57292 Toises as I think You put it in Your Table. The two extreme numbers are as the Cubes of 230 & 229, In y$^e$ rest the increment from y$^e$ Æquator is as the Versed Sine of y$^e$ doubled Latitude.

In Page 386 lin: penult. 1 $\frac{87}{1000}$ for 1 $\frac{92}{1000}$. Page 387 lin. 1 31 $\frac{7}{12}$ for 31 $\frac{1}{3}$. Line 11$^{th}$ I have put other numbers for y$^e$ semidiameters of the Earth, which I desire You would examine, since there are different ways of coming at those numbers & I may not possibly have taken that which You like best. Line 21$^{st}$ I put 95 Miles for 94. Line 27$^{th}$ 2'. 46'' for 2'. 45. Line 32 *Norwoodus noster in regionibus magis borealibus*, the word borealibus or something to y$^t$ effect was omitted in Your copy

In Page 389: line 26$^{th}$ I have put 8°. 24 for 9°. 34. In the last Period of y$^e$ same xxiii$^d$ Proposition I have made an alteration which You will see.

I think You have much improved the Method of the whole, but there seemes to be a mistake in y$^t$ Section of Prop xxxvii which begins with *Prœterea diametri Orbis in quo Luna &c.* The Moons force in her Syzygies & Quadratures should be increased & diminished in the triplicate proportion of those distances to her mean distance reciprocally, Your correction is nearly according to y$^e$ duplicate proportion. I am streightned in time at present, & will explain myself more fully in my Next

Your most humble Serv$^t$

R C

---

* Slip of the pen for 228.

S$^r$.                                   Cambridge April 15$^{th}$ 1712.

I hope You have received the sheets sent You by the
Carrier for Your examination, with my Letter. I come
now to the xxv$^{th}$ Proposition which I think were better to
end thus ...... ad dies 365. 6$^h$. 9'. id est, ut 1000 ad 178725
seu 1 ad 178 $\frac{29}{40}$. Unde ex proportione linearum *TM*, *ML*,
datur etiam vis *TM* : & hæ sunt vires Solis quibus Lunæ
motus perturbantur. Q.E.I. The two Periods which are left
out may be removed to Prop: xxxvi for I think they are
of no use till we come to that Proposition. If You remove
them I suppose You will at the same time alter them, by
putting in line 14$^{th}$ instead of y$^e$ proportion of 60$\frac{1}{4}$* to 60
the proportion of 40$\frac{1}{5}$ to 39$\frac{1}{5}$, if this be the Proportion
which may at last stand in Corol. 6$^{th}$ of Prop. xxxvii$^{th}$.
Now because the Proportion of 40$\frac{1}{5}$ to 39$\frac{1}{5}$ is made out in
y$^e$ xxxvii$^{th}$ Proposition, the xxxvi & xxxvii$^{th}$ Propositions
ought to change places, but this they cannot do because
the xxxvii$^{th}$ does in other respects depend upon y$^e$ xxxvi$^{th}$.
Whence it appeares that there ought to be a fur⁺her
alteration in y$^e$ Form of these Propositions, that the former
may not depend upon the latter. This may easily be done
& I think the whole would be clearer & more Methodical
if in y$^e$ former Proposition the Problem were to find
neither y$^e$ force of y$^e$ sun nor the force of y$^e$ moon, but
only their proportion to each other, & in y$^e$ latter the
Problem were to find the proportion of both forces to y$^e$
force of Gravity. And thus y$^e$ 3$^d$, 4$^{th}$, 5$^{th}$, 6$^{th}$, 7$^{th}$, & 8$^{th}$
Corollarys of y$^e$ xxxvii$^{th}$ will belong to y$^e$ former, & the
Corollary of y$^e$ xxxvi$^{th}$ together with the 1$^{st}$ & 2$^d$ corollarys
of y$^e$ xxxvii$^{th}$ will belong to y$^e$ latter. There will be this

---

* It should be $\frac{1}{3}$.

further advantage in the change, That in $y^e$ 7$^{th}$ Corollary of $y^e$ xxxvii$^{th}$ which will then be annex'd to $y^e$ former Proposition a good foundation may be laid for making out $y^e$ latter. In my Letter which I yesterday wrote to You I was somewhat in haste, I just mention'd a difficulty in Prop : xxxvii. Let $ST$ be the Moons distance from $y^e$ Earth when she is $17°\frac{1}{2}$ from her Syzygies & $QT$ be her distance at $17°\frac{1}{2}$ from her Quadratures & $MT$ her mean distance in $y^e$ Octants.

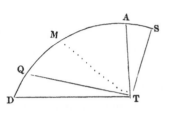

I think the force of $y^e$ Moon must be increased at $S$ in the proportion of $MT$ cub to $ST$ cub, & diminished at $Q$ in the proportion of $MT$ cub to $QT$ cub. Your last corrections increase it at $S$ in $y^e$ proportion of $QT$ to $ST$, which is nearly in the proportion of $MTq$ to $ST$ quad, & diminish it at $Q$ in $y^e$ same proportion. I could wish when the whole is settled that the proportion of $4\frac{1}{2}$ to 1 may be retain'd for the sake of Proposition xxxix.* I think there is no Proposition in Your Book which does more deserve Your care.

---

### LETTER XLVI.
#### NEWTON TO COTES·

S$^r$                                        London Apr. 22. 1712.

I have run my eye over the two proof sheets & approve yo$^r$ corrections. The sheets may be printed off. The xxv$^{th}$ Proposition may end thus.—ad dies 365. 6$^h$ 9′, id est ut 1000 ad 178725 seu 1 ad $178\frac{29}{40}$. Invenimus autem in Propositione quarta quod, si Terra et Luna circa commune gravitatis centrum revolvantur, earum distantia mediocris

---

* "Invenire Præcessionem Æquinoctiorum."

ab invicem erit $60\frac{1}{2}$ semidiametrorum mediocrium Terræ quamproxime. Et vis qua Luna in Orbe circa Terram quiescentem ad distantiam semidiametrorum 60 revolvi posset {est ad vim qua eodem tempore ad distantiam semidiametrorum 60 revolvi posset *,} ut $60\frac{1}{2}$ ad 60 & hæc vis ad vim gravitatis apud nos ut 1 ad 60 × 60. Ideoq: vis mediocris $ML$ est ad vim gravitatis in superficie Terræ ut $1 × 60\frac{1}{2}$ ad $60 × 60 × 60 × 178\frac{29}{40}$, seu 1 ad 638092,6. Unde ex proportione linearum $TM$, $ML$, datur etiam vis $TM$. Et hæ sunt vires Solis quibus motus Lunæ perturbantur. Q. E. I.

I here referr the summ of $y^e$ forces upon† the Sun upon the earth & Moon to the Moon alone & therefore consider the earth as resting & referr its motion to the Moon.

I am satisfied that the force of the Moon upon the Sea is in a triplicate ratio of her distance reciprocally & have altered the calculations accordingly, $w^{ch}$ I send you in the inclosed paper together with the emendation of the 39$^{th}$ Proposition.

I am

Yo$^r$ most humble Servant

For the R$^{nd}$ M$^r$ COTES, *Professor of* Is. NEWTON.
*Astronomy, at his Chamber in*
*Trinity College in Cambridge.*

The "inclosed paper" mentioned at the end of this letter is a folio sheet (Nos. 202,203,208), and contains Newton's further corrections of Prop. XXXVII. called for by the two preceding letters, and also those of the 39th Prop. which he had promised in his letters of Apr. 3d and 8th, (compare letters of March 18, Feb. 19, Feb. 16, and Feb. 12). It is not necessary to copy the whole of what relates to Prop. XXXVII. Every useful end will be answered by giving only those parts of it where it differs from the copy which Newton had recently sent (Apr. 8), leaving blanks to represent what is common to the two. The paper begins as follows: "In Prop XXXVII read Cæterum ob aquarum reci-

---

* I have added the words between braces from the 1st Ed. The identity of termination of the two clauses with "revolvi posset," combined with a little hurry in transcribing, will readily account for their omission.

† This should be "of."

procos motus…seu proxime sequuntur tertium Lunæ…vel potius (ut a *Sturmio* notatur) sunt tertii post diem novilunii vel plenilunii, seu post horam a novilunio vel plenilunio plus minus duodecimam, adeoq : incidunt in horam a novilunio vel plenilunio plus minus quadragesimam tertiam.

Incidunt vero in hoc portu in horam septimam circiter ab appulsu Lunæ ad meridianum loci, ideoq : proxime sequuntur appulsum Lunæ ad meridianum ubi Luna distat a Sole vel oppositione Solis gradibus plus minus octodecim vel novendecim in consequentia. Æstas … Sol distat a solstitiis decima circiter parte totius circuitus seu gradibus plus minus 36 vel 37 … a Sole decima circiter parte motus totius ab æstu ad æstum. Sit distantia illa graduum plus minus 18½. Et vis Solis in hac distantia Lunæ a syzygiis & quadraturis, minor erit ad augendum et ad minuendum motum … seu anguli graduum 37, hoc est, in ratione 10000000 ad 7986355. Ideoq : …debet 0,7986355 *S*. … in gradu 18½ post Quadraturas, in Declinatione graduum plus minus 22.13′ versatur…… est tantum 0,8570328 *L*. Est igitur *L* + 0,7986355 *S* ad 0,8570328 *L* − 0,7986355 *S* ut 9 ad 5….. ut 69 ad 70; ideoq : distantia…. cæteris paribus. Et distantiæ ejus in gradu 18½ a syzygiis … maximus generatur, & in gradu 18½ a quadraturis ubi æstus minimus generatur, sunt ad mediocrem ejus distantiam ut 69,100682 & 69,899318 ad 69½. Vires autem Lunæ ad mare movendum sunt in triplicata ratione distantiarum inverse, ideoq : vires in maxima et minima harum distantiarum sunt ad vim in medi{o}cri distantia ut 0,9828616 et 1,017342 ad 1. Unde fit 1,017342 *L* + 0,7986355 *S* ad 0,9828616 × 0,8570328 *L* − 0,7986355 *S* ut 9 ad 5. Et *L* = 4,4824 *S*.

Corol. 1 & 2, as before.

Corol. 3. … ut 4,4824 ad 1…ut 4,4824 ad 1 … sint 31′ 16″½… ut 4892 ad 1000 … ad densitatem    Terræ ut 4892 ad 3960 seu 21 ad 17. Est igitur …

Corol. 4. … ad massam Terræ ut 1 ad 39,363.

Corol. 5. … erit quasi triplo minor…

Corol. 6. … ut 40,363 ad 39,363.

Corol. 7. Et mediocris distantia centri Lunæ a centro Terræ erit semidiametrorum maximarum Terræ 60¼ quam proxime. Nam semidiameter maxima Terræ fuit pedum Parisiensium 19767630, et mediocris distantia centrorum Terræ et Lunæ ex hujusmodi semidiametris 60¼ constans, æqualis est pedibus 1190999707. Et hæc distantia (per Corollarium superius) est ad distantiam … centro Terræ et Lunæ ut 40,363 ad 39,363, quæ proinde est pedum 1161492740. Et cum Luna … centrum Terræ et Lunæ describit est 1275235, existente … pedes 1161492740 ad pedes 14,811762. Luna … in Orbe, cadendo in Terram, tempore minuti unius primi describet pedes 14,811762.

Et si hæc vis augeatur in ratione $177\frac{29}{40}$ ad $178\frac{29}{40}$ habebitur vis tota gravitatis in Orbe Lunæ per Corol. Prop. III. Et hac vi Luna cadendo, tempore minuti* unius primi describere deberet pedes 14,89513. Et ad sexagesimam partem hujus distantiæ, id est, ad distantiam pedum 19849995 a centro Terræ corpus grave cadendo, tempore minuti unius secundi describere deberet etiam pedes 14,89513. Diminuatur hæc distantia in subduplicata ratione pedum 14,89513 ad pedes 15,12028, et habebitur distantia pedum 19701651 a qua grave cadendo, eodem tempore minuti unius secundi describet pedes 15,12028, id est pedes 15, dig 1, lin 5,32. Et hac vi ... urbis *Lutetiæ Parisiorum*, ut supra ostensum est. Est autem distantia pedum 19701651 paulo minor quam semidiameter globi huic Terræ æqualis et paulo major quam Terræ hujus semidiameter mediocris ut oportet. Sed differentiæ sunt insensibiles. Et propterea vis qua Luna ... ad distantiam maximarum Terræ semidiametrorum $60\frac{1}{4}$, ea est quam vis gravitatis in superficie Terræ requirit.

Corol. 8. .... centrorum Terræ et Lunæ est mediocrium Terræ semidiametrorum $60\frac{1}{2}$ quam proxime. Nam semidiameter mediocris quæ erat pedum 19688725 est ad semidiametrum maximam pedum 19767630, ut $60\frac{1}{4}$ ad $60\frac{1}{2}$ quamproxime.

In his computationibus attractionem magneticam Terræ non consideravimus, cujus utiq : quantitas perparva est et ignoratur. Siquando vero hæc attractio investigari poterit, et mensura graduum in meridiano, ac longitudines ... parallelis, legesq : motuum maris, & parallaxis Lunæ cum diametris apparentibus Solis et Lunæ ex phænomenis ... ”

The following are the corrections of the 39th Prop. " In the xxxix$^{\text{th}}$ Proposition pag 470 lin 23 write —— id est (cum Terræ diameter minor $PC$ vel $aC$ sit ad diametrum majorem $AC$ ut 229 ad 230,) ut 52441 ad 459 ; si annulus iste Terram secundum Æquatorem cingeret & uterq : simul circa diametrum annuli revolveretur, motus annuli esset ad motum globi interioris (per hujus Lemma III) ut 459 ad 52441 et 1000000 ad 925275 conjunctim, hoc est, ut 4590 ad 485223, ideoq : motus annuli esset ad summam motuum annuli ac globi ut 4590 ad 489813. Vnde si annulus globo adhæreat, & motum suum quo ipsius Nodi seu puncta æquinoctialia regrediuntur, cum globo communicet : motus qui restabit in annulo erit ad ipsius motum priorem ut 4590 ad 489813 ; et propterea motus punctorum æquinoctialium diminuetur in eadem ratione. Erit igitur motus annuus punctorum æquinoctialium corporis ex annulo et globo compositi

---

* "minuti" here & "quam" p. 98. lin. 20 have been added by Cotes, who has made a number of other alterations in the MS., the principal of which are mentioned in Letter XLVIII.

ad motum 20$^{gr}$ 11′ 46″, ut 1436 ad 39343 et 4590 ad 489813 conjunctim, id est, ut 100 ad 292368.   Vires autem quibus &c.

Pag. 471 lin 19 write —— atq : adeo ad movenda puncta æquinoctialia evaderet minor quam prius in ratione 2 ad 5.   Ideoq : annuus Æquinoctiorum regressus jam esset ad 20$^{gr}$ 11′ 46″ ut 10 ad 73092, ac proinde fieret 9″ 56‴ 50″″.

Cæterum hic motus ob inclinationem plani Æquatoris ad planum Eclipticæ minuendus est, idq : in ratione sinus 91706 (qui sinus est complementi graduum 23½) ad Radium 100000.   Qua ratione motus iste jam fiet 9″. 7‴. 20″″.   Hæc est annua Præcessio Æquinoctiorum a vi Solis oriunda.

Vis autem Lunæ ad mare movendum erat ad vim Solis ut 4,4824 ad 1 circiter.   Et vis Lunæ ad Æquinoctia movenda est ad vim Solis in eadem proportione.   Indeq: prodit annua Æquinoctiorum Præcessio a vi Lunæ oriunda 40″ 53‴ 22″″, ac tota Præcessio annua a vi utraq : oriunda 50″. 00‴. 42″″.   Et hic motus cum phænomenis congruit.

Nam Præcessio æquinoctiorum ex Observationibus Astronomicis est minutorum secundorum plus minus quinquaginta

Si altitudo Terræ ad Æquatorem superet altitudinem ejus ad Polos milliaribus plus quam 17⅙, materia ejus rarior erit ad circumferentiam quam ad centrum : et Præcessio Æquinoctiorum ob altitudinem illam augeri, ob raritatem diminui debet.

Descripsim usjam Systema Solis, Terræ, Lunæ, et Planetarum : superest ut de Cometis nonnulla adjiciantur."

---

### LETTER XLVII.
#### COTES TO NEWTON.

S$^r$.

I have received Your last, but have not yet had time to try the Calculations of the inclosed sheet. I am satisfied as to the xxv$^{th}$ Proposition, upon reconsidering it.

In Page 441, lin: 25, the first & last numbers are 368682 & 362046: they should be 368676 & 362047.   The Æquation* which results from hence will be

$$88487,19 - 12307251,44x + 75578,14xx - 5082017,44x^3 + 42456,19x^4 = 0,$$

---

* The following is on a separate piece of paper, (No. 209):

Æquatio fit 88487,19 − 12307251,44 r + 75578,14 rr − 5082017,44 r³ + 42456,19 r⁴ = 0.

Inde r = 0,00719,   CT = 1,00719,   AT = 0,99281 adeoq: CT ad AT ut 70,041 ad 69,041, sive ut 70$\frac{1}{24}$ ad 69$\frac{1}{24}$ vel 70$\frac{3}{73}$ ad 69$\frac{3}{73}$.

Vera Radix iterato examine est, 0071900057 ter exam :

of which I find the Root to be 0,0071900057. If You approve of it I would alter the bottom of the Page thus [obtinetur $x$ æqualis 0,00719, & inde semidiameter $CT$ fit 1,00719 & semidiameter $AT$ 0,99281, qui numeri sunt ut $70\frac{1}{24}$ & $69\frac{1}{24}$ quam proxime. Est igitur distantia Lunæ a Terra in Syzygiis ad ipsius distantiam in Quadraturis (seposita scilicet Eccentricitatis consideratione) ut $69\frac{1}{24}$ ad $70\frac{1}{24}$ vel numeris rotundis ut 69 ad 70] This will cause an alteration in the xxix$^{th}$ Proposition & in the xxxi$^{st}$, page 450.

I have not computed the alterations for the xxix$^{th}$*, not knowing whether You will chuse the whole numbers 69 and 70 or the fractions $69\frac{1}{24}$ & $70\frac{1}{24}$.

As for the other place in page 450$^{th}$ I took the numbers 69 & 70 that I might find what alteration would arise in the conclusion of y$^e$ xxxii$^d$ Proposition. The result of my computation is as follows. Pag: 450. lin: 18 [69 ad 70] Lin: 20, [si capiatur angulus 16″. 21‴. 3′ᵛ. 30ʳ] Page 452$^d$. lin: 5, [erat 32″. 42‴. 7′ʳ] Lin: 8, [illud est 17‴: 43′ᵛ. 11ʳ] Lin: 10, [relinquit 16″. 16‴. 37′ᵛ. 42ʳ] Page 453, Lin: 22, [fit 39°. 38′. 7″. 50‴] Lin: 23, [19°. 49′. 3″. 55‴] Lin: ult: [seu 39,6355] Page 454, Lin: 3, [id est, ut 9,0827646 $ATq$] Page 455, Lin: 4 [prodibit 0,1188502]† Lin: 6, [est 1°. 29′. 58″. 3‴] Lin: 7 [subductis relinquit 18°. 19′. 5″. 52‴] Lin: 9 [relinquit 341°. 40′. 54″. 8‴] Lin: 12 [qui propterea erit 19°. 18′. 1″. 22‴]

In finding the Number 0,1188502, I supposed y$^e$ ordinate $eZ$ to bisect y$^e$ base $NT$ by which meanes the series for y$^e$ Area $TZeF$ converged quicker than the other for the Area $NeZ$, so y$^t$ on account of this Latter I would not depend upon the last figure 2, I think the other are right.

---

* These alterations of Prop. xxix. form the subject of Letter L.

† This correction, though approved by Newton, was subsequently modified (as also the four following corrections which depend upon it). The result which is substituted for it in the 2nd Ed. leads to the value .1188496 for the area of the curve $NeFn$: in the 1st Ed. it is .1188478.

In Line 14<sup>th</sup> You have 19°. 20'. 31''. 1''' from Flamsteeds
Tables. By Your Theory in D<sup>r</sup> Gregory tis 19°. 21'. 22''. 3''' *.
So in the following Proposition, page 456. Lin 13 You
have 9°. 10'. 40'' ; by Your Theory tis 9°. 11'. 3''.

There will need some other alterations in Prop. xxxiii<sup>d</sup>
& its Corollary upon account of those in the preceding
Proposition. You seem to depend too much upon Your
Readers quickness when you say [ut rem perpendenti con-
stabit] I hope when You review the whole You will make
it easier to apprehend the agreement of the two Constructions.

I do not rightly understand line 12<sup>th</sup> of page 458
[Inclinationis autem Variatio tantum augebitur per decrementum sinus $IT$, quantum diminuitur per decrementum
motus Nodorum]

I think I had observed nothing further before we come
to y<sup>e</sup> xxxvi<sup>th</sup> Proposition.

I am, S<sup>r</sup>,

Your most Humble Servant

Trinity College Apr. 24<sup>th</sup> 1712                    Roger Cotes.

*For* S<sup>r</sup> Isaac Newton *at his House*
*in St Martin's Street in Leicester*
*Fields London.*

———

LETTER XLVIII.

COTES TO NEWTON.

S<sup>r</sup>.                                             April 26<sup>th</sup> 1712.

I have examin'd your last Emendations† of the xxxvii<sup>th</sup>
Proposition. I am very glad to see the whole so perfectly

---

* Newton, in his next letter, adopts this correction and the following one. Afterwards, however, (Letter LII.) apparently forgetting that he had already given directions about them, he orders 19°. 21'. 20''. 45''' to be written in p. 455, and 9°. 11'. 3'' in
p. 456. Cotes, in his reply, (Letter LIII.) proposes to write 19°. 21'. 21''. 50''' in
p. 455, which Newton approves, (Letter LV.)

Flamsteed's *Tables* here referred to, are printed at the end of his *Doctrine of the
Sphere*, London, 1680.

† Sent in the Letter of Apr. 12.

well settled & fairly stated, for without regard to the conclusion I think $y^e$ distance of $18\frac{1}{2}$ degrees ought to be taken & is much better than $17\frac{1}{2}$ or $15\frac{1}{4}$ & the same may be said of $y^e$ other changes in $y^e$ principles from which the conclusion is inferr'd.

In examining Your Numbers I found it necessary to alter most of them, I here send you others {instead of them} for your approbation.

Præterea diametri Orbis in quo Luna......sunt ad mediocrem ejus distantiam ut 69,098747 & 69,897345 ad $69\frac{1}{2}$. Vires autem Lunæ...ad vim in mediocri distantia ut 0,9830427 et 1,017522 ad 1. Unde fit 1,017522 $L$ + 0,7986355 $S$ ad 0,9830427 × 0,8570327 $L$ -0,7986355 $S$ ut 9 ad 5. Et $S$ ad $L$ ut 1 ad 4,4815. Itaq: cum vis Solis sit ad vim gravitatis ut 1 ad 12868200 vis Lunæ erit ad vim gravitatis ut 1 ad 2871400.

Corol. 1. Cum igitur* aqua vi Solis agitata ascendat ad altitudinem pedis unius & undecim digitorum cum octava parte digiti, eadem vi Lunæ ascendet ad altitudinem octo pedum & digitorum octo. Tanta autem vis—

Corol. 2. Cum vis Lunæ ad mare movendum sit ad vim gravitatis ut 1 ad 2871400—

Corol: 3. Quoniam vis Lunæ ad mare movendum est ad Solis vim consimilem ut 4,4815 ad 1......et 32'.12") ut 4891 ad 1000. Densitas autem Solis......ad densitatem Terræ ut 4891 ad 3960 seu 21 ad 17. Est igitur......

Corol: 4......ad massam Terræ ut 1 ad 39, 371.

Corol: 6......ut 40,371 ad 39,371.

---

* The word "igitur" is omitted in the 2nd Ed., neither does it appear in Newton's first copy of the Prop. which is given at the end of Letter XXXIX.

After the words "digitorum octo," the sentence is continued as follows in the 2nd Ed., "& vi utraque ad altitudinem pedum decem cum semisse, & ubi Luna est in Perigæo ad altitudinem pedum duodecim cum semisse & ultra, præsertim ubi Æstus ventis spirantibus adjuvatur. Tanta autem vis......" corresponding to Newton's copy just referred to. Cotes's omission of these words in this draught of his letter probably arose from the fact of Newton's having omitted the passage in the emendations sent in his Letter of Apr. 8.

Corol: 7......ut 40,371 ad 39,371, quæ proinde est pedum 1161498340......ita sunt pedes 1161498340 ad pedes 14,811833......Et hac vi Luna cadendo, tempore minuti unius primi describere deberet pedes 14,89517......et habebitur distantia pedum 19701678 a qua grave cadendo, eodem tempore minuti unius secundi describet pedes 15,12028...

In the xxxix$^{th}$ Proposition. Vis autem Lunæ ad mare movendum erat ad vim Solis ut 4,4815 ad 1 circiter...... Præcessio a vi Lunæ oriunda 40″. 52‴. 52″. ac tota Præcessio annua a vi utraq: oriunda 50″. 00‴. 12″. Et hic motus......

The xxxvi$^{th}$ Proposition depends upon the latter part of the xxv$^{th}$, & must therefore stand as in the former Edition. I have altered the Corollary of it thus

Corol. Cum vis......ad vim gravitatis ut 1 ad 289..... mensura pedum Parisiensium 85820, vis Solaris de qua egimus, cum sit ad vim gravitatis ut 1 ad 12868200 atq: adeo ad vim illam centrifugam ut 289 ad 12868200 seu 1 ad 44527, efficiet ut......mensura tantum pedis unius Parisiensis & digitorum undecim cum octava parte digiti. Est enim hæc mensura ad mensuram pedum 85820 ut 1 ad 44527.

I have altered the xxxviii$^{th}$ Proposition thus. Pag: 467. lin: 10 [id est, ut 39,371 ad 1 & 100 ad 365 conjunctim, seu 1079 ad 100. Unde cum mare nostrum vi Lunæ attollatur ad pedes 8⅔, fluidum Lunare vi Terræ attolli deberet ad pedes 93½......excessu pedum 187

Your very Humble Servt.

R Cotes.

## LETTER XLIX.

### NEWTON TO COTES.

S$^r$

The corrections made in yo$^r$ last of Apr. 24$^{th}$ may all stand. In y$^e$ xxix$^{th}$ you may use either y$^e$ whole numbers 69 & 70 or the fractions 69$\frac{1}{24}$ & 70$\frac{1}{24}$. In pag 455 lin 14 & pag 456 I have put the motion of the Nodes of Moon from y$^e$ Equinox & should have put it from y$^e$ fixt starrs. In y$^e$ first place therefore for 19$^{gr}$ 20′ 31″ 1‴ write 19°.21′.22″.3‴ In y$^e$ second for 9°.10′ 40″ write 9°.11′.3″.

In pag. 458 lin 11. write. [Et in eadem ratione minuetur etiam Inclinationis Variatio.] And strike out the rest to the end of the Paragraph.

In y$^e$ xxxiii$^d$ Proposition, pag 456, instead of y$^e$ words [ut rem perpendenti constabit] may be written [ut rem perpendenti & computationes instituenti constabit.] And the numbers in this Proposition are to be suited to y$^e$ alterations made in y$^e$ preceding Proposition as you mention.

I am

London Apr. 24$^{th}$*           Yo$^r$ most humble Servant

1712                                                   Is. NEWTON

*For the* R$^{nd}$ M$^r$ ROGER COTES *Professor*
*of Astronomy, at his Chamber in*
*Trinity College in Cambridge.*

---

## LETTER L.

### COTES TO NEWTON.

S$^r$

I have received Your last, & taking the whole numbers 69 & 70, the alteration in Pag: 442.† lin. penult. will be

---

* The post mark is Ap. 29.
† Prop. xxix. Invenire Variationem Lunæ.

[68,6877 ad numerum 69. Quo pacto tangens anguli *CTP* jam erit ad tangentem motus medii ut 68,6877 ad 70, & angulus *CTP* in Octantibus, ubi motus medius est 45ᵍʳ. invenietur 44ᵍʳ. 27'. 28": qui subductus de angulo motus medii 45° relinquit Variationem maximam 32'. 32". Hæc ita se haberent si ...... & Variatio maxima quæ secus esset 32 . 32".\* jam aucta in eadem ratione fit 35'. 10".†] You go on thus‡. Hæc est ejus magnitudo in mediocri distantia Solis a Terra, neglectis differentiis quæ a curvatura Orbis magni majoriq: Solis actione in Lunam falcatam et novam quam in gibbosā & plenam, oriri possint. In aliis distantiis Solis a Terra, Variatio maxima est in ratione quæ componitur ex duplicata ratione revolutionis Synodicæ Lunaris (dato anni tempore) directe, et ratione anguli *CTa* directe, & triplicata ratione distantiæ Solis a Terra inverse; id est, ex triplicata ratione revolutionis synodicæ Lunaris directe et triplicata ratione distantiæ Solis a Terra inverse. Ideoq: in Apogæo Solis Variatio maxima est 33'. 11" & in ejus Perigæo 37'. 24", si modo eccentricitas Solis sit ad Orbis magni semidiametrum transversam ut 16 5⁄16 ad 1000.

Hactenus Variationem investigavimus in Orbe non eccentrico in quo utiq: Luna in Octantibus suis semper est in mediocri sua distantia a Terra. Si Luna propter eccentricitatem suam, magis vel minus distat a Terra quam si locaretur in hoc Orbe, Variatio paulo major esse potest vel paulo minor quam pro Regula hic allata: sed excessum vel defectum ab Astronomis per Phænomena determinandum relinquo.

I was going to diminish § Your numbers 33'. 11", &

---

\* 32'. 34" in Newton's MS.
† 35'. 12" in Newton's MS.
‡ Nos. 149, 150.
§ In the margin of Newton's MS. (No. 149,) Cotes has actually made this diminution, as he has done above, in the case of the numbers 32'. 34" and 35'. 12" at the end of the extract inclosed within brackets.

37'. 24" by 2" which is nearly the diminution if those numbers are right, which I am forc'd to take upon trust not knowing how to state the proportion of the Moon's Periodical Revolutions nor consequently of her Synodical in the Apogee & Perigee of $y^e$ Sun. But I cannot fully satisfy my self about Your Rule. As I take it, the duplicate ratio of $y^e$ Synodical revolution of $y^e$ Moon & $y^e$ simple ratio of $y^e$ angle $CTa$ compose not the triplicate ratio of $y^e$ Synodical revolution alone but this triplicate ratio directly & $y^e$ simple ratio of $y^e$ periodical revolution inversly : the angle $CTa$ being as $y^e$ Synodical revolution directly & $y^e$ Periodical revolution inversly. I have besides some scruple about introducing $y^e$ ratio of $y^e$ angle $CTa$, I have not throughly considered the thing, but I quæry whether it will not be sufficient to make the compounded ratio consist only of $y^e$ duplicate ratio of $y^e$ Synodical revolution directly & $y^e$ triplicate ratio of $y^e$ Sun's distance inversly according to $y^e$ 16$^{th}$ Corol : of Prop: LXVI$^{th}$ Lib. 1. I have transcribed $y^e$ whole $y^t$ You may review it and order it as You think it should stand.

Your &c.

May day 1712.                                          R C.

In his answer to this, (May 10,) Newton adheres to the statement that the Variation is proportional to

$$\frac{(\text{Moon's synodical period})^2 \; dato \; anni \; tempore \times \angle CTa}{(\text{distance between Sun \& Moon})^3}.$$

Cotes then (May 13) further explains his reasons for thinking that the $\angle CTa$ should be cancelled. Not receiving an answer, he writes again (May 25) to draw his attention to the point, and has the gratification of finding (see letter of May 27) that Sir Isaac has been convinced by his arguments.

## LETTER LI.

### COTES TO NEWTON.

S$^r$.

I fear I give You too much trouble with my Letters, but I think this will be my last till we come to the Theory of Comets. In the Corollary of the xxxiii$^d$ Proposition I put 16″.19‴.27′′. instead of 16″.18‴.41″ $\frac{1}{2}$. I am not certain how You would compute that motion, & therefore I mention it to You, I found it by this Proportion: As 19$^0$.18′.01″.23‴ to 19$^0$.21′.22″.3‴ so 16″.16‴.37′′.42$^v$ to 16″.19‴.26′.56′.

In Your last letter You order page 458. lin 11. thus. [Et in eadem ratione minueter etiam Inclinationis Variatio] This will cause some alteration in the following Corollarys & in the xxxv$^{th}$ Proposition unless You design to consider the Moons Inclination only as moving in Orbe circulari.

At the bottom of Page 461 You make use of 5$^0$.17′.46″ & 5$^0$ for the extream Inclinations; In D$^r$ Gregorys Astronomy You have 5$^0$.17′.20″ & 4$^0$.59′.35″. Which I suppose You find to be more agreeable to observations.

In the first Paragraph of y$^e$ New Scholium* to Prop: xxxv$^{th}$ You have [ad 11′.50″ circiter ascendit, & *additur* medio motui Lunæ ubi Terra pergit a Perihelio suo ad Aphelium & in opposita orbis parte *subducitur*] As I take it, the words *additur* & *subducitur* should change places. You have not mention'd how to find this Æquation in every place.

In the second Paragraph concerning the Annual Æquations of the Moon's Apogee & Node You have forgotten to mention when they must be added & when substracted.

In the third Paragraph You say [Per Theoriam gravitatis constitit etiam quod actio Solis in Lunam paulo major

---

* See the remarks which follow the Letter.

sit ubi transversa diameter Orbis Lunaris transit per Solem
quam ubi eadem ad rectos est angulos cum linea Terram
& Solem jungente & propterea Orbis Lunaris paulo *minor*
est in priore casu quam in posteriore] I think it should be
[paulo *major* est in priore]

In the fourth Paragraph concerning $y^e$ Æquation of $y^e$
Moon arising from $y^e$ position of her Nodes which You call
*Semestris secunda*, You have [*additur* vero medio motui
Lunæ dum Nodi transeunt a Solis Syzygiis ad proximas
Quadraturas & *subducitur* in eorum transitu a Quadraturis
ad Syzygias] As I apprehend it $y^e$ words *additur* & *subducitur* should change places.

The sixth Paragraph I do not understand. The Æquation which You there describe seems to be established not
so much from Observations as from the Theory of gravity,
but I cannot perceive how it answers Your design express'd in these words. In Perihelio Terræ propter majorem vim Solis Apogæum Lunæ velocius movetur in epicyclo circum centrum $D$ (I suppose it should be *centrum C*)
quam in Aphelio, idq: in triplicata ratione distantiæ Terræ
a Sole inverse. Ob æquationem centri Solis in argumento
annuo comprehensam Apogæum Lunæ velocius movebitur
in epicyclo in duplicata ratione distantiæ Terræ a Sole
inverse. Ut idem adhuc celerius moveatur in ratione simplici distantiæ inverse, sit &c.* Now the Æquation which

---

* We will add the remainder of the paragraph from Newton's MS. (No. 170):
"sit $TD$ excentricitas primò æquata, et producatur $TD$ ad $E$ ut sit $DE$ ad $TD$ ut

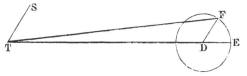

duplum excentricitatis Solis ad radium Orbis magni seu 33⅔ ad 1000. Capiatur angulus
$EDF$ æqualis argumento annuo, vel quod perinde est, agantur parallelæ $TS$ ac $DF$
solem versus, et sit $DF$ ipsi $DE$ æqualis, et erit $DTF$ æquatio annua apogæi Lunæ
& $FTS$ distantia Solis ab apogæo Lunæ ter æquata, & $TF$ excentricitas Lunæ bis
æquata in apogæum Lunæ ter æquatum tendens."

You describe in what follows does not in the least, as I see, depend upon the Sun's Anomaly but intirely upon $y^e$ Annual Argument of the Apogee. You will perhaps more easily perceive my difficulty if I tell You how I think the Æquation should be stated to answer what was propos'd. Let $CTD$ be $y^e$ *Æquatio Semestris* describ'd in $y^e$ preceding Paragraph; produce $CD$ to $E$, so $y^t$ $DE$ may be to $CD$ as $33\frac{7}{8}$ to 1000; make the angle $EDF$ equal to the Sun's Anomaly, & the line $DF$ equall to $DE$, & joyn $TF$: then will $DTF$ be the second annual Æquation of $y^e$ Apogee & $TF$ be the Eccentricitas Lunæ bis æquata in Apogæum Lunæ ter æquatum tendens.

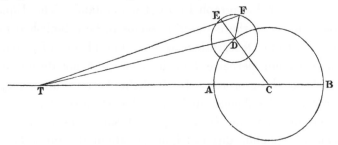

The following Paragraph concludes thus*.    Ducantur

---

* The former part of this paragraph is as follows, (No. 170): " Per eandem gravitatis Theoriam Sol fortius agit in Lunam annuatim ubi apogæum Lunæ et perigæum Solis conjunguntur quam ubi opponuntur. Et inde oriuntur æquationes duæ periodicæ, una medii motus Lunæ, altera apogæi ejus: quæ quidem æquationes nullæ sunt ubi apogæum Lunæ vel conjungitur cum perigæo Solis vel eidem opponitur, et maximæ in apogæorum quadraturis. In aliis apogæorum positionibus datam habent proportionem ad invicem, suntq: ut sinus distantiæ apogæorum ab invicem. Æquatio prior subducitur et posterior additur ubi apogæum Lunæ minus distat a perigæo Solis in consequentia quam gradibus 180; prior verò additur & posterior subducitur ubi distantia illa fit major. Harum æquationum quantum sentio, Æquatio maxima apogæi ascendit ad 15' vel 20' circiter, sed æquatio maxima motus medii Lunæ vix ascendit ad 30'', et ob parvitatem negligi potest donec quantitas ejus ex observationibus determinetur. Producatur excentricitas Lunæ bis æquata $TF$ ad $G$ ut sit $FG$ sinus æquationis maximæ periodicæ apogæi Lunæ 15' vel 20' ad radium $TF$. Ducantur," &c.

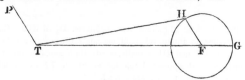

rectæ duæ parallelæ *TP*, *FH* in Perigæum Solis tendentes, vel quod perinde est, capiatur angulus *GFH* æqualis distantiæ Perigæi Solis ab Apogæo Lunæ, & sit *FH* ipsi *FG* æqualis ; et angulus *FTH* erit æquatio Periodica Apogæi Lunæ, & angulus *PTH* distantia Apogæi Lunæ quarto æquati a Perigæo Solis et *TH* eccentricitas tertio æquata in apogæum quarto æquatum tendens. Instead of which I propose the following alteration, leaving out yᵉ line *TP* in the Figure. Capiatur angulus *GFH* æqualis distantiæ Apogæi Lunæ a Perigæo Solis in consequentia et sit *FH* ipsi *FG* æqualis, & angulus *FTH* erit Æquatio periodica Apogæi Lunæ & *TH* eccentricitas tertio æquata in Apogæum quarto æquatum tendens. This Alteration will agree with what You lay down a little before in the same Paragraph, where speaking of this Periodical Æquation of yᵉ Apogee You say additur ubi Apogæum Lunæ minus distat a Perigæo Solis in consequentia quam gradibus 180 & subducitur ubi distantia illa fit major, Which Rule I think is right but not agreable to the conclusion of the Paragraph which I therefore propose to alter.

In the last Paragraph but one You say [pono mediocrem distantiam centri Lunæ a centro Terræ in Octantibus æqualem esse 60⅔ semidiametris maximis Terræ] I desire to know whether You will here retain 60⅔ or put instead of it 60¼ as in Corol 7ᵗʰ of Prop xxxvɪɪᵗʰ

Your &c.

May. 3ᵈ. 1712                                        R C.

The "New Scholium to Prop. xxxv." which forms so large a part of the subject of the preceding letter is a Scholium on the Lunar Theory, containing a statement of the origin and quantity of various Lunar Inequalities, and occupying the place of a short Scholium in the 1st Ed. relative to the motion of the Moon's Apogee. It is written on three sides of a sheet of foolscap (Nos. 169—171) which seems to have been doubled up and placed loosely between the pages of Newton's interleaved copy of the 1st Ed. It was probably sent to Cotes with the third and last division of Newton's copy the first week in July

1711. (Letter xxviii and note). The reason why the Scholium appears on *folio* paper is, no doubt, that there was not room for it on the quarto leaf in the interleaved copy : that quarto leaf is still preserved, and its first page (No. 190) is headed " Scholium" and is devoted to the opening words of it followed by an " &c." thus : " Hisce motuum Lunarium computationibus ostendere volui quod motus Lanares per Theoriam gravitatis &c." indicating that the Scholium was to be found written out on another paper. The second page of the leaf contains some supplementary matter to be added to Prop. xxxvi. These minutiæ are mentioned for the purpose of limiting the date of the composition of the Scholium, as the circumstance of its being written on a folio sheet might have led one to suppose that it was sent down to Cambridge not as part of the copy, but as an emendation of copy previously sent. The quarto leaves of Newton's handwriting in the Newtonian Volume all formed part of his interleaved copy of the Principia : those in folio were sent down in letters as corrections. The only exception to this remark that I have noticed is the sheet now referred to, which contains the Scholium on the Lunar Theory.

A distinct idea of the contents of this Scholium (or " first draught of the Moon's theory," as it is afterwards called), as it stood before undergoing the alterations which Newton made in it in consequence of the above letter from Cotes, may be obtained from the following outline of it. It consists of twelve paragraphs, which, for convenience of reference, I will number in the order in which they present themselves.

1. " Hisce motuum Lunarium computationibus...æquatio maxima erit 11′. 52″." (Annual Equation).

2. "Inveni etiam...æquatio maxima medii motus nodorum 9′.27″." (Annual Equations of mean motion of apogee and nodes.)

3. " Per theoriam gravitatis...quadratura ad radium." (Æquatio semestris, the argument of which is = twice the distance of apogee from Sun, i. e. twice the annual argument).

4. " Per eandem gravitatis theoriam...ad 49″ circiter ascendit." (Æquatio semestris secunda, the argument of which is = twice the distance of node from Sun).

The four preceding paragraphs stand as they are printed in the 2nd Ed. with the exception of the modifications introduced in conformity with Cotes's suggestions in the above letter. (See Letter lviii). In the 2nd the word " inverse" is also omitted after " si motus Solis esset in triplicata ratione distantiæ."

5. " Per eandem gravitatis Theoriam apogæum Lunæ...in apogæum secundo æquatum tendens". (The Equation of the centre and Evection combined, giving the æquatio semestris of the apogee and first correction of the eccentricity).

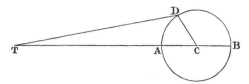

Same as in 2nd Ed. with two exceptions: (1) the *upper focus* of the moon's orbit (and not its *centre* according to Newton's subsequent correction in his paper of alterations, see letter LVII.) is represented as describing the epicycle *BDA* : by a similar inadvertence in paragraph 6 the *apogee* is made to move in that epicycle. (2) In the 2nd Ed. there is a clause " Habitis autem...per methodos notissimas" added at the end of the paragraph, which is in part transferred from paragraph 9.

Cotes has drawn two other lines in the figure (no doubt on receiving Newton's paper of alterations) viz. *DE* to the right, parallel to *AB*, and *DF* making an acute angle with it (not an obtuse angle as in the figure in the 2nd Ed.)

6. "In perihelio terræ...in apogæum Lunæ ter æquatum tendens." (Third correction of the place of the apogee and second of the excentricity by an " annual equation" whose argument = annual argument).

This paragraph is given in the preceding letter and note. It was completely remodelled in Newton's paper of alterations. Two paragraphs were substituted for it explanatory of what he says may be called " æquatio centri secunda" depending on the argument " distance of moon from sun + dist. of moon's apogee from sun's apogee." The latter of them merely contains an approximation to its value. Newton's mode of determining the position of the centre of the moon's orbit in a secondary epicycle with centre *D* became the subject of an active correspondence between him and Cotes (letters LVII-LXVI).

7. " Per eandem gravitatis Theoriam Sol fortius agit...in apogæum quarto æquatum tendens. (Fourth correction of the place of the apogee and third of the excentricity by a " periodical equation" whose argument = distance of apogee from Sun's perigee. Mention is also made of a " periodical equation" of the Moon's mean motion depending on the same argument having barely 30″ for its maximum: Damoiseau gives it 2″, Plana 0″.466, Pontécoulant 1″.496−1″.108 = 0″.388, Burckhardt 0″.7. See Pontécoulant, tom. IV. pp. 451-465, 580, 604, 626 : the two terms of which it is composed are of the fourth and fifth orders.)

This paragraph will also be found in the preceding letter and proper note.

8. " Si tres anguli *CTD, DTF* & *FTH* ad singulos gradus angulorum *BCD, EDF* et *GFH* computentur & in Tabulas referantur,

et si logarithmi quoq: trium distantiarum *TD*, *TF* & *TH* ad radios
*TC TD* et *TF* in partes 100000 divisos simul computentur & in
Tabulas referantur: aggregatum trium angulorum sub signis suis + &
– erit æquatio tota apogæi, et aggregatum trium Logarithmorum erit
Logarithmus excentricitatis veræ." This and the preceding paragraph were not given in Newton's
paper of alterations, where another paragraph (" Si computatio accu-
ratior...non multum errabitur") appeared relating to the " variatio
secunda," which was omitted in the 3rd Ed.

9.   " Habitis autem Lunæ motu medio & apogæo et excentricitate
ultimum æquatis, ut et Orbis diametro transversa partium 200000 ; ex
his eruetur verus Lunæ locus in orbe, et distantia ejus a Terra, idq : per
methodos notissimas. Deinde per Variationem et Reductionem ad
Eclipticam dabitur ejus longitudo et latitudo vera."

10.   " Diximus orbem Lunæ a viribus Solis per vices dilatari et
contrahi & æquationes quasdam motuum Lunarium inde oriri. Inde
etiam oritur variatio aliqua parallaxeos Lunæ, sed quam insensibilem
esse judico ; ideoq : in computationibus motuum Lunæ, pro mediocri
ejus distantia a centro Terræ semper usurpo numerum 100000, & pro
Orbis diametro transversa numerum 200000, et ad parallaxim inves-
tigandam pono mediocrem distantiam centri Lunæ a centro Terræ in
Octantibus æqualem esse $60\frac{2}{9}$ semidiametris maximis Terræ. Semidiametrum ejus maximam voco quæ a cent│r│o ad æquatorem ducitur,
minimam quæ a centro ad polos. Et hinc fit Lunæ parallaxis horizontalis mediocris apparens in Octantibus 57' 5", in Syzygiis 57' 30" in
quadraturis 56' 40". Lunæ vero diameter mediocris apparens in
Syzygiis 31.30 in Quadraturis 31.3 usurpari potest & Solis diameter
mediocris 32.12."

11.   " Et cum atmosphæra Terræ ad usq : altitudinem milliarium
35 vel 40 refringat Lucem Solis et refringendo spargat eandem in
umbram Terræ, & spargendo lucem in confinio umbræ dilatet umbram:
ad diametrum umbræ quæ per parallaxim prodit, addo minutum unum
primum in eclipsibus Lunæ, vel minutum unum cum triente."

12.   " Theoria verò Lunæ primò in Syzygiis, deinde etiam in quadraturis" &c. as in the 2nd Ed. except (1) as regards the changes in
some of the figures mentioned in Letter LXVI., and (2) the addition of
the clause " & differentiam meridianorum Observatorii hujus & Observatorii Regii Parisiensis 0$^{hor.}$ 9$^{min.}$ 20$^{sec.}$." at the close of the paragraph,
which does not appear here.

rectæ duæ parallelæ *TP*, *FH* in Perigæum Solis tendentes, vel quod perinde est, capiatur angulus *GFH* æqualis distantiæ Perigæi Solis ab Apogæo Lunæ, & sit *FH* ipsi *FG* æqualis ; et angulus *FTH* erit æquatio Periodica Apogæi Lunæ, & angulus *PTH* distantia Apogæi Lunæ quarto æquati a Perigæo Solis et *TH* eccentricitas tertio æquata in apogæum quarto æquatum tendens. Instead of which I propose the following alteration, leaving out y$^e$ line *TP* in the Figure. Capiatur angulus *GFH* æqualis distantiæ Apogæi Lunæ a Perigæo Solis in consequentia et sit *FH* ipsi *FG* æqualis, & angulus *FTH* erit Æquatio periodica Apogæi Lunæ & *TH* eccentricitas tertio æquata in Apogæum quarto æquatum tendens. This Alteration will agree with what You lay down a little before in the same Paragraph, where speaking of this Periodical Æquation of y$^e$ Apogee You say additur ubi Apogæum Lunæ minus distat a Perigæo Solis in consequentia quam gradibus 180 & subducitur ubi distantia illa fit major, Which Rule I think is right but not agreable to the conclusion of the Paragraph which I therefore propose to alter.

In the last Paragraph but one You say [pono mediocrem distantiam centri Lunæ a centro Terræ in Octantibus æqualem esse $60\frac{2}{9}$ semidiametris maximis Terræ] I desire to know whether You will here retain $60\frac{2}{9}$ or put instead of it $60\frac{1}{4}$ as in Corol 7$^{th}$ of Prop xxxvii$^{th}$

Your &c.

May. 3$^d$. 1712 R C.

The "New Scholium to Prop. xxxv." which forms so large a part of the subject of the preceding letter is a Scholium on the Lunar Theory, containing a statement of the origin and quantity of various Lunar Inequalities, and occupying the place of a short Scholium in the 1st Ed. relative to the motion of the Moon's Apogee. It is written on three sides of a sheet of foolscap (Nos. 169—171) which seems to have been doubled up and placed loosely between the pages of Newton's interleaved copy of the 1st Ed. It was probably sent to Cotes with the third and last division of Newton's copy the first week in July

1711. (Letter xxviii and note). The reason why the Scholium appears on *folio* paper is, no doubt, that there was not room for it on the quarto leaf in the interleaved copy: that quarto leaf is still preserved, and its first page (No. 190) is headed "Scholium" and is devoted to the opening words of it followed by an " &c." thus: "Hisce motuum Lunarium computationibus ostendere volui quod motus Lanares per Theoriam gravitatis &c." indicating that the Scholium was to be found written out on another paper. The second page of the leaf contains some supplementary matter to be added to Prop. xxxvi. These minutiæ are mentioned for the purpose of limiting the date of the composition of the Scholium, as the circumstance of its being written on a folio sheet might have led one to suppose that it was sent down to Cambridge not as part of the copy, but as an emendation of copy previously sent. The quarto leaves of Newton's handwriting in the Newtonian Volume all formed part of his interleaved copy of the Principia: those in folio were sent down in letters as corrections. The only exception to this remark that I have noticed is the sheet now referred to, which contains the Scholium on the Lunar Theory.

A distinct idea of the contents of this Scholium (or "first draught of the Moon's theory," as it is afterwards called), as it stood before undergoing the alterations which Newton made in it in consequence of the above letter from Cotes, may be obtained from the following outline of it. It consists of twelve paragraphs, which, for convenience of reference, I will number in the order in which they present themselves.

1. "Hisce motuum Lunarium computationibus...æquatio maxima erit 11′.52″." (Annual Equation).

2. "Inveni etiam...æquatio maxima medii motus nodorum 9′.27″." (Annual Equations of mean motion of apogee and nodes.)

3. "Per theoriam gravitatis...quadratura ad radium." (Æquatio semestris, the argument of which is = twice the distance of apogee from Sun, i. e. twice the annual argument).

4. "Per eandem gravitatis theoriam...ad 49″ circiter ascendit." (Æquatio semestris secunda, the argument of which is = twice the distance of node from Sun).

The four preceding paragraphs stand as they are printed in the 2nd Ed. with the exception of the modifications introduced in conformity with Cotes's suggestions in the above letter. (See Letter lviii). In the 2nd the word "inverse" is also omitted after " si motus Solis esset in triplicata ratione distantiæ."

5. "Per eandem gravitatis Theoriam apogæum Lunæ...in apogæum secundo æquatum tendens". (The Equation of the centre and Evection combined, giving the æquatio semestris of the apogee and first correction of the eccentricity).

LETTER LII.

NEWTON TO COTES.

\* " Prop. De Variatione Lunæ p. 402." {2ᵈ Ed.}.

Sʳ

I have received three letters from you since my last.
And the corrections wᶜʰ you send me in the two first
of them may all stand. In the second of them dated
May 1ˢᵗ, you cite my words. In aliis distantiis Solis a
Terra Variatio maxima est in ratione quæ componitur ex
duplicata ratione [temporis] revolutionis sy{n}odicæ Lunaris
(dato anni tempore directe, et ratione anguli $CTa$ directe,
et triplicata ratione distantiæ Solis a Terra inverse. Ideoq:
in Apogæo Solis Variatio maxima est 33′. 11″ et in ejus
Perigæo 37′ 24″ si modo excentricitas Solis sit ad Orbis
magni semidiametrum transversam ut 16$\frac{15}{16}$ ad 1000. Here
33 11 & 37 24 may be diminished by 2″ & the word temporis may be inserted where you see it wᵗʰin the brackets.
The Variatio maxima is composed of the ratios of the
time, the angle $CTa$, & the sun's force, as above ; because
if any one of the three ratios be enlarged while the rest
remain given, the variation will be enlarged. If the time
alone be enlarged the Variation will be enlarged in a
duplicate proportion, as may be gathered from the descent
of falling bodies in a greater or less time. If the angle be
enlarged the Variation wᶜʰ is a proportional part of yᵉ
Angle will be inlarged in the same simple proportion, &
the force also wᶜʰ is reciprocally as the cube of yᵉ Suns
distance enlarges the Variation in proportion to it self.

In pag 445 write. Idem per Tabulas Astronomicas est
19. 21. 20. 45†. Differentia minor est parte fere quadringentesima motus totius, et ab Orbis &c.

---

\* In Cotes's hand.

† This is the mean motion of the Moon's nodes in a *Julian* year. But it is the
mean notion in a *sidereal* year that is required in the place referred to. See Cotes's
answer.

Pag 456 lin 13 write $9^{gr}$. $11'$. $3''$. & lin 28 in Quadraturis autem regrediuntur motu horario $16''$ $19'''$ $51'^V$. I compute it thus. As $AB$ to $AD + AB$ so is the mean horary motion of the Node to $16''$. $19'''$. $51'^V$.

I am $S^r$

Yo$^r$ most humble Servant

London $10^{th}$ May 1712                    Is. NEWTON.

At the bottom of pag 461 you may put the numbers $5^{gr}$. $17'$. $20''$ & $4^{gr}$ $59'$ $35''$

Pag 456 lin 1 instead of $38\frac{1}{3}$ write $38\frac{3}{10}$.

The Lunar systeme must be altered *

*To* M$^r$ COTES *Professor of Astronomy*
*at his Chamber in Trinity College*
*in Cambridge*

---

### LETTER LIII.

#### COTES TO NEWTON.

S$^r$.                                    May $13^{th}$ 1712

I have received Your last, but I am not yet clear that the ratio of y$^e$ angle $CTa$ ought to be introduced in y$^e$ xxix$^{th}$ Proposition, though I do fully understand the reasons You give for it. As I apprehend it the duplicate ratio of y$^e$ Synodical time does itself account for the dilatation of the Angle, & therefore it ought not to be again accounted for. According to the reasoning of the $16^{th}$ Corollary of Prop: LXVI. Lib. I, the Variatio maxima which is the angular Error of y$^e$ moon whilst she describes the half of y$^e$ Arch $Cpa$, is as the Square of y$^e$ time imploy'd

---

* This is all the notice that Newton at present takes of Cotes's remarks upon the Scholium on the Lunar Theory. The necessity of an "alteration" in "the Lunar Systeme" points to the 6th and 7th paragraphs of the Scholium, especially the former. About the end of June, we are told, he intended to send down his corrections "very soon," but even with the stimulus of a letter from Cotes (July 20), it is only a little before Aug. 10 that they are despatched to Cambridge, (Letters LVI., LVII.)

in describing that half Arch directly & $y^e$ Cube of $y^e$
distance from $y^e$ Sun inversly: Or as the Square of $y^e$
Synodical time directly & $y^e$ Cube of $y^e$ distance inversly.
Now I think the dilatation is accounted for by taking the
angular Error which arises in the time of describing half
$y^e$ arch $Cpa$, instead of $y^e$ Error which would arise in $y^e$
time of describing half $y^e$ arch $CPA$. The thing may be
considered another way which perhaps will give more light
to $y^e$ understanding of my difficulty  The true Variatio
maxima $35'$. $10''$ arises from $y^e$ arch $Cpa$, but the Variatio
maxima $32'$. $32''$ arises from the arch $CPA$. Now this latter
by $y^e$ 16$^{th}$ Corollary of Prop LXVI Lib 1 must be altered
with $y^e$ Square of $y^e$ Periodical time directly & the Cube
of $y^e$ distance inversly, & so it will be more correct; after
it is thus corrected, the corrected true Variatio maxima
will be deduc'd from it, by enlarging or dilating it in $y^e$
proportion of $y^e$ Angle $CTa$ to $y^e$ Angle $CTA$ or in the
proportion of $y^e$ Synodical to $y^e$ Periodical time.   Therefore the corrected true Variatio max{i}ma will be as the
Square of $y^e$ Periodical time directly, the Cube of the
distance inversly, the Synodical time directly & the Periodical time inversly: that is, as the Periodical & Synodical
times directly & the cube of $y^e$ distance inversly.   In this
latter way I scruple not to account for the dilatation, but
in the former I think it is already accounted for by taking
the Square of the Synodical time instead of the Square of
$y^e$ Periodical.  If You find the Objection to be of any
moment, I desire you to send me other numbers instead
of $33'$. $11''$. & $37'$. $24''$.  If You choose to let the place
stand, yet still there must be a further alteration of those
numbers besides $y^e$ diminution by $2''$, for the Square of $y^e$
Synodical time compounded with $y^e$ ratio of $y^e$ angle $CTa$,
makes not the triplicate ratio of $y^e$ Synodical time (upon
which those numbers were computed but that triplicate

ratio directly & y$^e$ ratio of y$^e$ Periodical time inversly as I observ'd in my former Letter.

In Page 455 You direct me to write. Idem per Tabulas Astronomicas est 19°. 21′. 20″. 45‴. Differentia minor est parte fere quadringentesima motus totius &c. I would choose to put it thus. Idem per Tabulas Astronomicas est 19°. 21′. 21″. 50‴. Differentia minor est parte trecentesima &c. For according to Flamsteed's Tables the motion of y$^e$ Nodes from y$^e$ Fix't stars in 20 Yeares or 7305 Days is 1$^{rev}$. 0$^{sig}$. 27°. 6′. 53″, and therefore in 365$^d$. 6$^h$. 9$^m$ it is 19°. 21′. 21″. 50‴.

The mean horary motion of y$^e$ Nodes by the same Tables is 7″. 56‴. 56′$^V$ and as $AB$ to $AD + AB$ or as 373 to 766 so is 7″. 56‴. 56′$^V$ to 16″. 19‴. 26′$^V$. Therefore in Pag: 456, lin: 28, I would write 16″. 19‴. 26′$^V$. Unless You find other reason for writing 16″. 19‴. 51′$^V$ as You put it in Your Letters.

---

### LETTER LIV.

#### COTES TO NEWTON.

S$^r$.                              Trin: College May 25$^{th}$ 1712

I have not yet received an answer to my last of May 13$^{th}$ concerning the xxix$^{th}$ Proposition; I am therefore afraid it has miscarried.

I sent You by D$^r$ Bentley a small Treatise* of my own

---

* This was afterwards published in the *Philosophical Transactions,* (Jan—March, 1714), and subsequently formed the first part of Cotes's *Harmonia Mensurarum,* Cantab. 1722, edited by his cousin Rob. Smith. There is prefixed to it a short address to Halley as Secretary of the Royal Society, the first sentence of which is: "Mitto tibi, hortatu Illustrissimi Præsidis Newtoni, quæ aliquot abhinc annis conscripseram de Rationibus dimetiendis." Cotes had succeeded in integrating some general expressions, the integrals of which involve logarithms. His *Logometria* contains the application of the results to the solution of a variety of problems. Compare Letter CX. *fin.*

concerning Logarithms, of which the Title is, *Elementa Logometriæ* together with the Figures belonging to it. I desire the favour of You to deliver 'em to M$^r$ Livebody to be cut in Wood & to give him Your directions if he meets with any difficulty. I fear You are at this time taken up with other buisness, otherwise I would beg of You to peruse the Treatise. You will find I am there proposing a new sort of Constructions in Geometry which appear to me very easy, simple & general. But I am fearfull of relying upon my own Judgment alone, which possibly in this matter may be too much byass'd. What I think to be right, may to others appear whimsical & of no use & I would not willingly give them the satisfaction of laughing at my Dreams. If You think I may venture to publish it, I shall be glad to know what may want to be corrected or altered either in the Matter or Expression. I have been forc'd to use some new Terms, as *Modulus, Ratio modularis*, &c. If others more proper occur to You upon reading the Papers, I shall be very willing to make any alteration. I hope You will pardon this Trouble I give You. I am Sir

Your most Obliged & Humble Servant

*For* S$^r$ Isaac Newton *at his House in* S$^t$ *Martin's Street in Leicester-Feilds London.*           Roger Cotes.

---

LETTER LV.

NEWTON TO COTES.

S$^r$

I have reconsidered what you write about the Variation & agree to it. You may leave out the words [et ratione anguli $CTa$ directe] & instead of the numbers

$33'$ $11''$ & $37'$ $24''$ diminished by $2''$, write $33'$ $14''$ & $37'$ $11''$. ffor so I found them upon computing them anew.

Also in pag 455 lin 14 you may write. Idem per Tabulas Astronomicas est $19^{gr}$. $21'$. $21''$. $50'''$. Differentia minor est parte trecentesima &c And pag 456 lin. 28 you may write $16''$. $19'''$. $26''$.

I received yo$^r$ papers by D$^r$ Bently & have run my eye over them. I intend to read them over again & get the cuts done for you as soon as I can find out M$^r$ Livebody.

I am Yo$^r$ most humble Servant

London May 27 1712                               Is. NEWTON.

*For the* R$^{nd}$ M$^r$ ROGER COTES *Professor of*
*Astronomy at his Chamber in Trinity*
*College in Cambridge*

Brought probably by Bentley.

---

LETTER LVI.

COTES TO NEWTON.

S$^r$.                                Cambridge July 20$^{th}$ 1712

It is now about three Weeks since D$^r$ Bentley return'd from London. He told me, You then intended to send down Your Emendations of the Lunar Theory very soon. I have not received any thing from You since that time, & am therefore apprehensive of some miscarriage. He inform'd me, You had thoughts of adding something further upon the Subject of Comets*, & besides a small Treatise concerning the Methods of Infinite Series & Fluxions. I hope You will go on with Your design: it were better that the publication of Your Book should be deferr'd a little, than to have it depriv'd of those additions. I thank

---

* This was done (see Letter LXVIII.), but the project with respect to series and fluxions was abandoned.

You for the Picture which I have received of him: 'tis much better done than the former; but I could have wish'd it had been taken from the first of M$^r$ Thornhill's.

I am Sir Your most Humble Servant

*For* S$^r$ ISAAC NEWTON *at his House*        ROGER COTES. *in* S$^t$ *Martin's Street in Leicester* *Feilds London*

On the back of Cotes's draught of Apr. 26, there is the draught of a letter from him to Newton, which, from the allusion to the intended treatise on series and fluxions, seems to have been written about the same time as the letter we have just been reading. He probably never sent it, but replaced it by the above, suppressing the suggestions and remarks which, upon second thoughts, he may have considered as out of place. We need not, however, withhold it here. It is as follows:

"I am glad to understand by D$^r$ Bentley that You have some thoughts of adding to this Book a small Treatise of Infinite Series & the Method of Fluxions. I like the design very well, but I beg leave to make another proposal to You. When this Book shall be finished I intended to have importun'd You to review Your Algebra for a better Edition of it & to have added to it those things which are published by M$^r$ Jones & what others You have by You of the like nature. These together will make a Volume nearly of y$^e$ same size with Your Principia & may be printed in the same Character. Your Treatise of y$^e$ Cubick Curves should be reprinted, for I think the Enumeration is imperfect, there being five cases of Æquations viz: $xyy + ey = |$ $yy + gxxy = | xxy + ey = | xy = | y = |$ I should have acquainted You with this before M$^r$ Jones's book was published, if I had known any thing of the Printing of it, for I had observed it two or three yeares ago. I think there are some other things of less moment amiss in the same Treatise.

I am S$^r$ Your most Humble Serv$^t$

R. COTES"

Here we miss two communications from Newton, one of which accompanied the MS. of the "Elementa Logometriæ" on its return to its author, conveying his opinion of the tract in terms, the gist of which may be perceived, though more dimly than one could wish, through Cotes's *litotes* of "I am glad you are not displeased with it." (Next letter). The other contained his corrections of the Scholium on the Lunar Theory, (see note on the postscript of Letter LII). The nature of this lost paper may be easily collected from the correspondence that passed relative to parts of its contents.

Newton overlooked Cotes's suggestions on the first four paragraphs of the Scholium, and commenced his paper of alterations with paragraph 5, probably with the words "*Horroxius* noster... *Halleius* superiorem Ellipseos umbilicum," &c. The three last words are inadvertently copied from his first draught; they ought to be "centrum Ellipseos," as Cotes points out in the next letter. The diagram belonging to this and two following paragraphs, (the "new figure" mentioned in the next letter) seems to have been as represented in the annexed.

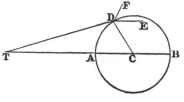

Next came paragraph 6, ("In perihelio Terræ" &c.) as it stands in the 2nd Ed. with the exception of the clerical errors rectified in the letter of Aug. 12, and the further correction (Aug. 26), in the mode of determining the ∠ *EDF*.

After that there was a new paragraph beginning "Computatio hujus motus difficilis est" &c. containing an approximation of the preceding paragraph. (The "æquatio centri secunda," whose argument is dist. of Moon from Sun + dist. of Moon's apogee from Sun's apogee). This paragraph, in consequence of the difficulties which Cotes found in it, was afterwards rendered more perspicuous in the paper of Aug. 26.

Then followed another new paragraph describing the "Variatio secunda," as it is printed in the 2nd Ed. except that "Aphelii" was twice written by mistake for "Apogæi." The Variatio secunda $= - (2'\overline{1 - \cos PE} + 1')$ sin $D$, if $PE$ = dist. of ☽'s apogee from ☉'s perigee and $D$ = dist. of ☽ from ☉.

And lastly, (omitting paragraphs 9, 10, 11, the first of which was partly removed to the end of paragraph 5) came the concluding paragraph "Theoria vero Lunæ" &c. as printed in 2nd Ed.

Compare the account of the first draught of the scholium which we have given after Cotes's letter of May 3. pp. 110—112.

## LETTER LVII.

### COTES TO NEWTON.

S^r                                    Cambridge August 10^{th} 1712

I thank You for Your care of the Wooden Cutts which I received of the Carrier together with the Manuscript *. I am glad You are not displeased with it, & I wish You had signified what Emendations might be made in it.

In my Letter of May the 3^d, I mentioned some alterations in the former part of Your Lunar Theory. You have left me uncertain as to Your resolution about them, by taking no notice of them in Your Last in which Your correction of the latter part of the Theory is set down.

I observe in the beginning of it, You have chang'd [et circulus *BDA* centro *C* intervallo *CB* descriptus erit Epicyclus ille in quo superior Ellipseos umbilicus locatur] for [Epicyclus ille in quo centrum Orbis Lunaris locatur]. I quæry whether [Halleius superiorem Ellipseos umbilicum in Epicyclo locavit] should not be also chang'd into [Halleius centrum Ellipseos] I have not D^r Halley's little Treatise by me concerning the Lunar Theory.

I do not yet understand the Paragraph beginning with [In Perihelio Terræ, propter majorem vim Solis &c.] As I apprehend it, the angle *EDF* in Your new Figure, should be equall to the excess of y^e doubled annual argument of the Apogee above the Sun's mean Anomaly as I had suppos'd it in my Letter of May y^e 3^d. Your Rule concerning that angle is this; [Et capiatur angulus *EDF* æqualis excessui argumenti annui supra distantiam Aphelii Lunæ ab Aphelio Solis.] I am uncertain how You understand the words [argumenti annui]; they may signify either the Annual argument of y^e moons apogee or the annual argument of the Sun, i. e, the Sun's mean Anomaly. I am also uncer-

---

* Of the *Elementa Logometricæ*.

tain about $y^e$ words [Aphelii Lunæ ab Aphelio Solis] I suppose it should be wrote [Apogæi Lunæ ab Apogæo Solis]. About the end of this Paragraph You say [Et concipe centrum orbis Lunæ......interea revolvi dum punctum $D$ revolvitur circum centrum $C$] I do not perceive why it should be thus.

The following Paragraph* is rather more obscure to me. I find I cannot form any conceptions of it, unless You will be pleased to give some further light to it. The Æquation which You here call *Æquatio centri secunda* is I perceive the same with that which in $D^r$ Gregories Astronomy You call *Æquatio loci Lunæ sexta* I shall be very glad to learn from You more distinctly the reasoning by which it is established.

I am $S^r$ Your obleged Freind

& most Humble Servant

<hr>

## LETTER LVIII.

### NEWTON TO COTES.

$S^r$

Upon the receipt of yo$^{rs}$ of Aug. $10^{th}$ I have looked back upon yo$^e$ of May $3^d$ w$^{ch}$ I had forgotten. In the first paragraph of $y^e$ new Scholium to Prop xxxv, where I have [ad $11'$ $50''$ circiter ascendit & *additur* medio motui Lunæ ubi Terra pergit a Perihelio suo ad Aphelium et in opposita Orbis parte *subducitur*] the words additur & subducitur should change places, & after the word *ascendit* let these words be added [in aliis locis æquationi centri solis proportionalis est,]

In the end of the second Paragraph add these words. Additur vero æquatio prior & subducitur posterior ubi

<hr>

* Beginning "Computatio hujus motus," &c.

Terra pergit a Perihelio suo ad Aphelium, & contrarium fit in opposita Orbis parte.

In the third Paragraph the words [paulo minor est in priore casu] are in my copy [paulo major est in priore casu] & should be so in yours.

In the fourth Paragraph the words additur & subducitur should change places.

In the beginning of the correction of the latter part of the Moons Theory you may write [Halleius centrum Ellipseos in Epicyclo locavit.]

In the next Paragraph beginning w$^{th}$ the words [In Aphelio* Terræ &c] after the first sentence of the Paragraph the word Aphelium is written five times erroneously for the word Apogæum. Write therefore [recta *DE* versus Apogæum Lunæ......excessui Argumenti annui Apogæi Lunæ supra distantiam Apogæi Lunæ ab Apogæo Solis, vel forte æqualis excessui Argumenti annui & 360$^{gr}$ supra distantiam Apogæi Lunæ ab Apogæo Solis.......Solis ab Apogæo Lunæ......Solis ab Apogæo proprio conjunctim. The Equation described in this Paragraph I had first from observations of Lunar Eclipses, & afterwards found that it answered the Theory of gravity in the manner here described. Its quantity when greatest came to about 2′ 10″† by Eclipses. By y$^e$ Theory tis 2′ 25″. I suppose you understand that the force of y$^e$ Sun for disturbing the Moons motions is reciprocally as the cube of the distance of the earth from y$^e$ Sun. The motion of the center of the Moons Orb in y$^e$ cycle *BDAB* arises from the force of the Sun, & as this force varies, the motion of the center of y$^e$ Moons

---

* Apparently a slip of the pen for " Perihelio."

† This is the value given in the *Lunæ Theoria Newtoniana*, in Gregory's *Astronomy*. In Mayer (modified by Lalande) it is 2′ 9″ ; Clairaut gives it only $-26″,8$ ; Damoiseau $-28″,67$ ; Plana $-28″,811$ ; Pontécoulant $-28″,511$ ; Burckhardt $-27″,6$. The terms which compose it are of the 3rd and higher orders (Pontécoulant iv. pp. 577, 602),the first term being $-\dfrac{15}{4} m . ee' = -53″,174$. See Letter LXV.

Orb should vary in this cycle both as to the length of the radius $DC$ & as to $y^e$ velocity of the rotation of this radius about the center $C$, supposing the suns annual motion to be always equal & uniform, & that his distance from the earth only changed. But because the suns annual motion accelerates & retards in a duplicate proportion of the Suns distance reciprocally, & this acceleration & retardation is allowed for in the angle $BCD$ so as to make the point $D$ accelerate & retard in the same proportion in $y^e$ cycle $BDAB$, here is a variation of the motion of the center of the Moons Orb in the cycle $BDAB$ in a duplicate proportion of the suns distance reciprocally & this without altering the length of the radius $CD$. Had this variation been in a triplicate proportion there would have been no need of any further æquation, but because it is only in a duplicate proportion, there wants a further allowance in a single proportion. And this allowance must be made $w^{th}$ respect to the Sun's motion & true place. If the suns true motion could be accelerated & retarded in this proportion, I would accelerate & retard the motion of the point $D$ in $y^e$ Epicy{c}le $BDAB$ in the same proportion. But because this cannot be done, I make the allowance by the rotation of the line $DF$ about $y^e$ center $D$, so that the center of the Moons orb may revolve about the center $D$ in an Epicycle described by the point $F$, & about $y^e$ center $C$ in a curvilinear Orb with a velocity reciprocally proportional to the cube of the distance of the earth from the Sun, or directly as the force of $y^e$ Sun $w^{ch}$ causeth this velocity; or that the velocity of the point $F$ in the said curvilinear Orb be to the velocity of the point $D$ in the Orb $BDAB$ reciprocally as the distance of the earth from the Sun. And this will come to pass quam pro{x}ime by determining $y^e$ length $DF$ & the angle $EDF$ as in the Theory.

The next Paragraph beginning with the words [Computatio motus hujus difficilis est] conteins only an approxi-

mation of the former paragraph, by computing the angle
at $y^e$ earth $w^{ch}$ the line $DF$ subtends at the Moon in her
mean distance from the earth. For the translation of the
center of the Moons Orb from $D$ to $F$, creates the same
translation of the whole orb of the Moon & of the Moon in
its Orb from the place in $w^{ch}$ they would otherwise be, &
so makes an equation or angle at the Earth $w^{ch}$ the line
$DF$ subtends at the Moon.

If the Sun did not act upon the Moon the center of
the Moons orb would be in the point $C$. By the action of
Sun it is transferred from the center to the circumference
of the Epicycle $BDAB$. If the earth moved uniformly in
a concentric circle about the Sun so that $y^e$ action of the
Sun upon the Moons Orb might be uniform, the center of
her Orb would move uniformly in $y^e$ Epicy{c}le $BDAB$. By
the inequality of the Suns action the center of the Moons
orb is transferred from the center to the circumference of
a secondary epicycle described with $y^e$ radius $DC*$ about
the point $D$. If the inequality of the Suns force of action
on $y^e$ Moons orb arose only from the variation of the distance of the earth from $y^e$ Sun & the angular motion of
the earth about the Sun was uniform, the point $D$ would
move uniformly in the epicycle $BDAB$, the angle $BCD$
$w^{ch}$ is double to the argumentum annuum increasing uniformly & the center of the Moons orb would move uniformly about the point $D$ in an Epicycle whose radius is $3\,DF$.
But the angular motion of the earth about the Sun not
being uniform, the angular motion of the radius $CD$ about
the Center $C$ is not uniform. If the angular motion of the
earth about the Sun was as the cube of the distance of the
earth from the Sun reciprocally, that is as the force of the
Sun upon the Moons Orb, the angular velocity of the
Radius $CD$ about the center $C$ would be in the same pro-

---

\* A slip of the pen for $DF$.

portion, & the center of the Moons orb being placed in the point $D$ would have a velocity in the Orb $BDAB$ proportional to the force of the Sun w$^{ch}$ causeth it, & there would be no need of a secondary Epicycle about the center $D$. But because the angular motion of the earth about the Sun is but in a duplicate proportion of the distance of the Sun reciprocally, the motion of the point $D$ in the epicycle $BDA$ will {be} but in a duplicate proportion & for making up this proportion a triplicate one, the center of the Moons Orb must be placed not in the point $D$ but in an Epicycle about the point $D$, & the radius of the Epicycle must be but a third part of such a Radius as would make the epicycle alone answer to a triple proportion, so that the motion of the center of the Moons orb in this Epicycle & of the point $D$ about the center $C$ may together compound a motion in a triplicate proportion of the distance of the earth from the Sun reciprocally.

In yo$^r$ papers* I met w$^{th}$ nothing w$^{ch}$ appeared to me to need correction.

I am

Yo$^r$ most humble Serv$^t$

London Aug. 12. 1712.                                    Is. NEWTON.

For the R$^{nd}$ M$^r$ ROGER COTES Professor
of Astronomy at his Chamber in
Trinity College in Cambridge.

LETTER LIX.

NEWTON TO COTES.

S$^r$                                        London. 16† Aug. 1712.

In the Letter I wrote to you two days ago, the words [Apogæi Lunæ] were interlined after the words [excessui

---

* The Elementa Logometriæ.
† The post mark is Aug. 14.

Argumenti annui.]* Its better to strike out the interlined words, & at the end of the Paragraph to add this sentence. [Per Argumentum annuum intelligo excessum qui relinquitur subducendo medium locum Apogæi Lunæ semel æquatum a vero loco Solis, vel a summa veri illius loci et 360$^{gr}$.

Yo$^r$ humble Servant

*For the* R$^{nd}$ M$^r$ Cotes *Professor*                Is. NEWTON
*of Astronomy at his chamber in*
*Trinity College in Cambridge.*

The directions given in this billet were superseded by the communication of Aug. 26.

---

## LETTER LX.

### COTES TO NEWTON.

S$^r$                                    Cambridge August. 17$^{th}$ 1712

I have received two Letters from You by the last Post & the foregoing. I thank You for the trouble You have given Your self to make the thing clearer to me, but am sorry to find You had mistaken my difficulty. I was very well satisfied as to the design of introducing a secondary Epicycle about y$^e$ point $D$: the motion which You had given the point $F$ in that Epicycle was what I stuck at, & consequently Your manner also of determining the angle $EDF$. By making the angle $BCD$ equal to the doubled annual argument of y$^e$ Moons Apogee the motion of the point $D$ in the primary Epicycle $BDAB$ was not yet enough accelerated in the Earths Perihelium nor enough retarded in the Earths Aphelium : the secondary Epicycle was therefore added that the velocity might be in a triplicate instead of a duplicate proportion, & an increase of velocity be made in y$^e$ Earths Perihelium & a decrease be

---

* *All* these five words are interlined.

made in its Aphelium. Hence it seem'd evident to me, that the motion of $y^e$ point $F$ in the secondary Epicycle ought to be such that it might arrive at $y^e$ place of its nearest distance from $y^e$ point $C$ in $y^e$ earths Perihelium & there by its motion conspiring with $y^e$ motion of the point $D$ might render the compound of both the swiftest & again that it might arrive at $y^e$ place of its furthest distance from the point $C$ in $y^e$ earths Aphelium & there by its motion contrary to $y^e$ motion of $y^e$ point $D$ might render the compound of both the slowest. Wherefore* if $CD$ be produced to $G$ so that $DG$ be equal to $DF$ & on the other side between $D$ & $C$, $DH$ be also taken equal to $DF$: tis evident that in the Earths Aphelium $DF$ will coincide with $DG$ & in $y^e$ Earths Perihelium $DF$ will coincide with $DH$ so revolving about $y^e$ centre $D$ $y^t$ the angle $GDF$ may always be equal to the suns mean Anomaly. Hence the angle $EDF$ or $EDG - GDF$ or $BCD - GDF$ will be equal to the excess of $y^e$ doubled Annual argument above $y^e$ suns mean Anomaly as I observ'd in my last. This is the only way according to which I can apprehend the motion of $y^e$ point $F$ in the secondary Epicycle to be regulated; but I cannot perceive how it may be reconcil'd with Your way of determining the angle $EDF$† or with the time You Assign for its

---

* Cotes does not give any figure: the annexed is added for the convenience of the reader.

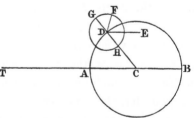

† i. e. by making it = annual argument − dist. of Moon's apogee from Sun's apogee, = twice annual argument − Sun's anomaly.
Cotes himself afterwards (letter of Sept. 7) contends for this mode of determining the ∠ EDF, taking the ∠ GDF = Sun's true anomaly, not its mean, as he makes it in this and former letters.

revolution by making it equal to $y^e$ time in which $y^e$ point $D$ revolves about $y^e$ centre $C$.

What I have here said will also affect the following Paragraph beginning with [Computatio motus hujus difficilis est &c.] But besides this there were two other difficultys containd in this Period [Et hæc recta [$DF$] subtendit angulum ad Terram quem translatio centri Orbis Lunæ a loco $D$ ad locum $F$ generat, & cujus duplum propterea dici potest Æquatio centri secunda.] The angle at the Earth which $DF$ subtends is $y^e$ angle $DTF$ comprehended by $y^e$ lines $TD$, $TF$. I understood You thus, but I perceive by Your Letter that You do not mean the angle $DTF$, but an angle at $y^e$ Earth which is subtended by a line at the Moon equal & parallel to $DF$; so $y^t$ I can now understand what follows [Et hæc æquatio est ut sinus anguli quem recta illa $DF$ cum recta a puncto $F$ ad Lunam ducta continet quam proxime] which I could not before. However I am still at a loss to understand why You take the double of that angle for the *Æquatio centri secunda*.

The following Paragraph describes the *Variatio secunda*. I suppose it was deriv'd from Observations. In it the word Aphelium is twice used instead of Apogæum.

I am $S^r$ Your &c.

---

LETTER LXI.

NEWTON TO COTES.

$S^r$                     London Aug 26. 1712.

For removing the difficulties in the Theory of the Moon mentioned in yo$^{rs}$ of Aug. 17 I have sent you the inclosed paper conteining some alterations in the description of the latter part of that Theory. I had by mistake writ [Aphelio Solis] & changed it to [Apogæo Solis\*] & should have changed it to [Perigæo Solis,] as I have done in this paper inclosed. By considering that the angle *CDF* is the complement of y$^e$ Suns Anomaly to a circle (as I have exprest it in the paper inclosed) you may perceive that whenever the Sun is in his Apoge the point *F* will fall between the points *D* & *C* & so will be in its slowest motion in the Curve line w$^{ch}$ it describes about the center *C*. If the line *DF* kept parallel to it self the points *F* & *D* would have equal motions: but by the revolving of the point *F* about the point *D* according to the order of the signes this motion of the point *F* is subducted from the motion of the point *D*, & the difference is the motion of the point *F* in the said curve line, w$^{ch}$ motion is therefore the slowest that it can be. And on the contrary, in the Sun's Perige the line *DF* will lye in directum with the line *DC*, & the motion of the point *F*† in the said curve line will be at the swiftest being the † summ of the two motions. By the inclosed paper you will understand also why I took the double of the angle subtended by a line at the Moon equal & parallel to *DF*, for the *Equatio centri secu{n}da*. The line must be doubled at the superior focus of the Moon's Orb & carried thence to the Moon.

I am Yo$^r$ most humble Servant

*For the* R$^{nd}$ M$^r$ Cotes *Professor of*        Is. Newton.
*Astronomy at his Chamber in*
*Trinity College in Cambridge*

Paper inclosed in the above.

...... Capiatur angulus *BCD* æqualis duplo argumento annuo, seu duplæ distantiæ veri loci Solis ab Apogæo Lunæ semel æquato, et erit *CTD* æquatio secunda‡ Apogæi

---

\* In his letter of Aug. 12, adopting the conjecture thrown out by Cotes in his letter of Aug. 10.

† The " *F* " and part of " the " are covered by the wax.

‡ In the fair copy of the Scholium which Cotes made for the printer (No. 173), he

Lunæ et *TD* excentricitas Orbis ejus. Habitis autem
Lunæ motu medio et Apogæo et excentricitate, ut et Orbis
axe majore partium 200000; ex his eruetur verus Lunæ
locus in Orbe et distantia ejus a Terra idq: per methodos
notissimas.

In perihelio Terræ, propter majorem vim Solis centrum
Orbis Lunæ velocius movetur in epicyclo *BDA* circum
centrum *C* quam in Aphelio, idq: in triplicata ratione distantiæ Terræ a Sole inverse. Ob æquationem centri Solis
in argumento annuo comprehensam, centrum Orbis Lunæ
velocius movetur in Epicyclo illo in duplicata ratione distantiæ Terræ a Sole inverse. Vt idem adhuc velocius
moveatur in ratione simplici distantiæ inverse; ab Orbis
centro *D* agatur recta *DE* versus Apogæum Lunæ seu
rectæ *TC* parallela, et capiatur angulus *EDF* æqualis excessui Argumenti annui prædicti supra distantiam Apogæi
Lunæ a Perigæo Solis in consequentia; vel quod perinde
est, capiatur angulus *CDF* æqualis complemento Anomaliæ
veræ Solis ad gradus 360. Et sit *DF* ad *DC* ut dupla excentricitas Orbis magni ad distantiam mediocrem Solis a
Terra et motus medius diurnus Solis ab Aphelio* Lunæ
ad motum medium diurnum Solis ab Apogæo proprio conjunctim, id est, ut 33⅞ ad 1000 et 52′. 27″. 16‴ ad 58′. 8″. 10‴
conjunctim, sive ut 3 ad 100. Et concipe centrum Orbis
Lunæ locari in puncto *F*, et in Epicyclo cujus centrum est
*D* et radius *DF* interea revolvi dum punctum *D* progredi-
tur in circumferentia circuli *DABD*. Hac enim ratione
velocitas qua centrum orbis Lunæ circum centrum *C* in
linea quadam curva movebitur, erit reciproce ut cubus distantiæ Solis a Terra quamproxime, ut oportet.

Computatio motus hujus difficilis est, sed facilior red-

---

has altered "secunda" into "semestris", and added the words " in Apogæum secundo
æquatum tendens" after "Orbis ejus", in both instances returning to the phraseology
of the first draught from which Newton had, probably without intending it, departed.

* Altered by Cotes to Apogæo.

detur per approximationem sequentem. Si distantia mediocris Lunæ a Terra sit partium 100000, et excentricitas *TC* sit partium 5505 ut supra: recta *CB* vel *CD* invenietur partium 1172¾, et recta *DF* partium 35¼. Et hæc recta ad distantiam *TC* subtendit angulum ad Terram quem translatio centri Orbis a loco *D* ad locum *F* generat in motu centri hujus; et eadem recta duplicata in situ parallelo ad distantiam superioris umbilici Orbis Lunæ a Terra, subtendit eundem angulum, quem utiq: translatio illa generat in motu umbilici, et ad distantiam Lunæ a Terra subtendit angulum quem eadem translatio generat in motu Lunæ, quiq: propterea æquatio centri secunda dici potest. Et hæc æquatio in mediocri Lunæ distantia a Terra est ut sinus anguli quem recta illa *DF* cum recta a puncto *F* ad Lunam ducta continet quamproxime, et ubi maxima est evadit 2′ 25″. Angulus autem quem recta *DF* et recta a puncto *F* ad Lunam ducta comprehendunt, invenitur &c.

In the next Paragraph but one* write *Apogœi* twice for *Aphelii*.

---

## LETTER LXII.

### COTES TO NEWTON.

Sr            Cambridge Aug: 28th 1712

I receiv'd Yours with the inclosed paper, but cannot yet agree with You. In my former Letters I had suppos'd the point *F* to come the nearest to *C* in ye Suns Perigee & to be the furthest from *C* in the Suns Apogee: You on the contrary suppose it to be ye the nearest in ye Suns Apogee & the furthest in the Suns Perigee. According to your supposition the motion of ye point *F* in its curvilinear Orb

---

* The words "but one" are added by mistake. They led Cotes to suspect that Newton's copy contained an additional paragraph which was not in his.

will then be the swiftest when that point is at its greatest
distance from y<sup>e</sup> Centre C, & slowest at its least distance
from the same, for we agree that tis the swiftest in the
Suns Perigee & slowest in his Apogee: whereas according
to my supposition the swiftest motion accompanys the least
distance & y<sup>e</sup> slowest the greatest, as I think it ought to
do.

By considering that the angle CDF is the complement
of y<sup>e</sup> Suns Anomaly to a circle, You say, I may perceive
that whenever the Sun is in his Apogee, the point F will
fall between the points D & C, & so will be in its slowest
motion in the Curve line which it describes about the centre C.   I do indeed perceive that y<sup>e</sup> point F will fall between y<sup>e</sup> points D & C, but I think it will then be in its
swiftest motion not its slowest.   For since y<sup>e</sup> angle CDF
is, by supposition, the complement of the suns Anomaly to
a circle; it follows, that as that Anomaly is continually
increasing its complement must be continually decreasing.
Therefore the line DF does so revolve to the line DC as
by its motion to diminish continually the angle CDF:
Whence it appeares that in respect of y<sup>e</sup> line DC the line
DF does revolve with a motion contrary to y<sup>e</sup> order of y<sup>e</sup>
signes I say in respect of y<sup>e</sup> moveable line DC, not in
respect of y<sup>e</sup> Fixt Stars & it is in respect of y<sup>e</sup> line DC
that its motion must be estimated in order to compound it
with the motion of y<sup>e</sup> point D in the circle ABD.   The
motion then of y<sup>e</sup> point F in its passage over y<sup>e</sup> line DC
or, by supposition, in the Suns Apogee does conspire with
y<sup>e</sup> motion of y<sup>e</sup> point D & therefore the sum of y<sup>e</sup> two
motions renders the motion of y<sup>e</sup> point F in its Curvilinear
Orb the swiftest in the Suns Apogee, which ought not
to be.

I think I apprehend Your meaning very well where
You say, The line DF must be doubled at y<sup>e</sup> superior
Focus of the Moons Orb, & carried thence to the Moon:

but I cannot see any reason why y^e doubled line at y^e superior Focus rather than the single line at y^e centre, should be carried to the Moon, excepting that Observations may require it.

Your &c. R. C.

By Your Letter I suspect that in Your copy there is a Paragraph between that beginning with *Computatio motus hujus difficilis* &c. & that beginning with *Si computatio accuratior desideretur;* they immediately follow one the other in my Copy.

---

### LETTER LXIII.

#### NEWTON TO COTES.

S^r

The reason why the doubled line at the superior focus rather then the single one at the center should be carried to the Moon is this. The angles about the superior focus are (quamproxime) proportional to the times, those about y^e Center are not. And therefore if the superior focus be translated, the line drawn from it to y^e Moon will keep its parallelism, & by doing so will make the same translation in the Moon.

As for your other difficulty, if the line *DF* kept parallel to it self, so as being produced to cut the line *TB* in a given angle the motion of the points *D* & *F* would be always equal to one another. I do not speak of the angular

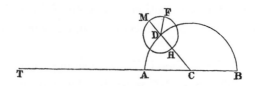

motion of the lines *CD* and *CF* about the center *C* but of the local motion of the points $D$ & $F$ in their curvilinear Orbs w^{ch} in this case will be two equal circles. Let the circle $FMN^*$ be described w^{th} the center $C$† & radius $DF$ & be cut by the line $CD$ in the point $H$ & by the line $CD$ produced in the point $M$. And if the line $DF$ keep parallel to it self, the increase of the angle $MDF$ will be equal to the increase of the angle $BCD$. I meane that y^e two angles will increase w^{th} equal swiftness or have equal augmentations in equal times. And in this case the motions of the points $D$ & $F$ will be equal. But if the angle $MDF$ increase but half so fast (w^{ch} is the case of the Theory), the motion of the point $F$ will be accelerated neare $M$ & retarded neare $N^*$. When the line $DF$ keeps parallel to it self & has no angular motion, its motion in it{s} orb will be equal to that of the point $D$. But if it has an angular motion according to the order of the letters $FMHF$ (as in the Theory) that angular motion will accelerate the point $F$ neare $M$ & retard it neare $N^*$. You seem to consider the angular revolution of the line $DF$ or $CF$ in respect of the line $DC$. I consider not the relative angular motion of the line $DF$ or $CF$ but the absolute linear motion of the point $F$ in its linear orb described about the point $C$ in the unmoved plane of the Moons orb w^{th}out any relation to the angular motion of the line $CD$.

There is no Paragraph between that w^{ch} begins w^{th} *Computatio motus hujus difficilis* &c & that w^{ch} begins w^{th} *Si computatio accuratior desideretur* &c If the words of the paper inclosed in my last are not right, pray correct them. After these two Paragraphs there is or should be a Paragraph concerning the refraction of the Atmosphere whereby the Diameter of the earths shadow is enlarged in Lunar

---

\* The " $N$ " should be " $H$ " if we follow the figure, as it is also in Cotes's figure, (Letter LX.) It would naturally drop from the pen after " $M$."

† A slip of the pen for $D$.

Eclipses.    That Paragraph was (I think) in the first
draught I sent you of the Moons Theory*.

I am Yo$^r$ most humble Servant

London Sept 2$^d$ 1712.                    Is. NEWTON.

For the R$^{nd}$ M$^r$ ROGER COTES *Professor of*
*Astronomy at his Chamber in Trinity*
*College in Cambridge*

---

LETTER LXIV.

COTES TO NEWTON.

S$^r$

I received Your last, by which I do at length perceive,
that You consider the absolute linear motion of the point
$F$ in its linear Orb described about the centre $C$, & not
the angular revolution of the line $CF$ about the same
centre, which I had before suppos'd You to do.

I am satisfied that this linear motion of the point $F$ will be
accelerated near $M$ & retarded near $N$ & therefore if it be
the linear motion which ought to be considered in Your
Theory & not the angular You do rightly in making the
angle $CDF$ equal to the complement of the Suns Anomaly
to a Circle, or which is the same thing, in making the
angle $EDF$ equal to the excess of the Annual Argument
above the distance of the Moons Apogee from the Suns
Perigee.

But I am of opinion that You ought rather to consider
the angular motion of the point $F$ than the linear.   And if
so, because the angular revolution of y$^e$ line $CF$ about the
centre $C$ in the unmoved plane of the Moons Orb, is
accelerated near $N$ & retarded near $M$; the angle $MDF$
must be taken equal to the suns Anomaly, or which is the
same thing, the Angle $EDF$ must {be} taken equal to the

---

* It is paragraph 11.   See p. 112.

excess of the Annual Argument above the distance of the Moons Apogee from the Sun's Apogee*

I will not set down other reasons for considering the Angular motion rather than the linear, which may admit of dispute. What I offer is as follows. I suppose these words at $y^e$ end of the Paragraph answer to observations [—subducendam si summa illa sit minor semicirculo, addendam si major. Sic habebitur—] But these words are not true by the Theory if the angle $EDF$ be taken equal to the excess of the annual Argument above the distance of the Moons Apogee from the Suns Perigee, as it must be taken if the linear motion be considered. And they are true by the Theory if the angle $EDF$ be taken equal to the excess of $y^e$ Annual Argument above $y^e$ distance of the Moons Apogee from the suns Apogee, as it must be taken if the angular motion be considered. Therefore the angular motion ought to be considered rather than the linear, that the Theory may answer to the Observations.

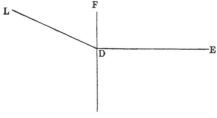

Let $DL$ be a line drawn from the point $D$ to the Moon, then will the *Æquatio centri secunda* be as the sine of the angle $FDL$. I suppose You agree with me that the Æquation must be substracted whenever the angular distance of $y^e$ line $DL$ from the line $DF$ taken according to the order of the signs is less than a semicircle & be added whenever $y^t$ distance is bigger, or in other words, that it

---

* This is precisely the value which Newton gave to the ∠ $EDF$ by mistake in his Letter of Aug. 12, (see his Letter of Aug. 26), and *against* which Cotes argues in his Letter of Aug. 17, where he takes $MDF$ = Sun's mean anomaly, not its true, as here.

must be substracted whenever the excess of the Moons Anomaly above the angle $EDF$ is less $y^n$ a semicircle & be added whenever that excess is bigger.

If then the angle $EDF$ be taken equal to $y^e$ excess of the Annual argument above the distance of the Moons Apogee from the Suns Perigee : the excess of the Moons Anomaly above the angle $EDF$ will be equal to the sum of distances of the Moon from the Sun & of the Moons Apogee from the Suns Perigee, & therefore the Æquation must be substracted when this sum is less $y^n$ a Semicircle & added when it is greater. Now this sum is less than a Semicircle when the sum of the distances of the Moon from the Sun & of the Moons Apogee from the Suns Apogee is greater than a Semicircle, and on the contrary the first sum is greater than a Semicircle when the second is less. Therefore the Æquation must be substracted when the second sum is greater than a semicircle & added when it is less. But this Rule deriv'd from the Theory is contrary to Your Rule at the end of the Paragraph derived from Observation. From which contrariety I think it is evident that the angle $EDF$ ought not to be taken equal to $y^e$ excess of the Annual Argument above the distance of the Moons Apogee from the Sun's Perigee & consequently the linear motion of the point $F$ ought not to be considered but its angular motion.

I am Your &c.

Cambridge Sep$^t$. 7$^{th}$ 1712                    R C

## LETTER LXV.

### NEWTON TO COTES.

S$^r$
London Sept. 13$^{th}$ 1712.

If it could be supposed that the force of the sun upon the Moon for disturbing her motions could be increased w$^{th}$out altering the periodical times of the sun & Moon, & that the Orb of the earth was concentric to the Sun : the line *DF* would vanish & the radius *DC* would be increased in proportion to the Sun's force without altering its angular motion about the center *C*. By the increase of the Suns force, the linear motion of the point *D* would be increased by its moving in a larger orb, but its angular motion about the center *C* would remain the same as before. But the earths orb being excentric & the excentricity causing a variation of the Suns force upon the Moon greater then in proportion to the variation of the Suns velocity, I compensate the excess or defect of the force by a secondary epicycle described w$^{th}$ the radius *DF* about the center *D*, so that the distance *CF* may increase or decrease accordingly as there is an excess or defect of the suns force & by increasing or decreasing cause the linear motion of the point *F* in the plane of the Moons Orb to be greater or less then the linear motion of the point *D* in the circle *BDA* in proportion to the said excess or defect of the suns force.

I thank you for putting me upon examining the words [—*subducendam si summa illa sit minor semicirculo, addendā si major. Sic habebitur &c.*] I have compared them with my calculations of the Moons place in Eclipses & find that they must be corrected & put [—*addendam si summa illa sit minor semicirculo, subducendam si major. Sic habebitur &c.*] The Equation* I gathered from Observations many years ago & put it when greatest, to be 2′ 10″. The last

---

* Compare p. 123.

year I gathered its quan{ti}ty from observations to be 2′ 25″ when greatest, but in describing it, committed the mistake w<sup>ch</sup> I have now corrected by reviewing my old calculations.

I am S<sup>r</sup>

Yo<sup>r</sup> most humble Servant

*For the* R<sup>nd</sup> M<sup>r</sup> ROGER COTES *Professor*           Is. NEWTON.
*of Astronomy, at his Chamber in Trinity*
*College in Cambridge*

---

LETTER LXVI.

COTES TO NEWTON.

S<sup>r</sup>

I have received Your last Letter. & am now sufficiently satisfied as to the *Æquatio centri secunda.* I hope the description of the *Variatio secunda* is accurate. The Paragraph concerning the refraction of the Atmosphere in Eclipses was in Your first draught, but was left out in Your Alteration* of it. There was also another Paragraph before it describing the dimensions of the Sun's & Moon's Diameters & Parallaxes which was also omitted in Your Paper of Alterations. I am uncertain whether You would have both of them inserted or that only concerning the Effect of y<sup>e</sup> Atmosphere. They stood thus†.

Diximus Orbem Lunæ a viribus Solis &c.

Et cum Atmosphæra Terræ &c.

I suppose You would omit the first of these Paragraphs since the substance of it is in other parts of Your Book, excepting that You have 60¼ semidiameters in Corol. 7. Prop. XXXVII. Lib. III instead of 60⅔. Be pleased to send what You would have inserted.

---

* See the introduction to Letter LVII. p. 120.

† These form paragraphs 10 and 11 in the first draught of the Lunar Theory, and will be found in the account which we have given of it, (p. 112.) This being only the draught of his letter, Cotes has not copied them at full length.

In the last Paragraph I suppose You have designedly altered Your first draught by putting ♑ 20°. 43′. 40″ for ♑ 20°. 43′. 50″, and ♒ 15°. 20′. 00″ for ♒ 15°. 19′. 50″, and ♓ 8°. 20′. 00″ for ♓ 8°. 18′. 20″.

Sept. 15. 1712                    Your &c. R C.

## LETTER LXVII.
### NEWTON TO COTES.

Sr

I beleive it will be sufficient to insert only the last of the two Paragraphs wᶜʰ you have copied in your last, vizᵗ that wᶜʰ concerns the refraction of the Atmosphere. The alterations made in the last Paragraph of the Scholium were advisedly. The description of the Variatio secunda is derived only from phænomena & wants to be made more accurate by them that have leasure & plenty of exact observations. The public must take it as it is. It brings the Moon nearer to the Sun in both the Quadratures.

I am Yoʳ most humble Servant

London. Sept. 23 1712.                    Is. NEWTON.

*For the* Rⁿᵈ Mʳ COTES *Professor of Astronomy in the University of Cambridge At his chamber in Trinity College.*

## LETTER LXVIII.
### NEWTON TO COTES.

Sr

I send you the conclusion* of the Theory of the Comets to be added at yᵉ end of the book after the words

---

* Nos. 252—255, beginning "Cæterum Cometarum revolventium, &c." and ending "primus omnium quod sciam deprehendit," (pp. 476—481 of 2nd Ed.)

[Dato autem Latere transverso datur etiam tempus periodicum Cometæ *Q. E. I.*]

There is an error* in the tenth Proposition of the
second Book, Prob III, w$^{ch}$ will require the reprinting of
about a sheet & an half. I was told of it since I wrote to
you, & am correcting it. I will pay the charge of reprinting it, & send it to you as soon as I can make it ready.
With my service to D$^r$ Bentley

I remain Yo$^r$ most humble Servant

London 14 Octob. 1712.                           Is. NEWTON.

*For the* R$^{nd}$ M$^r$ ROGER COTES *Professor of*
*Astronomy at his Chamber in Trinity*
*College in Cambridge*

---

* This error in finding the value of the resistance to the motion of a projectile in the air
(see Letter LXXIV.) was pointed out to Newton by Nicolas Bernoulli (John's nephew),
who was on a visit to England during the months of September and October, 1712.
"Monente tandem D. Nic. Bernoulli quod error aliquis admissus fuisset in Prop. x.
Lib. II. constructionem propositionis correxi et correctam ei ostendi, et imprimi curavi
non subdole sed eo cognoscente." Letter of Newton in Macclesfield *Corr.* II. 437.
Newton's result, when the curve described is a circle, had been previously shewn to be
erroneous by John Bernoulli, in a Letter to Leibniz, in August, 1710, (see their Correspondence, II. 231), and in a communication made to the French Academy, in Jan.
1711, (see *Memoires* for 1711, pp. 50—56, not published until 1714,) in an appendix to
which his nephew corrects two others. of Newton's examples, and professes to explain
the origin of the mistake (en examinant avec soin sa solution generale, j' en ay trouvé
l'origine). John afterwards resumed the inviting subject in the Leipsic *Acts* for Feb. and
March, 1713, (see Letters LXXXII., LXXXVII.) It is remarkable that both of these
mathematicians mistook the source of the error. They imagined that Newton had
taken the coefficients of the successive powers of *h* in the expansion of $(x + h)^n$ for the
successive fluxions of $x^n$. This was one of the points upon which Keill was subsequently
engaged in controversy with John Bernoulli or his partisans, who worked their crotchet
with wearisome pertinacity in the Leipsic *Acts.* Keill informs us that Newton told
Nicolas that the mistake did not arise from the use of series. Newton, through Nicolas,
thanked the sturdy professor of Basle for the timely notification of the error, sent him a
copy of his *Analysis*, &c., published by Jones in 1711, and nine days after the date of
this letter, proposed him as a member of the Royal Society, into which he was accordingly elected on the 1st of December following.

## LETTER LXIX.

### NEWTON TO COTES.

S$^r$

I sent you last tuesday a sheet inclosed in a Letter. It concerned the * The Theory of Comets to be added to y$^e$ end of the book. I should be glad to hear that it came to your hands. I mentioned also an error that I was lately told of & w$^{ch}$ wants to be set right. I have heard nothing from you this month or above & should be glad of a line to know in what forwardness the Press is.

I am Yo$^r$ most humble Servant

London. Octob. 21. 1712.      Is. NEWTON

*For the* R$^{nd}$ M$^r$ ROGER COTES *Professor of
Astronomy at his Chamber in Trinity
College in Cambridge*

---

## LETTER LXX.

### COTES TO NEWTON.

S$^r$                     October. 23. 1712.

I received both Your last Letters, together with the Sheet to be added at the end of the Book, which was inclosed in the former. You mention'd an Error in the x$^{th}$ Proposition of the $\mathrm{II}^d$ Book, which will require the reprinting of about a Sheet & an half. I have not revis'd that Proposition to see if I might find it out, but shall stay for Your corrections. The sheet which is now under the Press, ends in Page 492 of y$^e$ old Edition, and Page 456 of the new Edition. I have not observ'd anything of moment which may be altered in the Theory of Comets. In the new fourth Corollary† of Prop. XL I have

---

* sic.          † No. 245.

inserted after the first line [& quadratum radii illius ponatur esse partium 100000000]. Pag. 490, lin. 5, I have put [in subduplicata ratione $SQ$ ad $St$] instead of [in subduplicata ratione $St$ ad $SQ$] In the last Page of the Book, lines 8 & 9, I design to put $2G - 2C$ & $2T - 2S$ for $G - C$ & $T - S$, unless You forbid it. I suppose the Astronomical computations relating to the Comets are exact, having been examined both by Your self & by D$^r$ Halley.

I should have given You notice sooner, that I had received Your additional Sheet at the end of the Book, but that I expected D$^r$ Bentley would have seen You before this time, for he once intended to have been at London a week sooner. I am S$^r$.

---

### LETTER LXXI.

#### COTES TO NEWTON.

S$^r$

I here send You the Sheets as far as they are Printed off, that Your self or some freind may revise them, in order to see what Errata may be put in a Table. I know not whether You have got the Copper-plate of the Comet yet done. The Printer tells me there will be 750 requisite. The next week I shall be in the Countrey, when I return I suppose You will have the corrections ready which You mention'd for the Sheet to be reprinted

I am Sir

Your most Humble Serv$^t$

Nov. 1$^{st}$. 1712 ROGER COTES

*For* S$^r$ ISAAC NEWTON *at His*
*House in* S$^t$ *Martin's-street*
*Leicester feilds London*

## LETTER LXXII.

### COTES TO NEWTON.

Sᵣ

I hope You have received the Sheets which I sent last, ending in Page 456 of the New Edition. We have since printed off 3 Sheets more, which take in the whole Book with the Additional Sheet, excepting about 20 lines. To fill up the following Sheet may be added a Table of the Contents of each Section, if You think fit. Dᵣ Bentley was proposing to have subjoyned an Index to the whole, but particularly to the Third Book. If You approve of it, such an Index may soon be made. If Your alterations in the Second Book are finished I desire You will be pleased to send 'em.

I am Sir, Your most

Humble Servant

Cambridge Novᵇʳ. 23ᵈ. 1712          ROGER COTES

*For Sᵣ ISAAC NEWTON at his*
*House in Sᵗ Martin's Street*
*Leicester-Feilds London*

---

## LETTER LXXIII.

### NEWTON TO COTES.

Sᵣ

I send you enclosed* the tenth Proposition of the Second book corrected. It will require the reprinting of a sheet & a quarter from pag 230 to pag. 240. There is wooden cut belonging to it wᶜʰ I intend to send you by the next Carrier. I think this Proposition as it is now done will take up much the same space as before. If not, the

---

* Nos. 262—265.

10

space about the cuts may be made a little wider or a little narrower, or the number of lines in a page may be increased or diminished by a line. When this sheet & a quarter is printed off I hope your trouble of correcting will be at an end. As for making a Table to the book I leave it to you to do what you think. I beleive a short one will be sufficient. I shall send you in a few days a Scholiu{m}* of about a quarter of a Sheet to be added to the {end}* of the book: & some are perswading me to add an Appendix concerning the attraction of the small particles of bodies. It will take up about three quarters of a Sheet, but I am not yet resolved about it. I am

Yo<sup>r</sup> humble & obedient

Servant

London. Jan. 6. 171$\frac{2}{3}$.                                    Is. NEWTON

*For the* R<sup>nd</sup> M<sup>r</sup> COTES *Professor of*
*Astronomy at his Chamber in*
*Trinity College in Cambridge.*

LETTER LXXIV.

COTES TO NEWTON.

S<sup>r</sup>.                                    Cambridge Jan. 13<sup>th</sup> 1713.

I have considered Your alteration of Prop. x, Lib. II. and am well satisfied with it. I observe that You have increased the Resistance in the proportion of 3 to 2, which is the only change in Your Conclusions, arising from hence (as I apprehend it) that in the new Figure *LH* is to *NI* as $Roo$ to $Roo + 3So^3$, whereas in y<sup>e</sup> former Figure *kl* was to *FG* as $Roo$ to $Roo + 2So^3$. Some things in Your Paper I have altered, they are not worth Your

---

* These four letters within { } have disappeared with the wax.

notice, being only faults in transcribing*. I have this day received the Wooden Cut. I shall expect the Scholium at y$^e$ end of the Book & the Appendix at Your leasure.

I am Sir

Your Obliged Freind

& Humble Servant

ROGER COTES.

*For* S$^r$. ISAAC NEWTON *at his*
*House in* S$^t$ *Martin's Street*
*Leicester. Feilds London*

---

LETTER LXXV.

NEWTON TO COTES.

S$^r$

The inclosed† is the Scholium w$^{ch}$ I promised to send you, to be added to the end of the book. I intended to have said much more about the attraction of the small particles of bodies, but upon second thoughts I have chose rather to add but one short Paragraph about that part of Philosophy. This Scholium finishes the book. The cut for the Comet of 1680 is going to be rolled off. I am

Yo$^r$ most humble & obedient Servant

London 2$^d$ March‡ 171$\frac{2}{3}$. ISAAC NEWTON.

*For the* Rev$^{nd}$ M$^r$ ROGER COTES *Professor of*
*Astronomy, at his Chamber in Trinity College*
*in Cambridge.*

---

* Cotes, however, besides making the alterations alluded to here, has (perhaps from want of room) omitted a paragraph at the beginning of the Scholium of the Prop. (p. 269, Ed. 1, p. 240, Ed. 2.) in which Newton points out another mode of viewing the problem which is the subject of the Proposition. The paragraph runs as follows :

"Fingere liceret projectilia pergere in arcuum *GH, HI, IK* chordis & in solis punctis *G, H, I, K* per vim gravitatis & vim resistentiæ agitari, perinde ut in Propositione prima Libri primi corpus per vim centripetam intermittentem agitabatur, deinde chordas in infinitum diminui ut vires reddantur continuæ. Et ſ ɔlutio Problematis hac ratione facillima evaderet."

† Nos. 269, 270, 272.

‡ The Post mark is March 3, (Tuesday.)

## LETTER LXXVI.

### NEWTON AND BENTLEY TO COTES.

S$^r$

I sent you by last tuesdays Post the last sheet of y$^e$ Principia, & told you that the cut for y$^e$ Comet of 1680 was going to be rolled off.  But we want the page where it is to be inserted in the book.  I think y$^e$ page is 462 or 463.  Pray send me w$^{ch}$ it is, that it may be graved upon the Plate for directing the Bookbinder where to insert it.

I am Yo$^r$ most humble Servant

London 5 March 171$\frac{2}{3}$.        Is. Newton

I have S$^r$ Isaac's Leave to remind you of what You and I were talking of, An alphabetical Index, & a Preface in your own Name; If you please to draw them up ready for y$^e$ press, to be printed after my Return to Cambridg, You will oblige

Yours

*For the* R$^{nd}$ M$^r$ Roger Cotes *Professor of*       R Bentley.
*Astronomy, at his Chamber in Trinity*
*College in Cambridge*

---

## LETTER LXXVII.

### COTES TO NEWTON.

S$^r$.

I received both Your Letters with the last sheet of the Book inclosed in the former of them.  The Paragraph beginning with *Cæterum Trajectoriam quam Cometa descripsit* &c., which is in the 497$^{th}$ page of the former Edition, falls in the 465$^{th}$ page of the new Edition.  This is the place to which I suppose You would refer the Cut for the Comet.

I intend in a day or two to set about the Alphabetical Index.  I will write to D$^r$ Bentley concerning the Preface by y$^e$ next Post.

March. 8. 171$\frac{12}{13}$        I am S$^r$. Your &c.

## LETTER LXXVIII.

### COTES TO BENTLEY.

To D$^r$ Bentley

S$^r$.

March. 10$^{th}$. 171$\frac{2}{3}$

I received what You wrote to me in S$^r$ Isaac's Letter. I will set about the Index in a day or two. As to the Preface I should be glad to know from S$^r$ Isaac with what view he thinks proper to have it written. You know the book has been received abroad with some disadvantage, & the cause of it may easily be guess'd at. The Commercium Epistolicum lately publish'd by order of the R. Society gives such indubitable proof of Mr Leibnitz's want of candour that I shall not scruple in the least to speak out the full truth of the matter if it be thought convenient There are some peices of his looking this way which deserve a censure, as his *Tentamen de Motuum Cœlestium causis**. If S$^r$ Isaac is willing that something of this nature may be done, I should be glad if, whilst I am making the Index, he would be pleas'd to consider of it & put down a few notes of what he thinks most material to be insisted on. This I say upon supposition that I write the Preface my self. But I think it will be much more adviseable that You or He or both of You should write it whilst You are in Town. You may depend upon it that I will own it & defend it as well as I can if hereafter there be occasion.

I am S$^r$ &c.

---

* Newton had himself drawn up some strictures upon this piece, which were made use of by the editors of the *Commercium Epistolicum* (p. 97). See the paper entitled "Ex Epistola cujusdam ad Amicum," printed in the Appendix to this work.

## LETTER LXXIX.

### BENTLEY TO COTES.

Dear Sir,                              At S[r] Isaac Newton's March 12.

I communicated your Letter to S[r]. Isaac, who happend to make me a visit this morning, & we appointed to meet this Evening at his House, & there to write you an Answer. For y[e] Close of your Letter, w[ch] proposes a Preface to be drawn up here, and to be fatherd by you, we will impute it to your Modesty; but You must not press it further, but go about it your self. For y[e] subject of y[e] Preface, you know it must be to give an account, first of y[e] work it self, 2[dly] of y[e] improvements of y[e] New Edition; & then you have S[r]. Isaac's consent to add what you think proper about y[e] controversy of y[e] first Invention. You your self are full Master of it, & want no hints to be given you: However when it is drawn up, You shall have His & my Judgment, to suggest any thing y[t]. may improve it. Tis both our opinions, to spare y[e] *Name* of M. Leibnitz, and abstain from all words or Epithets of reproch: for else, y[t] will be y[e] reply, (not that its untrue) but y[t] its rude & uncivil. S[r]. Isaac presents his service to you.

I am Yours

*For* M[r]. ROGER COTES *Professor of*                    R. BENTLEY*
*Astronomy at Trinity College in*
*Cambridg.*

---

* The original of this Letter, which has been already printed in the Bentley Correspondence (p. 460), is in the possession of Dawson Turner, Esq., who has kindly furnished me with a new transcript of it.

---

## LETTER LXXX.

### COTES TO NEWTON.

S^r

I have received D^r Bentlys Letter in answer to that which I wrote to him concerning the Preface. I am very well satisfied with the directions there given, & have accordingly been considering of the Matter. I think it will be proper besides the account of the Book & its improvements, to add something more particularly concerning the manner of Philosophizing made use of & wherein it differs from that of Descartes and Others, I mean in first demonstrating the Principle it employs. This I would not only assert but make evident by a short deduction of the Principle of Gravity from the Phænomena of Nature in a popular way that it may be understood by ordinary readers & may serve at y^e same time as a specimen to them of the Method of y^e whole Book. That You {may} y^e better understand what I aim at I think to proceed in some such manner. [Tis one of y^e primary Laws of Nature, that all bodys persevere in their state &c. Hence it follows that Bodys which are moved in curve-lines & continually hindred from going on along the tangents to those curve-lines must incessantly be acted upon by some force sufficient for that purpose. The Planets (tis matter of fact) revolve in Curve-lines, therefore. &c. [Again, tis Mathematically demonstrated that *Corpus omne, quod movetur &c. Prop.* 2 *Lib* 1, & *corpus omne, quod radio &c. prop.* 3 *Lib* 1. Now tis confess'd by all Astronomers that the Primary Planets about y^e Sun & the Secondary about their respective primary doe describe areas proportional to the times. Therefore y^e force by which they are continually diverted from the tangents of their Orbits is directed & tends towards their central Bodies; which force (from what cause soever it proceeds) may therefore not improperly be

call{ed} Centripetal in respect of the revolving Bodies &
Attractive in respect of yᵉ central ones. [Furthermore tis
Mathematically demonstrated that. Cor. 6, Prop. 4. Lib. 1 &
Cor. 1, Prop. 45, Lib. 1. But tis agreed upon by Astronomers that &c. or &c. Therefore the centripetal forces
of the Primary Planets revolving about the Sun & of the
Secondary Planets revolving about their Primary ones, are
in a duplicate proportion &c. In this manner I would proceed to the 4ᵗʰ Prop of Lib. ɪɪɪ & then to the 5ᵗʰ. But
in the first corollary of this 5ᵗʰ Proposition I meet with
a difficulty *, it lyes in these words [Et cum attractio
omnis mutua sit] I am persuaded they are then true when
the Attraction may properly be so called, otherwise they
may be false. You will understand my meaning by an
Example. Suppose two Globes *A* & *B* placed at a distance
from each other upon a Table, & that whilst yᵉ Globe *A*
remaines at rest the Globe *B* is moved towards it by an invisible Hand; a by-stander who observes this motion but
not the cause of it, will say that yᵉ Globe *B* does certainly
tend to the centre of yᵉ Globe *A*, & thereupon he may call
the force of the invisible hand the centripetal force of
*B* & the Attraction of *A* since the effect appears the same
as if it did truly proceed from a proper & real Attraction
of *A*. But then I think he cannot by virtue of this Axiom
[Attractio omnis mutua est] conclude contrary to his sense
& Observation that the Globe *A* does also move towards
the Globe *B* & will meet it at the common centre of Gravity
of both bodies. This is what stops me in the train of

---

* The difficulty raised by Cotes here affords an instance of the temporary haze
which may occasionally obscure the brightest intellects. Compare the story told of
Lagrange by Biot (*Journal des Savants*, 1837, p. 84): "Lagrange tira un jour de sa
poche un papier qu'il lut à l'Académie, et qui contenait une démonstration du fameux
*Postulatum* d'Euclide, relatif à la théorie des parallèles. Cette démonstration reposait
sur un paralogisme évident, qui parut tel à tout le monde; et probablement Lagrange
aussi le reconnut pour tel pendant sa lecture. Car, lorsqu'il eut fini, il remit son papier dans sa poche, et n'en parla plus. Un instant de silence universel suivit, et l'on
passa aussitôt à d'autres objets."

reasoning by which I would make out as I said in a popular way Your 7<sup>th</sup> Proposition of y<sup>e</sup> III<sup>d</sup> Book. I shall be glad to have Your resolution of the difficulty, for such I take it to be. If it appeares so to You also, I think it should be obviated in the last Sheet of Your Book which is not yet printed off or by an *Addendum* to be printed with y<sup>e</sup> Errata Table. For till this objection be cleared I would not undertake to answer any one who should assert that You do *Hypothesim fingere*, I think You seem tacitly to make this supposition that y<sup>e</sup> Attractive force resides in the Central Body

After this Specimen I think it will be proper {to} add somethings by which your Book may be cleared from some prejudices which have been industriously laid against it. As that it deserts Mechanical causes, is built upon Miracles, & recurrs to Occult qualitys. That You may not think it unnecessary to answer such Objections You may be pleased to consult a Weekly Paper called *Memoires of Literature* & sold by Ann Baldwin in Warwick-Lane. In the 18<sup>th</sup> Number of y<sup>e</sup> second Volume of those Papers which was published May 5<sup>th</sup>, 1712* You will find a very extraordinary Letter of Mr Leibnitz to Mr Hartsoeker which will confirm what I have said. I do not propose to mention Mr Leibnitz's name, twere better to neglect him, but the Objections I think may very well be answered & even retorted upon the maintainers of Vortices. After I have spoke of Your Book it will come in my way to mention the Improvements of Geometry upon which Your Book is built, & there I must mention the time when those improvements were first made & by whom they were made. I intend to say nothing of Mr Leibnitz, but desire You will give me leave to appeal to the Commercium Epis-

---

* p. 137. Leibniz. Opp. Tom. II. Pars II. p. 60. The letter is dated, Hanover, Feb. 10, 1711. Leibniz does not mention Newton's name.

tolicum to vouch what I shall say of Your self & to insert into my Preface the very words of the Judgment of the Society (page 120<sup>th</sup> Com. Ep) that foreigners may more generally be acquainted with the true state of the Case.

Feb. * 18. 171⅔

The plan of the Preface sketched in the above letter was afterwards modified. The Indices compiled by Cotes supplied the place of "an account of the book", and the short preface which Newton sent him in his letter of March 31 made it unnecessary to enter into a detail of "its improvements." The intended notice of the method of fluxions and of the dispute relative to its discovery was abandoned, whether in consequence of Newton's declaration at the close of the letter just quoted that he "must not see it," or from a feeling that it was better to leave the evidence in the Commercium Epistolicum to work its own way, we have no precise information. Cotes's Preface therefore is confined to an exposition of "the manner of philosophizing made use of" in the work, and to an examination of the objections of Leibniz (without mentioning his name) and of the system of Vortices.

Leibniz in a letter (Apr. 9, 1716. N.S.) written under excitement, (it is his reply to Newton's raking fire of Feb. 26.) calls this Preface "pleine d'aigreur," an expression which may be taken as a measure of that extraordinary man's sensitiveness at the time.

---

### LETTER LXXXI.

#### NEWTON TO COTES.

S<sup>r</sup>

I had yo<sup>rs</sup> of Feb 18<sup>th</sup>, & the Difficulty you mention w<sup>ch</sup> lies in these words [Et cum Attractio omnis mutua sit] is removed by considering that as in Geometry the word Hypothesis is not taken in so large a sense as to include the Axiomes & Postulates, so in Experimental Philosophy it is not to be taken in so large a sense as to include the

---

* It is clear that this is a mistake for *March*, though Newton himself in his answer to this letter speaks of it as "yo<sup>rs</sup> of Feb. 18."

first Principles or Axiomes w$^{ch}$ I call the laws of motion. These Principles are deduced from Phænomena & made general by Induction: w$^{ch}$ is the highest evidence that a Proposition can have in this philosophy. And the word Hypothesis is here used by me to signify only such a Proposition as is not a Phænomenon nor deduced from any Phænomena but assumed or supposed w$^{th}$out any experimental proof. Now the mutual & mutually equal attraction of bodies is a branch of the third Law of motion & how this branch is deduced from Phænomena you may see in the end of the Corollaries of y$^e$ Laws of Motion, pag. 22. If a body attracts another body contiguous to it & is not mutually attracted by the other: the attracted body will drive the other before it & both will go away together w$^{th}$ an accelerated motion in infinitum, as it were by a self moving principle, cōtrary to y$^e$ first law of motion, whereas there is no such phænomenon in all nature.

At the end of the last Paragraph but two now ready to be printed off I desire you to add after the words [nihil aliud est quam ffatum et Natura.] these words: [Et hæc de Deo: de quo utiq: ex phænomenis disserere, ad Philosophiam experimentalem pertinet.]

And for preventing exceptions against the use of the word Hypothesis I desire you to conclude the next Paragraph in this manner [Quicquid enim ex phænomenis non deducitur Hypothesis vocanda est, et ejusmodi Hypotheses seu Metaphysicæ seu Physicæ seu Qualitatum occultarum seu Mechanicæ in Philosophia experimentali locum non habent. In hac Philosophia Propositiones deducuntur ex phænomenis & redduntur generales per Inductionem. Sic impenetrabilitas mobilitas & impetus corporum & leges motuum & gravitatis innotuere. Et satis est quod Gravitas corporū revera existat & agat secundum leges a nobis expositas & ad corporum cœlestium et maris nostri motus omnes sufficiat.

I have not time to finish this Letter but intend to write to you again on Tuesday.

I am

Yo<sup>r</sup> most humble Servant

London. 28 March {Saturday} 1713.    Is. NEWTON

*For the* Reverend M<sup>r</sup> ROGER COTES *Professor
of Astronomy, at his Chamber in Trinity
College in Cambridge.*

---

LETTER LXXXII.

NEWTON TO COTES.

S<sup>r</sup>                                      London. 31 Mar. 1713.

On Saturday last I wrote to you, representing that Experimental philosophy proceeds only upon Phenomena & deduces general Propositions from them only by Induction. And such is the proof of mutual attraction. And the arguments for y<sup>e</sup> impenetrability, mobility & force of all bodies & for the laws of motion are no better. And he that in experimental Philosophy would except against any of these must draw his objection from some experiment or phænomenon & not from a mere Hypothesis, if the Induction be of any force.

In the same Letter, I sent you also an addition to the last Paragraph but two & an emendation to the last Paragraph but one in the paper now to be printed off in the end of the Book.

I heare that M<sup>r</sup> Bernoulli has sent a Paper\* of 40

---

\* Part of it appeared in the Number for Feb. 1713, pp. 77—95, the remainder in the March number, pp. 115—132. See *Comm. Epistol.* Leibn. and Bernoull. II. 299. Bernoulli afterwards (Letter to Leibniz, Feb. $\frac{17}{28}$, 1714), in consequence of his not receiving a copy of the *Commercium Epistolicum*, and of the 2nd Ed. of the *Principia*, which Demoivre, in Newton's name, had promised more than a year before to send him, fancied that Newton was offended at his animadversions, and seems to have stated his suspicions to Demoivre; but the tone of the article did not prevent the author of the *Principia* from expressing his sense of the merits of Bernoulli's solution of his problem.

pages to be published in the Acta Leipsica relating to
what I have written upon the curve Lines described by
Projectiles in resisting Mediums. And therein he partly
makes Observations upon what I have written & partly
improves it. To prevent being blamed by him or others
for any disingenuity in not acknowledging my oversights
or slips in the first edition I believe it will not be amiss to
print next after the old Præfatio ad Lectorem, the following Account of this new Edition.

In hac secunda Principiorum Editione, multa sparsim
emendantur & nonnulla adjiciuntur. In Libri primi Sect. ii,
Inventio virium quibus corpora in Orbibus datis revolvi
possint, facilior redditur et amplior. In Libri secundi
Sect. vii Theoria resistentiæ fluidorum accuratius investi-
gatur & novis experimentis confirmatur. In Libro tertio
Theoria Lunæ & Præcessio Æquinoctiorum ex Principiis
suis plenius deducuntur, et Theoria Cometarum pluribus
et accuratius computatis Orbium exemplis confirmatur.

28 Mar. 1713.                                    I. N.

If you write any further Preface†, I must not see it‡.
for I find that I shall be examined about it. The cuts for
yᵉ Comet of 1680 & 1681 are printed off & will be sent to
Dʳ Bently this week by the Carrier.

I am

Yoʳ most humble Servant

*For the* Rⁿᵈ Mʳ Cotes *Professor of Astro-*          Isaac Newton
*nomy in the University of Cambridge. At*
*his Chamber in Trinity College in Cambridge*

---

"J'ai vu Mr. Neuuton, qui m'a dit, qu'il avoit lu avec beaucoup de plaisir vôtre
methode de resoudre le probleme de la resistance, il vous rend justice en Homme, qui
n' est nullement offensé, il dit qu' elle est admirablement belle, & meme qu' elle est
commode pour des expressions finies." Extract from a Letter of Demoivre to Bernoulli
in Leipsic *Acts* for July 1716, p. 309.

† Newton seems to have particularly in his eye Cotes's proposed allusion to the
dispute about the invention of fluxions.

‡ Compare *Commerc. Epistol.* 2nd Ed. ad Lectorem pag. penult. "Quæ novæ

This is the last letter in the Trin. Coll. collection that passed between Newton and his editor while the work was in the press. The proof-sheet however of the Scholium Generale must have been sent up to Newton, as there is a paper (No. 271) in his handwriting containing some alterations of the Scholium, in which the pages and lines are referred to as we find them in the printed book.

The Index was finished in April (letter cxiii), and the Preface is dated May 12. In his letter of May 3 to Jones (letter cxiv), Cotes "hopes the whole book may be finished in a fortnight or 3 weeks:" "it might have been done by this time" but for indisposition. It was not however until about June 18 that the impression was finished. (See next letter).

It was probably about this time that the Cambridge Aristarchus made his emendations of Halley's verses prefixed to the Principia. See Rigaud's Essay, pp. 86, 87.

---

## LETTER LXXXIII.

### COTES TO Dʳ SAM. CLARKE.

Sʳ

Cambridge June 25ᵗʰ 1713.

I received Your very kind Letter. I return You my thanks for Your corrections of the Preface, & particularly for Your advice in relation to that place where I seem'd to assert Gravity to be Essential to Bodies. I am fully of Your mind that it would have furnish'd matter for Cavilling, & therefore I struck it out immediately upon Dʳ Cannon's mentioning Your Objection to me, & so it never was printed. The impression of the whole Book was finished about a week ago.

My design in that passage was not to assert Gravity to be essential to Matter, but rather to assert that we are ignorant of the Essential propertys of Matter & that in re-

---

Principiorum editioni præmissa sunt, Newtonus non vidit antequam Liber in lucem prodiit." Dalembert's misstatement on this point ("préface faite sous les yeux de l'auteur," *Encycloped.* i. 854) is noticed by Wilson (Robins's *Tracts*, Appendix, ii. 334).

spect of our Knowledge Gravity might possibly lay as fair a claim to that Title as the other Propertys which I mention'd. For I understand by Essential propertys such propertys without which no others belonging to the same substance can exist: and I would not undertake to prove that it were impossible for any of the other Properties of Bodies to exist without even Extension.

Be pleased to present my humble Service to S$^r$ Isaac when You see him next, & let him know that the Book is finished*

I am S$^r$

Your much Obliged Freind

& Humble Servant

To D$^r$ CLARK            R C

It appears from the above letter that a meaning has been given to expressions in Cotes's Preface which he did not intend them to convey. He has been understood to assert that gravity is an essential property of bodies: his words are "Inter primarias qualitates corporum universorum vel Gravitas habebit locum; vel Extensio, Mobilitas & Impenetrabilitas non habebunt." His supposed views are controverted by D$^r$ Whewell (Philosophy of the Inductive Sciences, I. 249, or 258 2nd Ed.), and are quoted with approbation in a recent work (Le Cartésianisme ou la véritable rénovation des sciences, par Bordas-Demoulin, Paris 1843,—a work less remarkable for accuracy than for liveliness of declamation). Though Newton, says this last writer, had not the true idea of attraction, "cette notion perce et triomphe déjà chez quelques-uns de ses disciples immédiats, tels que Roger Côtes." (I. 304). He also refers to Maupertuis and Lalande as holding the same opinion. "Pour moi, dit Lalande, je pense avec M. Maupertuis et la plupart des métaphysiciens anglais, que l'attraction dépend d'une propriété intrinséque de la matière." Astron. ed. 2. art. 3384."

---

* On Monday July 27 Newton waited on the Queen with a copy of the new edition of his book. (Baily's Flamsteed, p. 98.) Jones's letter of thanks for a presentation copy (letter cxv) is dated July 11. Compare Bentley's Correspondence, p. 465. Flamsteed gave 18$s.$ for a copy (Baily, p. 305). In Clare Hall Library are two copies of the book, one of which belonged to Cotes's friend Charles Morgan "Ex dono Clariss$^i$. Editoris Pr. 1$^{lb}$. 1$^s$. 0$^d$." and the other to Rob. Green "Pret. 15$s.$" In a catalogue of Keill's library in his own hand-writing among the Lucasian papers the price of a copy is put down at £1.

Newton was obliged on several occasions to protest against the doctrine of innate gravity being ascribed to him. See letters to Bentley, Jan. 17. Feb. 25. 169⅔. Advertisement to 2ᵈ Ed. (in English) of his Optics, July 16, 1717: " And to shew that I do not take Gravity for an essential Property of Bodies, I have added one question {the 21ˢᵗ} concerning its Cause, chusing to propose it by way of a Question, because I am not yet satisfied about it for want of Experiments*." & his letter in Macclesfield Corresp. II. 437.

## LETTER LXXXIV.

This is not, properly speaking, a letter, but a paper of Corrections and Additions sent by NEWTON to COTES through Cornelius Crownfield, the University Printer, six months after the publication of the book. See next letter.

### Corrigenda et Addenda in Lib. I.

Pag 7. lin. 8, *post* veriore tempore *adde* mensurent. P. 10, l. 6, *post* sed *adde* sunt P. 10. l. 17, *lege* difficillimum est.

P. 15 l 16 *lege* in plana, ut $pN$ ad $pH$. Ib. l. 20 veritatem ejus.

P. 17 l. 20 *pro* communis *lege* corporis. P 31, l. 38 $AD$ et $DB$. P. 36, l. 6 *lege* Cor. 5. P. 38 l 24 *lege* Corol. 2. l. 26 *lege* Corol. 4.

P. 41 l 5 *lege* $P$ et $Q$. P. 42, l. 8 *post* vel *adde* circulum concentrice tangit, id est.

P 44 l 23 *lege* $QR \times \overline{RN + QN}$. P. 45, l. 14 *post* hoc est *adde* (ob datam specie figuram illam) Ib. l. 21 *post* Spiralem *adde* concentrice. P. 46 l. 30 *post* intelligatur *adde* recta. Ib. l. ult. *post* $Pv \times uV$ *lege*, Adde rectangulum $uPv$ utrinq: et prodibit quadratum chordæ arcus $PQ$ æquale rectangulo $VPv$. P. 47 l 4 *post* conica in $P$, *lege*

---

* This declaration was probably drawn from him by the recent controversy between Leibniz and Clarke.

adeoq: ex natura Sectionum Conicarum, circuli hujus
chorda $PV$ æqualis erit $\frac{2DC^q}{PC}$. P. 52, l. 16 *dele* per.

P. 54, l. 4, *post* area $QT \times SP$ *adde* quæ dato tempore
describitur. P. 57, l. 25 *post* si ea *adde* sit. P. 59, l. 7 *post* axi
principali figuræ, *adde* id est axi in quo umbilici jacent. P. 61, l.
12 *lege* ita ut sit $GA$ ad $AS$ et $Ga$ ad $aS$ ut est
$KB$ ad $BS$, et axe $Aa$.

Ib. l. 15, 16 *lege*, et cum sit $GA$ ad $AS$ ut $Ga$ ad $aS$,
erit divisim $Ga - GA$, seu $Aa$ ad $aS - AS$ seu $SH$ in
eadem ratione. P. 86, l. 7, *post* biseca *adde* in $M$ et $N$.
P. 87 l. 7 *lege* per Prob. xiv.
P. 89 & 90 *in Figura jungatur FD.*
P. 92, *in Figura jungantur FG et HI.* P. 101, l. 6, 7,
8, *lege*, Nam centro $O$ intervallo $OA$ describatur semicircu-
lus $AQB$ rectæ $LP$ si opus est productæ, occurrens in $Q$,
junganturq: $SQ$, $OQ$, quarum $OQ$ producta occurrat arcui
$EFG$ in $F$, et in eandem $OQ$ demittatur perpendiculum
$SR$. Ib. l. 36 *post* quæ *adde* per punctum $P$ transit et.
P. 109, l. 1 post Hyperbola *adde* rectangula. Ib. *in Schemate pro litera O scribatur litera H.* P. 117. l 15 *lege* prioris in $I$. Et stantibus. P. 121 *in Schemate e regione literæ
p scribatur litera K in Orbe VPK.* P. 127, l. 7, 9 graduum.
P. 131 l. 17 *lege*, m æqualis 1 et n.

P. 136, l. 2 pro $Bp$ scribe $BP$. P. 137, l. 16 *post* sinus
versus *adde* est. P. 139, l. 10 *post* adeoq: ad *adde* globi
exterioris. l. 12 *post* habet ad *adde* globi interioris. P. 148,
l. 4 *post* distantiæ *adde* corporum. Ib. l. 7 *pro* terminos
suos communi *scribe* terminum suum communem. P. 151,
l. 8, 21 *scribe* ad primum duorum. P. 151, l. 18 *scribe* ut
primum duorum. P. 156 l 31 *scribe* maximo. Nam. P 158
l 32 *Post* atq: *adde* ut, *et post* proportionalitate *dele* ut. Ib.
l. 36 *post* non sit, *adde* reciproce.

P. 166, l. 9 *dele* quadratum temporis periodici *et scribe*

11

tempus periodicum. P. 169, l. 26, 33, 34, & P. 170 l 3 *pro*
*C* scribe *O, et in schemate inter P ac T scribe literam O.*
P. 184 l 21 *post* area *adde ABNA.* P. 187 l 4 *pro* duplo
ejus *scribe* ejus duplo. P. 190, l. 15 *pro* similia *scribe* continue proportionales *SI, SE, SP,* similia sunt. Ib. l. 19,
*post PE*ⁿ *adde,* (ob proportionales *IE* ad *PE* ut *IS* ad *SA*)
P. 191, l. 7, *lege* corpus *P* erit ut $\dfrac{DF \times O}{PF^{n-1}} - \dfrac{DF^q \times O}{2PF^n}$.

P 196 l. 25 *post* qua annuli *adde* centro *A* intervallo *AE*
in plano prædicto descripti. P. 197 l 24 *pro* diametro *lege*
semidiametro.

Corrigenda et addenda in Lib. II.

Pag. 213, lin. 10, 12 *Pro BC et BD scribe BACH* et
*BADE.* Ib. lin 14 *post* partes *adde* rectæ *AB.* Ib. l. 24
*pro BC scribe BACH.* Ib. l. 26 *pro AH scribe RACH*
P. 214, l. 33 *post* gravitatis qua *adde* corpus illud. P. 223
l. 20, 22 *pro* sesquialtera *scribe* sesquiplicata. P. 229, l. 7,
8 *lege* omne ascendendi ad locum summum ut Sector Circuli, et tempus omne descendendi a loco summo ut Sector
Hyperbolæ. Ib. l. 13, 14, 15 *post* Circularis *A t D* ut tempus *lege* omne ascendendi ad locum summum, & Sector
Hyperbolicus *ATD* ut tempus omne descendendi a loco
summo; si modo Sectorum. Ib. l. 21, *post* ut *lege*
$\dfrac{qDp \times tD^{\text{quad.}}}{pD^{\text{quad.}}}$, id est, ob datam *t D,* ut. Ib. l. 26 *post*
mento *adde* velocitatis. l. 30, *post* est ut *adde* tempus totum
ascendendi ad locum summum. Q. E. D.

P. 233, lin. ult. *pro* 2 *QRo lege* 2 *QRo*³. P. 240 l. 27 *pro*
*MX lege NX.* P. 241, l. 13 Parabolæ prædictæ.

P. 244, l. 22 *lege* $\dfrac{2nn - 2n}{n - 2}$ *VG.* Pag. 248, l. 2 *lege* sit.

Ib. l. 10, *pro* omnis futuri *lege* totius Ib. l. 23, 42 *pro* futuri
*lege* totius. P. 249, l. 20 *post* tempus *adde* totum P. 251,
l. 32 *post* et *AB* ut *adde* area. P. 255 l. 8 *pro* sit *lege* est.
P. 285 l. 17 *post* arcubus *adde* vel. P. 290 l. 31 *pro* aere

*scribe* aqua. Ib. l. 34 *pro* aqua *scribe* aere. P. 300 l. 11
*pro CB scribe AB.* P. 301, l. 7 *post* axis sui *adde* uniformiter progrediendo. Ib. l 9 *post* diametri suæ *adde* uniformiter progrediendo. Ib. l. 12 *pro* totum globi motum *lege* motum globi. Ib. l. 15, *post* diametri suæ *adde* uniformiter progrediendo
p. 317 l. penult. *pro* maximam *G lege* maximam *H.*

Corrigenda et Addenda in Lib. III.

Pag. 358, l. 3, 4 *lege* affirmatur. Corpora plura dura esse experimur; oritur autem.
P. 367, l. 14 *lege* foret. P. 378 l. 28 *pro* circa annum *lege* anno.
P. 379 l. 13, 23 *pro* centripetam *lege* centrifugam.
P. 387 l. 22 *lege* quam. P. 396, l. 17 *pro* erit *Kk* ad *lege* erit *FK* æqualis *TK* & *Kk* erit ad. Ib l 19 *post FKkf adde* erit. P. 399 l. 6 *post* Solem *adde* vel ab ea superatur.
P. 415 l. 12, 15 *pro* annua *et* annuæ *scribe* semestris *et* semestri. P 422 l. 34 *post* hæc æquatio *adde* maxima.
P. 425 l. 23 dilatet. P 444 l. 33 *dele* formata est, *et post* inter se *adde* formata sunt. P 450 l. 16 *lege* ad ejus velocitatem. P. 453 l 17 lege quorum *AM.* P 457 l. penult. & ult. *post* manentem *dele* parum diligenter definivit. Nam Cometa, *&* *scribe* ex observationibus definire neglexit. Cometa autem. P. 459. l. 3 *lege* partium 100000. P 459 *proxime post Tabulam lege* Apparuit etiam hic Cometa mense Novembri præcedende* in signis Virginis & Libræ ut stella secundæ vel tertiæ magnitudinis, & Florentiæ quidem ad horam octavam Italicam ea nocte quæ mensis hujus diem vigesimum & vigesimum primum intercessit, st. novo, id est, decimum & undecimum st. vet. visus fuit in signo Virginis sub stellis in sinistro pede [vel femure] Leonis cum Ascentione* recta graduum 165, referente Cassino. Erat igitur Cometa in ♍ 13½ circiter. Nam et

---

* sic.

Hillus quidam hora quinta matutina die 12 vel potius 10 Novembris, Cantuariæ in Anglia distantiam cæpit* hujus Cometæ a Corde Leonis graduum septendecim in Orientem et a Cauda Leonis paulo plusquam graduum undecim in austrum. Unde Cometa tunc erat in ♍ 12$^{gr}$ 24′ cum latitudine boreali 2$^{gr}$ circiter. Crassissimæ fuerunt hæ observationes; meliores sunt quæ sequuntur. Pag. 459 lin 35 post Galletius etiam scribe Avenioni. Ib. l. 39 Cellius in ♎ 13. 30′ Ib. l. 40 dele Romæ. P. 460 l. 33 post Australi 1$^{gr}$ 16′ adde Cellius in ♎ 28. Ib. l. 37 post, id est 2$^{gr}$ 2′ vice linearum quinq: sequentium adde. Eodem die ad horam quintam matutinam Ballasoræ in India Orientali, capta est distantia Cometæ a Spica ♍ 7$^{gr}$ 35′ in Orientem. In linea erat recta inter Spicam et Lancem australem, ideoq: versabatur in ♎ 26$^{gr}$. 58′, cum Latitudine australi 1$^{gr}$ 11′ circiter; et propterea post horas 5 & 40′, ad horam scilicet quintam matutinam Londini erat in ♎ 28$^{gr}$ 11′ cum Latitudine australi 1$^{gr}$ 16′ circiter. Pag. 462 lin 30 post factæ videntur adde Die 22 ubi Cometa ex observatione Montenari erat in ♏ 2$^{gr}$ 36′ Venetiis, & propterea in ♏ 2$^{gr}$ 48′ eadem hora matutina Londini: Hookius noster eundem locavit in ♏ 3. 30′ ut supra. Montenarus in defectu Hookius in excessu errasse videntur. Nam et Ballasoræ eodem die ante ortum Solis, Cometa observabatur in ♏ 1$^{gr}$ 50′, ideoq: eadem hora matutina Londini erat in ♏ 3$^{gr}$ 5′. Die 24 ad horam quintam matutinam Ballasoræ Cometa observabatur in ♏ 11$^{gr}$ 45′, ideoq: ad horam quintam Londini erat in ♏ 13$^{gr}$ circiter. Pag. 463 in Tabula priore pro ♎ 27. 52′, ♏ 2 56, ♏ 12. 58, lege ♎ 28. 0. ♏ 3. 5. ♏ 13. 0. Ib. initio secundæ Tabulæ addantur Novem. 9. 17 | 101551 | ♏ 12. 25. 50 | 0. 43. 30 Bor. Pag 472 lin 27 lege cadent. Pag 474 lin 23, inter Et et similis lege in Chronico Saxonico. Ib. dele 1101 vel. Ib. lin. 26 post habet adde etiam.

---

* sic.

P. 478, l. 25 *pro* prima *lege* secunda. P. 482 l. 2, *post* spatiis *adde* ob defectum aeris. Ib. lin 18 *lege* ut se mutuo quam minime trahant. Ib. l. 29 *lege* non in corpus proprium (uti sentiunt quibus Deus est anima mundi,) sed in servos. P. 483 l 36, *post* Fatum et Natura. *adde*, A necessitate Metaphysica, quæ utiq: eadem est semper et ubiq:, nulla oritur rerum variatio. Omnis illa quæ in mundo conspicitur pro locis ac temporibus diversitas a voluntate sola Entis necessario existentis oriri potuit. Dicitur autem Deus per Allegoriam videre, audire, loqui, ridere, amare, odio habere, cupere, dare, accipere, gaudere, irasci, pugnare, fabricare, condere, construere, & intelligentes (vitam infundendo) *generare. Nam sermo omnis de Deo a rebus humanis per similitu- <br> \* Job. 38. 7. <br> Luc. 3. 38. <br> dinem aliquam desumi solet. Et hæc de Deo; de quo utiq: ex phænomenis disserere ad Philosophiam experimentalem pertinet.

The following notes are in Cotes's hand : they are the elements of the next letter.

p. 3. l: 14

p. 41. l: 3

p. 47 l: penult.

p. 47. l: 4 non emend.

p. 109. in schem. non *H* pro *O*

p. 148. l. 7 *n*.

p. 151. l. 8, 18, 21 *n*

p. 191. l. 7 *n*.

℞ . $12^0.25'.50''$ *non* ℞

p. 230, l. penult. *post* incremento *adde* velocitatis

p. 460. p. 462 *n* intell.

LETTER LXXXV.

COTES TO NEWTON.

S<sup>r</sup>

I lately received from You by M<sup>r</sup> Crownfeild a Paper
of *Errata, Corrigenda & Addenda* to be printed* & bound
up with Your Principia.   I take leave to send You some
observations upon them.

By comparing Your Catalogue with my Table of
*Corrigenda,* I find you have omitted that of pag: 3.
lin : 14.  I think it convenient to make some such alteration,
that You may not seem to assert what is false.   You have
also omitted that of pag. 47. lin. penult. which I think is
requisite to determine Your meaning.   Whilst that Sheet
was printing I remember I did not understand what it was
that You there asserted, & not having then time to examine the thing to the bottom, I was forc'd to let it go.
Soon after I considered it, & found in what sense You{r}
words could be true & accordingly made the Alteration.
Since Your book has been published I have been ask'd the
meaning of that place by one who told me he knew not
what sense to put upon Y<sup>r</sup> words: I referr'd him to the
Table of Corrigenda & then I perceiv'd he understood
You.

Your addition of pag. 47 lin. 4 should I think be
omitted.   For if that addition be made the 8 preceding
lines are to no purpose & ought to be omitted.  Tis very
evident that $PV$ is equal to $\dfrac{2DCq}{PC}$ by  pag. 46 lin. antepenult.

In pag 109 You direct to put $H$ in the Figure instead
of $O$.   You mean instead of the lower $O$ which bisects
the transverse diameter of the Hyperbola.   If this be

---

* I am not aware that this table of Errata was ever printed.  Cotes does not seem
to have been altogether pleased at the receipt of so formidable a list.

done, then the Figure will not agree with the second line of this page, nor indeed with the whole Demonstration as it relates to the Hyperbola. In pag. 148: lin, 7. I think the alteration should not be made. There are three different *distantiæ*, & three different *termini* & one common angular motion.

Pag. 151. You change *prima* the Fæminine into *primum* y^e Neutre. Tis my Opinion that this alteration is not necessary. I understand the printed text thus: *prima duarum medie proportionalium quantitatum.* If it were adviseable to make an alteration, I would rather choose the Masculine & put it; *primus duorum medie proportionalium terminorum inter &c.*

Pag. 191. lin. 7 I think wants no correction. I cannot understand by what reasoning You make one; You will be pleas'd to reconsider it. If Your correction be true, it will be very necessary to explain it more fully.

Page 463 in the beginning of the second Table I suppose You intended to put m 12⁰. 25'. 50″ not m 12. 25. 50 as it is in Your written copy

You order the 3 last lines of page 460, & the 2 first of page 461 to be struck out; & in their room You place what follows. [Eodem die ad horam quintam matutinam Ballasoræ in India Orientali, capta est distantia Cometæ a Spica m 7^gr. 35' ———— Londini, erat in ♎ 28^gr. 11' cum Latitudine australi 1^gr. 16 circiter.] I suppose You intended to make this addition at the end of the Paragraph which begins with *Nov.* 21. *Ponthæus &c.* & would not have the 5 first lines of the following Paragraph struck out.

I observe You have put down about 20 Errata besides those in my Table. I am glad to find they are not of any moment, such I mean as can give the reader any trouble. I had my self observ'd several of them, but I confess to You I was asham'd to put 'em in the Table, lest I should appear to be too diligent in trifles. Such *Errata* the

Reader expects to meet with, and they cannot well be avoided. After You have now Your self examined the Book & found these 20, I beleive You will not be surpriz'd if I tell You I can send You 20 more as considerable, which I have casually observ'd, & which seem to have escap'd You: & I am far from thinking these forty are all that may be found out, notwithstanding that I think the Edition to be very correct. I am sure it is much more so than the former, which was carefully enough printed; for besides Your own corrections & those I acquainted You with whilst the Book was printing, I may venture to say I made some Hundreds, with which I never acquainted You

I am S<sup>r</sup>

Your very

Humble Serv<sup>t</sup>

Dec. 22<sup>d</sup> 1713.                                    R. Cotes


END OF CORRESPONDENCE ON THE PRINCIPIA.

# LETTERS OF NEWTON TO KEILL.

## LETTER LXXXVI*.

### NEWTON TO KEILL.

S<sup>r</sup>

Yo<sup>r</sup> Letter of Feb. 8<sup>th</sup> I delayed to answer till the Journal Literaire for November and December should come out. It is just come from Holland & I desired M<sup>r</sup> Darby to send you a copy w<sup>ch</sup> I doubt he has not done because he sent one to me this morning w<sup>ch</sup> I reccon to be for you & I designe to send it to you the first opportunity by the Carrier. M<sup>r</sup> Leibnitz in August last, by one of his correspondents published a paper† in Germany conteining the judgment of a nameless Mathematician‡ in opposition to the judgment of the Committee of the Royal Society, with many reflexions annexed. This paper hath been sent to M<sup>r</sup> Johnson with remarks prefixed to it. And the whole is printed in the journal Literaire pag. 445. And

---

* Letters LXXXVI., XCII., XCIII. were formerly among the papers belonging to the Lucasian Professor.

† A "charta volans," dated 29 Jul. 1713, without name of place, printer, or author.

‡ i.e. John Bernoulli in the letter of June 7, 1713, to Leibniz. There are two circumstances connected with this letter—one of them affecting the writer of it, the other his correspondent—which are not calculated to add lustre to either of these great names. To mention the latter first: Bernoulli accompanied the letter with the request that in any use that might be made of it, his name might not be mixed up with the controversy. Leibniz observed his friend's injunction of secrecy at the time, but between two and three years afterwards, without Bernoulli's permission or knowledge, he quoted the letter with Bernoulli's name, in letters to Count Bothmar and—(*quæ legat ipsa Lycoris*)—Madame la Comtesse de Kilmansegg. He had shortly before intimated the fact in the plainest terms in his letter of April 9, 1716, to Conti for Newton. The other point alluded to wears a more serious aspect. Though Bernoulli was confessedly the writer of the letter, (which accordingly appears in his Correspondence, published during his lifetime,) he afterwards (1719), in a letter which he sent to Newton, disavowed the authorship of it. The following references will be sufficient to enable any reader to form his own judgment upon these two points. Leibn. and Bernoull. *Commerc.* II. 311, 323, 330, 334, 378. Leibniz. *Opp.* III. 459, 462. Macclesfield *Correspondence*, II. 436. Des Maizeaux to Conti, MSS. Birch, 4284. fol. 222, Brit. Mus.

now it is made so publick I think it requires an Answer.
It is very reflecting upon the Committee of the Royal
Society, & endeavours to derogate from the credit of some
of the Letters published in the Commercium Epistolicum
as if they were spurious.  If you please when you have it,
to consider of what Answer you think proper, I will within
a Post or two send you my thoughts upon the Subject,
that you may compare them w^th your own sentiments &
then draw up such an Answer as you think proper.  You
need not set your name to it.  You may write either in
English or in Latine & leave it to M^r Johnson to get it
translated into F{r}ench.  M^r Darby will convey yo^r Answer to the Hague.

I am

Yo^r most humble Servant

London. 2 Apr. 1714.                          Is. NEWTON

*For* D^r JOHN KEILL, *Professor of
    Astronomy, at his house in Ox-
    ford.*

LETTER LXXXVII*.

NEWTON TO KEILL.

S^r

I am glad you have read both the pieces concerning
the Commercium inserted in the Journal Literaire & are of
opinion that they must be immediately answered & are
thinking of an Answer.  As to what you want to know
concerning things in the Principia contrary to the doctrine
of fluxions or differences I take it to be this.  In the
Scholium of y^e 10^th Proposition of the second book of the
Principia I have made use of y^e method of Infinite Series
for determining the Curves in w^ch Projectiles will move in

---

* This and the two following Letters were "the gift of Mr Watson, fellow of the
College, 1771," (afterwards Bishop of Llandaff).  They were formerly placed in a
folio volume, which is now marked R. 4.59.

a resisting Medium such as is air. John Bernoulli has published in the Acta Eruditorum for Febr. & March was a twelve month, a Paper upon that Scholium, in w<sup>ch</sup> he represents that the Method there used is the Method of fluxions, & that it appears thereby that I did not understand y<sup>e</sup> 2<sup>d</sup> ffluxions when I wrote that Scholium because (as he thinks) I take the second terms of the series for the first fluxions, the third terms for the second fluxions & so on*. But he is mightily mistaken when he thinks that I there make use of the method of fluxions. Tis only a branch of y<sup>e</sup> method of converging series that I there make uses of. The Acta Eruditorum for the last year are but just come to London, & I find thereby that John Bernoulli is the great Mathematician† who accuses me on this account. But I beleive it's better not to reflect upon him for it nor so much as to name him any otherwise then by the general name of the great Mathematician. They are seeking to pick a quarrell with me & its better to lett them begin it still more openly without a provocation.

There is another great Mathematician‡ to whom Leibnitz referred the examination of the Commercium Epistolicum. He makes use of two arguments against me. One

---

* See p. 142 note. An abortive attempt has been made to revive this delusion by M. Jean Trembley (Berlin Mémoires, 1798) in a paper which professes to overthrow Lagrange's explanation of the real source of the error in the expression for the resistance given in the 1st edition of the Principia. Lagrange has shewn (Théorie des Fonctions, Paris, 1813. pp. 339—349 : see also p. 6) that if powers of θ (the time of describing a small arc) above the square be neglected, we get Newton's first result, but that if we include terms involving θ³, we obtain the correct value. He has not, however, pointed out in what respect Newton's geometrical expression is erroneous, or at what step of the demonstration the fallacy is introduced. The error consists in substituting

$$FG \text{ (which} = Ro^2 + So^3 = \tfrac{1}{2}g\theta^2 - \tfrac{1}{6}g\frac{r}{u}\theta^3 \text{) for } fg \text{ (which} = Ro^2 + 2So^3 = \tfrac{1}{2}g\theta^2 + \tfrac{1}{3}g\frac{r}{u}\theta^3 \text{),}$$

where r = resist. and u = vel. I am fully sensible of the danger of dissenting from that great geometer on a point of mathematics, but I think that a remark to the effect just stated would have been less open to objection than his mode of arriving at the correct expression by substitution in an erroneous formula (p. 347. lines 15, 16, 17.)

Lacroix (Calc. Diff. et Int. tom. 3. p. 644. Paris. 1819) does not seem to have read the part of the Principia in question with much attention.

† i.e. the "eminens quidam Mathematicus," quoted in the *Charta Volans*. See next page, line 3.

‡ John Bernoulli. See preceding Letter and note.

that I made no use of the prickt letters till of late, the
other that when I wrote the Principia I understood not
the second fluxions as a certain great Mathematician (Bernoulli) has observed *. The Answer is that I use any notation for fluents & any other notation for fluxions, & an
unit for the fluxion of time or its exponent & the letter
$o$ for the moment of time or of its exponent, & the rectangles of the fluxions & the moment $o$ for the moments of
other fluent quantities. That in the Analysis per æquationes numero terminorum infinitas I represent fluents by the
areas of figures, time by the Abscissa flowing uniformly,
the fluxions of fluents by the Ordinates of curves, the
moments of fluents by the rectangles under the Ordinates
& $o$ the moment of the Abscissa : but do not confine my
self to any certain symbols for the Ordinates or fluxions.
That I do the same in the book of Quadratures & even to
this day. That where I use prickt letters they signify not
moments or differences w$^{ch}$ are infinite little quantities but
fluxions or the Ordinates of curves as the exponents of
fluxions w$^{ch}$ are finite quantities, unless they be multiplied by the symbol $o$ (either exprest or understood) to
make them infinitely little : but it is not necessary that the
Ordinates of curves should be represented by prickt letters
Such letters may be a convenient sort of notation but not
necessary to the method. That prick letters are older
symbols for fluxions then any used by M$^r$ Leibnitz : for he
has no symbols for fluxions to this day. That the rectangles under the Ordinates of curves & the moment $O$
are older symbols for moments or differences then any
used by M$^r$ Leibnits they being used by me in my Analysis abovementioned communicated by D$^r$ Barrow to M$^r$
Collins in the year 1669 & the symbols $dx$ & $dy$ being not
used by M$^r$ Leibnitz before the year 1677. And whereas
M$^r$ Leibnits præfixes the letter $\int$ to the Ordinate of a

---

* " Quemadmodum ab eminente quodam Mathematico dudum notatum est." These
words were inserted in Bernoulli's letter in the *Charta Volans* by Leibniz.

curve to denote the Summ of the Ordinates or area of the Curve, I did some years before represent the same thing by inscribing the Ordinate in a square as may be seen in the Analysis. My Symbols therefore (so far as I have used any particular symbols) are the oldest in the kind.

The other argument used by the great Mathematician, is that when I wrote my Principia I understood not the second differences, as a certain great Mathematician (viz$^t$ Bernoulli) has noted, meaning in the Scholium to y$^e$ 10$^{th}$ Proposition of y$^e$ second Book. But this great Mathematician is grosly mistaken in taking the method there made use of, w$^{ch}$ is a branch of the method of converging series to be the method of fluxions. The Elements of the method of fluxions are set down in y$^e$ 2$^d$ Lemma of the second Book & are very different from y$^e$ method made use of in this Scholium.

The author of the Remarks* cites D$^r$ Wallis as favouring M$^r$ Leibnitz & yet D$^r$ Wallis in the Preface to the first Volume of his works A.D. 1695 writes that in my two letters of June 13 & Octob. 24, 1676 I expounded my method of ffluxions to M$^r$ Leibnitz found by me ten years before.

In my Letter of 10 Decem. 1672 sent to M$^r$ Collins, in writing of a method whereof the method of Tangents of Slusius was but a Corollary, & which stuck not at surds, & w$^{ch}$ was therefore the method of fluxions, I represented that this method was very general & amon{g}st other things extended to the determining the curvature of Curves. Whence its manifest that I then understood the second fluxions or differences of differences.

I received yo$^r$ Letter this afternoon at three of the clock & have time to add no more but that I am

Yo$^r$ most humble Servant

London 20 April {Tuesday} 1714.　　　　Is. NEWTON

---

* In the Journal Literaire. See *antea*, p. 169.

In the book of Quadratures where I use prickt letters
for fluxions I solve some Problems in the Introduction to
yᵉ book without making use of such Letters & therefore
did not then confine the method of fluxions to such Letters.

*For the* Rⁿᵈ Dʳ JOHN KEILL *Professor*
*of Astronomy in the University of*
*Oxford.*

———————

## LETTER LXXXVIII.

### NEWTON TO KEILL.

Sʳ

I have read over your Letter & find it right. The
Marquess de L' Hospital in his Treatise de Infinitement
Petits teaches that if the Ordinates *AB*, *CD*, *EF* be at
Equal distances, & the chord *BD* be produced till it cuts
the Ordinate *EF* produced in *N*, the line *FN* shall be the

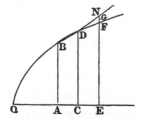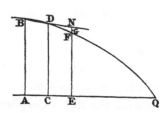

second difference of the three Ordinates.    And the points
*B*, *D* being infinitely neare, perhaps Burnoulli may take
*BD* for a tangent of the Curve at *D* & so reccon that the
distance between the Curve & yᵉ Tangent is the second
difference : whereas *BDN* is not a tangent but cuts the
Curve at *D*, & the tangent at *D* is parallel to the chord
*BF* & bisects the second difference *FN*, suppose in *G*.
So that the line *FG* wᶜʰ lies between the Curve & the
tangent, & is equal to the third term of the series, is but
half the second Difference, as I have put it.    Mʳ Burnoulli

therefore is mistaken in affirming that I put the third terme of the series equal to the second difference, & I am in the right in putting it equal to $y^e$ line between the Curve & the Tangent & by consequence to half the second difference as you observe. And I think $yo^r$ Demonstration is good.

I have corrected a paragraph in $y^e$ 11$^{th}$ page of $y^e$ papers you sent me & put it thus. , [3$^{dly}$ We do not dispute about the ,antiquity of the symbols of ffluents Fluxions & Moments, , Summs & Differences used by $M^r$ Newton & $M^r$ Leibnitz, ,they being not necessary to the method, but liable to ,change. And yet the symbol $\boxed{\dfrac{aa}{64x}}$ used by $M^r$ Newton ,in his Analysis for fluents or summs is much older then ,the symbol $\int \dfrac{aa}{64x}$ used by $M^r$ Leibnitz in the same sense. ,And some of the symbols of fluxions used by $M^r$ Newton ,are as old as his said Analysis, whilst $M^r$ Leibnitz has no ,symbols of fluxions to this day. And the rectangles under ,the fluxions & the letter $o$ used by $M^r$ Newton for mo- ,ments are much older then the symbols $dx$ & $dy$ used ,by $M^r$ Leibnitz for the same quantities. But these are only ,ways of Notation & signify nothing to $y^e$ method it self ,$w^{ch}$ may be without them]. I have made this alteration to avoyd quoting my Manuscripts $w^{ch}$ are not upon record. And for the same reason the last leaf of the papers you sent me must be altered. But I have time to add no more at present but that I am

S$^r$

Yo$^r$ most humble Servant

London May 11$^{th}$* 1714.  Is. NEWTON

*For the* R$^{nd}$ D$^r$ JOHN KEILL *Professor*
*of Astronomy in the University of*
*Oxford.*

---

* The post mark is 13 Ma.

## LETTER LXXXIX.
### NEWTON TO KEILL.

S$^r$                                    London May 15 1714.

I wrote to you on Tuesday that the last leafe of the papers you sent me should be altered because it refers to a Manuscript in my private custody & not yet upon Record. For setting right this leafe it is to be considered that altho I use prickt Letters in the first Proposition of the book of Quadratures, yet I do not there make them necessary to the method. For in the Introduction to that book I describe the method at large & illustrate it w$^{th}$ various examples without making any use of such letters. And it cannot be said that when I wrote that Preface I did not understand the method of fluxions because I did not there make use of prickt letters in solving of Problems. The book of Quadratures is ancient, many things being cited out of it by me in my Letter of 24 Octob 1676. A copy of the first Proposition where letters with pricks are used, was at the request of D$^r$ Wallis sent to him in the year 1692 & the next year published in the second Volume of his works. And in the Principia Pholosophiæ {sic} pag 254 the Notarum formulæ used in those days in explaining this Proposition are referred unto.

ffluxions & moments are quantities of a different kind. ffluxions are finite motions, moments are infinitely little parts. I put letters with pricks for fluxions, & multiply fluxions by the letter $o$ to make them become infinitely little & the rectangles I put for moments. And wherever prickt letters represent moments & are without the letter $o$ this letter is always understood. Wherever $\dot{x}, \dot{y}, \ddot{y}, \dddot{y}$ &c are put for moments they are put for $\dot{x}o, \dot{y}o, \ddot{y}oo, \dddot{y}o^3$. In demonstrating Propositions I always write down the letter $o$ & proceed by the Geometry of Euclide & Apollonius without any approximation. In resolving Questions or investigating truths I use all sorts of approximations w$^{ch}$ I think will create no error in the conclusion & neglect to write down the letter $o$, & this do for making dispatch. But where $\dot{x}$, $\dot{y}$, $\ddot{y}$, $\dddot{y}$ are put for fluxions without the letter $o$ understood to make them infinitely little quantities, they never signify differences. The great Mathematician therefore acts unskilfully in comparing prickt letters with the marks $dx$ & $dy$, those being quantities of a different kind. M$^r$ Leibnitz has no mark for fluxions & therefore prickt letters are older marks for fluxions then any used by him & so are others {sic} marks used by me for fluxions. The rectangles under fluxions & the moment $o$ being my marks for moments are to be compared with the marks $dx$ & $dy$ of M$^r$ Leibnitz & are much the older being used by me in the Analysis communicated by D$^r$ Barrow to M$^r$ Collins in the year 1669.

The Author of the Remarks represents that D$^r$ Wallis was for M$^r$ Leibnitz & yet the D$^r$ in the Preface to the first Volume of his works represents that I in my Letters of June 13 & Octob 24, 1676 explained to M$^r$ Leibnitz this method found out by me ten years before or above, that is in the year 1666 or 1665.

I am

Yo$^r$ most humble Servant

*For the* R$^{nd}$ D$^r$ JOHN KEILL *Professor*
*of Astronomy in the University of*
*Oxford.*

Is. NEWTON

Keill's " Answer" to the Leibnizian cartel, drawn up, as we see by the four preceding Letters, with Newton's assistance, appeared in the Journal Literaire, for July and August, 1714, (Tom. IV. p. 319), and produced an anonymous reply in the Leipsic *Acts* for July, 1716, under the title of *Epistola pro eminente Mathematico, Dn. Johanne Ber-*

*noullio, contra quendam ex Anglia antagonistam scripta\**. Among
the Lucasian papers (packet No. 5) are found the draught and fair
copy of an answer† to this " Epistola," by Keill, in French, probably
intended for insertion in the Journal Literaire, but, as far as I am
aware, never published. Newton's Letter of May 2, 1718, (q. v. p. 185.)
may have led to its suppression.

---

\* This was in reality Bernoulli's own production, though in a disguised form. In
its original shape it formed almost the entire contents of a letter to Christian Wolf (dated
Apr. 8, 1716), who, jointly with Leibniz, interpolated, abridged and otherwise altered it
(*e.g.* by changing the first person into the third, and writing *antagonista, Anglus iste* or
*antagonista audax* for *Keilius*) previous to its insertion in the Acts. See two papers by
a grandson of Bernoulli in the Berlin Memoirs for 1799—1800 and 1802, in the latter of
which a comparison is exhibited, in parallel columns, of the Epistola and the MS. copy
of Bernoulli's letter to Wolf. Bernoulli was extremely anxious to preserve a strict
*incognito*, "ingratum enim," he observes, "mihi valde foret a Keilio bile sua perfricari
et contumeliose traduci, ut solent ejus antagonistæ, postquam ille me hactenus satis
humaniter tractavit." Hermann suspected that he was the author, "quod tamen," says
Wolf, in announcing the fact, "hactenus constanter negavi." All the precautions,
however, that had been taken to elude detection were defeated by the unlucky "meam"
which had been overlooked in the process of transforming the letter (See p. 185 note and
p. 186). It was more than a year before Bernoulli's attention was directed to the oversight, when he desired Wolf (Sept. 18, 1717) to insert in the Errata " pro *meam* legen-
dam esse *eam*," adding " sed hoc tamen non satis quadrat ; vellem itaque ut invenires
modum commodiorem, quo culpa in typothetam plausibiliter rejici posset." But Wolf
was in no great hurry to meet his wishes, and ten months later we find Bernoulli em-
ploying his son Nicolas as his mouthpiece in an explanatory statement upon the subject,
in which he attempts to effect his escape under cover of the change which his letter
had undergone in the editorial hands of the friend to whom it was addressed. See
p. 185 note.

† The title of it is *Lettre de Mr. Jean Keill...à Jean Bernoulli.* This may, possibly,
be the piece alluded to by J. Bernoulli in his article on Keill's problem, about the
path of a projectile in the air, (Leipsic *Acts*, May, 1719, p. 218. *Opp.* II. 395):
" Taceo alia, ut rumor fert, dictu horrenda, ex quibus nuper conflavit libellum, (editum
an ineditum nescio) quem tum manuscriptum circumferebat prælo destinatum. Fue-
runt, ut mihi scribitur, inter ipsos adversæ partis sequaces, qui perlegendo cohor-
ruerunt."

# LETTERS OF COTES TO NEWTON.

## LETTER XC.

COTES TO {NEWTON.  After Apr. 25. 1715}.

S<sup>r</sup>.

I think it my duty to send You what Observations I could make of the late Eclipse

I beg Your pardon for troubling You with so large an account of my Method for correcting the Pendulum. I must confess to You, I have a design in it for the advantage of our yet imperfect Observatory. The Clock which I used was borrowed of a Clock-maker in this Town who took it for a very good one. Not expecting so great inæquality in its motion I was very much surpriz'd to find it by the Observations, & since I have found it I cannot think of making use of such ordinary workmanship again, unless in case of necessity. To speak plainly, I beg of You to let that excellent Clock* be now sent down to us which You order'd to be made for the use of our Observatory. I cannot think of a more accurate Instrument for the setting of it, than such an one as I have been describing :† having it therefore by me I think I am prepar'd to receive Your Noble gift. I have written to M<sup>r</sup> Street to wait upon You for Your resolution

I am Sir Your

Obliged Humble Serv<sup>t</sup>

ROGER COTES.

I will send You an account of what was observ'd at Cambridge during the total Obscuration in another Letter.

---

* See Letter XCVIII.

† The description of his mode of adjusting a telescope for the purpose of finding the time by the method of corresponding altitudes is wanting in the MS., which is only a rough draught of a letter: it will however be found in Smith's *Optics*, Vol. ii. p. 328.

On the opposite side of the leaf is the following :

        Day

1.   xxi.   $4^h.01'.21''$   pm.   Sun's upper limb observ'd
                                 at $y^e$ $3^d$ Pin

2   xxii.   $6^h.48'.41$   am.   Upper limb   $2^d$ Pin
           $6.52.09$   am.   Lower limb   $2^d$ Pin

3.   xxiii.   $6^h.47'.29''$ $\Big\}$ am.   Upper limb $\Big\}$ $2^d$ Pin
           $6.50.58$        Lower limb

4   xxiii.   $7^h.51'.10''$   am.   Upper limb.   $3^d$ Pin

5   xxv.   $6^h.44'.53''$ $\Big\}$ am   Upper limb $\Big\}$ $2^d$ Pin
          $6.48.22$        Lower limb

6   xxv.   $5^h.08'.18''$ $\Big\}$ pm   Lower limb $\Big\}$ $2^d$ Pin
          $5.11.47$        Upper limb

Allowing for the variation of Declination I find

By $y^e$ $2^d$ & $3^d$ the length of $y^e$ Solar day measured by the Clock was $24^h.00'.18''$.

By $y^e$ $3^d$ & $5^{th}$ the length of 2 Solar days measured by $y^e$ Clock was $48^h.00'.18''$ Which 2 deductions shew the Clock inequal. of motion

By $y^e$ $1^{st}$ & $4^{th}$ the Meridian of $y^e$ $xxii^d$ day was at $11^h.57'.32''$
By $y^e$ $5^{th}$ & $6^{th}$, the Meridian of $y^e$ xxv day was at $11.58'.02''$

And therefore the Meridian of $y^e$ xxii at     $11.57.26$
I put the correct Meridian of $y^e$ xxii day at $11.57.29$

The "Eclipse" of this and the following Letter is the total eclipse of the Sun which occurred Apr. 22, 1715. See letter cxvi.

In an account of this eclipse by Halley (Phil. Trans. March—May 1715 : see also Number for Sept. and Oct.) he states that Cotes "had the misfortune to be opprest by too much company, so that, though the Heavens were very favourable, yet he miss'd both the time of the Beginning of the Eclipse and that of total Darkness. But he observed the Occultations of the three spots...the End of total Darkness...and the exact End of the eclipse at $10^h.21'.57''$." Some of its popular effects are described by Mead in his "De Imperio Solis ac Lunæ in Corpora Humana" Lond. 1746. pp. 65, 66.

Rud in his diary under the date Apr. 11, after noticing the time of the middle of the eclipse as calculated by Whiston and Halley, adds "$M^r$ Robt. Smith T.C.C.S. says at 7 min: past 9. but I suppose He calculates for Cambridge; whereas they calculate for London. Observe who is nearest the truth." In the Memoirs of the French Academy for 1715 there are no fewer than seven papers on the subject of this eclipse, not to mention several others relating to the luminous ring round the Moon's disk during the time of total obscuration, which the writers endeavour to account for without having recourse to the hypothesis of a lunar atmosphere, to which Louville and Halley attributed the phenomenon. One of these papers by Maraldi commences with the remark that this eclipse "est mémorable par sa grandeur, par la rencontre d' une Tache qui s'est trouvée dans le Soleil, & par les Personages Augustes qui l'ont observée"—the King, the Duke of Orleans and a brilliant Court. It was the last eclipse that had the honour of being observed by the Grand Monarque. Louis died on the 21st of August following.

---

## LETTER XCI.

### COTES TO {NEWTON}.

Sʳ

Dʳ Bentley has told me, You have been pleas'd to give orders, that the Clock may be sent to Cambridge. I take this oportunity of returning You my hearty thanks for it, & of giving You an account of what was observ'd by Us during the time of the sun's total obscuration in the late Eclipse, so far as I judge it to be of any moment. The sky was perfectly clear all the Morning till about two or three minutes after the recovery of the suns light. It surpriz'd us to find so great a quantity {of} Light remaining in the middle of the Eclipse: I think it did very much exceed the brightness of the clearest Moon-light nights. A Freind assur'd me He could very easily & distinctly read the smallest letters engrav'd about Mr Whistons Scheme of the Heavens, which he had in his hands at that time. We saw the Planets Jupiter, Mercury, & Venus, with some fixed stars, but they appear'd with far less splendour & fewer in number than we expected, or than they might have done by Moon-light. I took the greatest part of this remaining light to proceed from the Ring

which incompass'd the Moon at that time. As nearly as
I could guess, the breadth of this Ring was about an eighth
or rather a sixth part of the Moons Diameter, the light of
it was very dense where it was contiguous to the Moon
but grew rarer continually as it was further distant, till it
became insensible: its colour was a bright clear white.
I saw this Ring begin to appear about five seconds before
the total immersion of the suns body, & it remain'd visible
to me as long after His emersion. I did not apply my self
to observe whether it was of the same breadth in all its
parts during the total Obscuration. Mr Walker* a Fellow
of our College whom I can very well depend upon assur'd
me He was very certain it was not. He says He took notice
with a great deal of attention that at first the Eastern part
was very sensibly broader & brighter than the Western,
afterwards they became equal, & some time before the
emersion the Western side was manifestly broader &
brighter than the Eastern. His design in attending so
diligently to such an Observation was this; He thought,
as he afterwards told me, that I might desire to note
the Time of the middle of the Obscuration; & being in
the same Room with me, He was willing to assist me in
judging of that Time, & beleiv'd the method which He
took to be the properest for it; accordingly I do remember
that I heard him call out to Me, *Now's the Middle*, though
I knew not at that time what he meant. I think this
Observation of M^r Walkers is of moment, I have therefore
been very particular in giving You the circumstances of it
that You may Your self judge how far it may be depended
upon, for my part I cannot see any reason to doubt of it.
Besides this Ring there appear'd also Rays of a much
fainter Light in the form of a rectangular Cross: I have
drawn You a Figure which represents it pretty exactly,

---

* Richard Walker, afterwards (in 1734) Vice-Master, Bentley's devoted adherent.
Though four years junior to Cotes, in academical standing, he was six years older,
having been entered at the mature age of 27.

as it appeard to Me. The longer & brighter branch

of this Cross lay very nearly along the Ecliptick, the light of the shorter was so weak that I did not constantly see it. The colour of the Light of both was the same : I thought it was not so white as that of the Ring even in it's fainter parts, but verg'd a little towards the colour of very pale copper. You may observe, that in my Figure the branches of the Cross are represented as bounded by parallel lines, for so it was they appear'd to me. But there are others here, who saw a very different form. I have therefore sent You another Figure

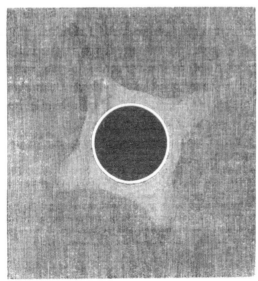

the most remote of any I have met with from my own,
This was drawn by a very ingenious Gentleman representing the appearance as seen by himself.  He differs also
from me in this particular, viz<sup>t</sup> that he takes the Cross
light to be only a continuation of the Ring whereas I make
'em to be intirely distinct from each other.    I am Sir.

May 13. 1715.

---

Edmund Halley born in London 1656, died 1742.

### LETTER XCII.

#### HALLEY TO KEILL.

Dear S<sup>r</sup>                          London Octob 3° 1715

We have printed a French translation of y<sup>e</sup> account of
the Commercium given in the Transactions*, in order to
send it abroad: S<sup>r</sup> Isaac is desirous it should be publisht
in the Journal Literaire, and M<sup>r</sup> Gravesant has promised
to gett it done, but cares not to do it as of his own head;
and therefore proposes that you would signifie to M<sup>r</sup> Johnson at the Hague, by a letter enclosed either to S<sup>r</sup> Isaac
or me, that you are desirous that the said French paper be
inserted in his Journal, as containing the whole state of y<sup>e</sup>
controversy between you and M<sup>r</sup> Leibnitz.   S<sup>r</sup> Isaac is
unwilling to appear in it himself, for reasons I need not
tell you, and therfore has ordered me to write to you
about it, who have been his avowed Champion in this
quarrell; and he hopes you will gratifie him in this matter
by the first opportunity†

I have rec<sup>d</sup> Cloaks Lady days rent, but hear not one

---

* For Jan. and Feb., 1715, pp. 173—224.  "An Account of the Book entituled
*Commercium Epistolicum......*"

† Keill, gladly enough, no doubt, complied with the request.  The French translation of the "Account" or Abstract, alluded to, was inserted in the 7th Vol. of the
*Journal Literaire*, pp. 114—158, and 344—365.  A Latin translation of the "Account"
was prefixed to the 2nd Ed. of the *Commercium Epistolicum*, (1722).

word of Spetty; Pray let me know what I shall say to him about the Lease, and I will endeavour to make him pay the Years rent due at Lady day, or at least the best part of it, before I come down to you, which will not be long.

I am

Dear S$^r$ your most faithfull Serv$^t$

EDM: HALLEY.

---

## LETTER XCIII.

### NEWTON TO KEILL.

D$^r$ Keill

I received about a month ago the inclosed Letter from M$^r$ Monmort*. It conteins some extracts of Letters to him from M$^r$ Bernoulli & his son. The chief point is that M$^r$ Bernoulli denies† that he is the author of y$^e$ Memoir entituled Epistola pro eminente &c that is inserted in the

---

* Born 1678, died Oct. 7, (N. S.) 1719. He acted as a sort of messenger between the Cocles of the Leibnizian bridge, as Fontenelle calls Bernoulli, and some of the English mathematicians. See his *Eloge* by Fontenelle. We see him here, and on another occasion (p. 187), in the amiable character of a peacemaker. The extracts from his letters, which were emulously published against each other after his death, by the belligerent parties, shew that he could go considerable lengths in adapting his language to suit the different tastes of his correspondents. His pen has left us an impassioned tribute to the beauty and accomplishments of Newton's niece, Miss Catharine Barton. Letter to Taylor, Apr. 1716, in *Contemp. Philos.* p. 93.

† In the Leipsic *Acts* for the following June, by way of Appendix to a paper on trajectories, Bernoulli's eldest son, Nicolas, then 23 years of age, took occasion to refer to the subject of the "Epistola pro eminente Mathematico," and to express his father's annoyance at the rumour which attributed it to him. He admits, says Nicolas, that at the request of a friend, he put down in writing, "sine ulla animi commotione," the main of the facts contained in the Letter, but his responsibility did not extend to the "modus scribendi" and form in which the Letter appeared. In confirmation of this, Nicolas, whose Latin, at this stage of his explanation, becomes somewhat obscure, points to the ludicrous oversight into which the *soi-disant* writer falls towards the close of his diatribe, where the mask drops and Bernoulli is found speaking in his own person. "Examinent etiam considerentque, quam brevi via quamque diversa a Newtoniana incesserit Bernoullius, {in the solution of the inverse problem of central forces}, dicantque postea, an alius quispiam præter antagonistam sibi persuadere possit, meam formulam ex Newtoniana esse desumtam." Leipsic *Acts* for July, 1716, p. 314.

Acts of Leipsic 1716. The Memoir it self lays it upon
M$^r$ Bernoulli by the words *meam solutionem*, & if M$^r$ Bernoulli is injured thereby it is not you but the author of
the Memoir who has injured him. The injury is public
& in justice requires a public satisfaction, not from you
but from him that has done the injury. The question is
therefore whether you will take notice of M$^r$ Bernoulli's
excusing himself in private or leave him to do it in publick.
I have not yet returned any Answer to M$^r$ Monmort, because I thought it best to stay till I had your sense upon
this matter. I think to discourse also your friends D$^r$
English * & D$^r$ Bower about it. I am

Your faithful friend &

humble Servant

London. 2 May. 1718. {Friday}.      Isaac Newton

I pray return M$^r$ Monmorts Letter by D$^r$ Halley because I am to answer it.

*For* D$^r$ John Keill, *Professor of*
*Astronomy at Oxford.*

This letter, as has already been observed, p. 178. may have been the
means of inducing Keill to suppress the answer which he had prepared
to the " Epistola pro eminente Mathematico." Fragments of it, however, may be discerned in a Latin dress in the first few pages of a subsequent publication, the origin of which may claim a notice here.

———

* Keill's cousin, John Inglis, M.D. Among the Lucasian MSS., (packet No. 3,)
there are two short letters from him to Keill. In the first of them, (Dec. 19, 1717),
after congratulating him on his marriage, the writer proceeds as follows : " Your papers
have been in Sir Isaac's hands ever since they came into mine, and as yet I have heard
nothing about them ; but as soon as I receive them, I shall endeavour to forward them
to Holland by the first sure hand." These " papers " were probably Keill's answer to
the *Epistola pro eminente Mathematico.* See *antea*, p. 178. The second Letter, (Jan.
14, 171$\frac{7}{8}$), also relates to the aforesaid "papers." " I acquainted Sir Is. Newton that
you was fully satisfyd with his corrections, and referr'd the whole to his judgement ;
which he received very kindly, though he had been impatient to hear from you. But
you have forgott to send me back his paper, as we had done to take a copy of it, and
therefore you must send it me, to free Sir Is. of the trouble of going over it again......
Doctor Bower is yours." Bower was M.D. and Professor of Mathematics at Aber-
deen. He and Inglis were Fellows of the Royal Society.

In the Journal Literaire for 1716 Keill had published an article * in defence of Newton against some remarks of John Bernoulli and his nephew relative to the inverse problem of central forces and the error in the 10th Prop. Book 2, of the 1st ed. of the Principia. An answer to this, framed under Bernoulli's eye by a pupil of the name of Crusius, appeared in the Leipsic *Acts* for October 1718, which had the effect of rousing Keill once more. He drew up a reply to it in the shape of a Latin letter to Bernoulli, but while the *brochure* was passing through the press, Newton shewed him a letter which he had received (July 1719) from Bernoulli through Monmort, disavowing the authorship of the famous letter of June 7, 1713. Upon talking the matter over, Keill seems to have consented to proceed no further with the publication of his pamphlet †. His pacific intentions, however, were scattered to the winds by the arrival of the May number of the Leipsic Acts (1719) containing a paper by Bernoulli ‡ in which that mathematician ushers

---

* There is a MS. copy of this among the Lucasian papers, (packet No. 5) : it is entitled "Apologie pour le Chevalier Newton, dans laquelle on repond aux remarques de Messieurs Jean et Nicolas Bernoully inserees dans les Mémoires de l'Academie Royale des Sciences pour les années, 1710 & 1711, par J. Keill..." It appears that on Jan. 19, 1716, Halley wrote to Fontenelle with a view to this *morceau* of Keill's being inserted in the *Mémoires de l'Academie*, where the papers against which it was directed had appeared. Monmort spoke in favour of the application, but the feeling of the majority of the members was adverse to it. (See *Contemplatio Philosophica*, p. 85.) Fontenelle in his answer, (dated March 8,) a copy of which, in Keill's hand, is extant in a folio book in the custody of the Lucasian Professor, says, "Nous ne cedons point ici aux Anglois meme en estime et en veneration pour Mr Newton. Et l'Academie voudroit fort qu' il fust possible " to insert Keill's paper in their Memoirs, but that it was their invariable rule to admit only articles written by members of their body.

† Quantum sentio, a litibus in posterum abstinebit, (draught of a letter of Newton in Macclesfield *Corres.* ii. 437.) I assume that the letter, of which the draught is printed in the work referred to, without date or address, was addressed to Monmort, (about the end of July, 1719,) though the editor (Preface, p. x) states that "it was found impossible clearly to make out the date." The point may be set at rest, if the letter to which this is an answer, should turn up among the Portsmouth Papers.

‡ *Joannis Bernoulli Responsio ad Non neminis Provocationem, ejusque solutio quæstionis ipsi ab eodem propositæ de invenienda Linea curva quam describit projectile in medio resistente.* Leipsic *Acts*, May, 1719, p. 216. Bernoull. *Opp.* ii. 393. The tone and language of this piece are such, that even Bernoulli's friends, the conductors of the *Acts*, thought it necessary to apologize for inserting it without modification.

In justice to Keill, it ought to be observed, that the problem which led to this explosion does not appear to have been sent as a challenge to Bernoulli, and still less to foreign mathematicians, as has been represented. It was mentioned incidentally in a private letter of his to Taylor, in which he expressed a wish that Bernoulli would apply his skill to questions of real utility (as, for instance, the one referred to, which Leibniz had attempted in vain), instead of wasting it upon such problems as that of Trajectories. An extract from this letter was (contrary to Keill's intention, and without his knowledge) sent by Taylor to Monmort, who forwarded it to Bernoulli. Keill seems to have intimated to Monmort, his dissatisfaction at the extract being communi-

in a construction which he gives of a generalization of Keill's projectile problem by a most violent attack upon its proposer. Forbearance was out of the question: Keill let loose his " Epistola ad Jo...Bernoulli," (London 1720) and gave further vent to his feelings in an " Additamentum" appended to it, which he closed with some stinging extracts from Monmort's letters to Taylor who kindly supplied them for the purpose,—a species of weapon which enabled Bernoulli afterwards to take ample revenge by turning it upon Taylor (Leips. *Act.* May 1721, p. 207 seqq. Bernoull. Opp. II. 493. seqq.).

There are rough draughts of Keill's letter in English and Latin among the Lucasian papers, and part of it was read by Halley (no doubt in the original English) at a meeting of the Royal Society May 28, 1719 at which Newton presided. Before publishing it, Keill laid a complaint before the Royal Society against his adversary " for affronting him with scurrilous language," and called upon the Society to take steps " to shew their dislike of such foul proceedings." " The President ordered that the consideration of this complaint be deferred till Dr Halley (Secretary) comes to town, & that enquiry be made into precedents for the better information & direction of the Society." *Journal Book*, May 26, 1720. The Society does not seem to have moved any further in the matter.

---

J. A. Arlaud or Arland, an eminent painter, born at Geneva 1668, died 1764. " Newton fut son ami, et lui fit présent de la version française de son Optique ; il était en correspondance avec lui." Biogr. Univ. At the age of 20 he went to live at Paris.

### LETTER XCIV.
### NEWTON TO ARLAND.

Vir celeberrime,

Gratias tibi debeo quam maximas quod Schema experimenti quo lux in colores primitivos & immutabiles separatur, emendasti, et longe elegantius reddidisti quam prius. Sed et me plurimum obligasti dum Schema illud in lamina

---

cated to Bernoulli, for among the Lucasian papers, (packet No. 2) we find a very civil letter from Monmort to Keill, (it is not dated, but bears the London post mark, " Nov. 5," probably in 1718), in answer to one from Keill to him, (dated Sept. 3) in which he states that he thought that the extract was intended to be sent on to Bernoulli, and protests that if he had had any idea of the offence that he should give, he would never have sent it.

ænea incisum & inter imprimendum obtritum, refici curasti, ut impressio libri* elegantior redderetur. Gratias itaque reddo tibi quas possum amplissimas. Quod inventa mea de natura lucis & colorum viris summis, D<sup>no</sup> Cardinali Polignac† & D<sup>no</sup> Abbati Bignon non displiceant, valde gaudeo. Utinam hæc vestratibus non minus placerent quam elegantissimæ vestræ & perfectissime delineatæ picturæ nostratibus placuerunt. Ut Deus te liberet a doloribus capitis & salvum conservet, ardentissime precatur

Servus tuus humillimus

& obsequentissimus

Dabam Londini 22 Oct. 1722.      Isaacus Newton‡.

*Celeberrimo Viro* D<sup>no</sup> Arland

---

* Peter Coste's French translation of Newton's *Optics*, Paris, 1722.

† Born 1661, died 1741. Author of *Anti-Lucretius* (a posthumous Latin poem). It is said that he took great pains to have Newton's fundamental experiments on light properly performed in France, and had the honour of receiving a letter of thanks from our philosopher in consequence.

‡ The original is in the Library at Geneva, to which institution Arlaud bequeathed several medals, paintings, &c.

# COTES'S CORRESPONDENCE WITH HIS UNCLE.

S$^r$.                                   {London Dec. 31*. 1698}

I am now very well recovered; and am I thank God in as good health as ever. As for y$^e$ works of Kepler, and Galilæo as far as I can learn they are dispersed in divers Volumes, put forth at different times. I have from severall choice Catalogues, as Draudius's Bibliotheca classica, A Catalogue of y$^e$ Mathematicall books in y$^e$ Savilian Library at Oxford. and y$^t$† immense one of D$^r$ Francis Bernard's Library which is now under y$^e$ Auctioners Mallet at London and is Like to continue so for many Months. and severall others collected what I could find of those t{w}o Learned Authors. I send 'em you here in y$^e$ latest Editions y$^t$ I could find there set down. You may from hence pitch upon those you most like of, & I shall be very glad to use my utmost endeavours to procure 'em for You—

{*Here follows in the MS a long list of Kepler's and Galileo's works, which it has not been considered necessary to print*}.

I suppose there might be added to each Catalogue especially to y$^t$ of Galilæus. Perhaps this is more than You expected of theire Works. The first Tome of Galilæus's Works translated into English came out some Yeares ago {in 1661}; but y$^e$ Second is as yet unpublished

---

* The day of the month is taken from the post mark.

† "A Catalogue of the Library of the late learned D$^r$ Francis Bernard, Fellow of the College of Physicians, and Physician to S. Bartholomew's Hospital....which will be sold by Auction at the doctor's late Dwelling House in Little Britain: the Sale to begin on Tuesday, Octob. 4. 1698."

and perhaps will never see light *. I have my self Galilæo's Nuncius Sidereus put out at London in 8$^{vo}$ together with Kepler's Dioptricks and Gassendus's Astronomy ; if you please I will send you 'em. You wrote of y$^e$ Quadrature of Curve's, as yet I cannot enquire of any Mathematician about 'em. S$^r$ Edw: Sherbourn in his Appendix to his Translation of Manilius's Astronom: {Lond. 1675} tell's us y$^t$ from M$^r$ Isaac Newton is expected a New general Analytical method by infinite Series for y$^e$ Quadrature of Curvilinear figures. I have D$^r$ Wallis's Algebra {London 1685} I think I bought it very cheape I am very well pleased w$^{th}$ y$^e$ Book. The D$^{rs}$. Buisness therein is to shew y$^e$ Original, Progress & Advancement of Algebra from time to time, and by what steps it hath attained to y$^t$ height at which it now is he give{s} us a full Account of y$^e$ Methods used by Vieta Harriot Oughtred De-Chartes and Pell & others and of y$^e$ several methods of exhaustions, Indivisibles, Infinites, Approximations &c. amongst other things he speak's of squaring Curves and after other ways of approximations shewed he show's you this of M$^r$ Newton† he determin's it impossible to do y$^e$ buisness exactly. In my mind there are many pretty things in y$^t$ book worth looking into. If you have a mind to see it, or have not seen it already I will send it w$^{th}$. Galilæo's Nuncius I thank you for your Directions about Instruments in your last letter dated December 21 You your self put me ofof y$^e$ Instrumentary way while I was with you but I meant In my Letter such Instrument's y$^t$ were not superseded by calculation or some more exact way ; as a Quadrant is

＊　　　＊　　　＊　　bigg as y$^e$　　＊　　themselves

＊　　　＊　　　＊　　＊　　＊　　＊

＊　　　＊　　　＊　　＊　　＊　　＊　　＊

---

* It was published in 1665, but nearly the whole impression was destroyed by the fire of London. See Macclesfield *Corresp.* i. 120.

† From the famous Letters of June 13 and Oct. 24, 1676, to Oldenburg, to be forwarded to Leibniz.

&ast;      &ast;      &ast;      &ast;      &ast;      &ast;

&ast;      &ast;      &ast;      &ast;      &ast;      &ast;      &ast;

&ast;      &ast;      &ast;      &ast;      &ast;      &ast;

sometimes be at a loss for    But I will not be so bold as to
ask my Grandfather for y$^e$ larger size.    I wi{ll}    *   *
*    *    *    little one in a concave case with y$^e$ Circles only which will serve y$^e$ end as well as y$^e$ largest size
it will als{o}    *    *    *    *    pocket and ready
upon all occasions.

I am
your very Obedient Servant and Nephew

*These For y$^e$ Reverend M$^r$ SMITH of Lea*          R. COTES.
*nere Gainsborough IN Lincolnshire*
*Newark Bagg*

The lower part of the second leaf of the letter has been torn off.

---

## LETTER XCVI.
### JOHN SMITH TO COTES.

Dear Cos: Roger                    Aug: 30, 1701.

I was very glad to hear of your welfare by your Father
who befriended us w$^{th}$ his company about a fortnight ago;
he showed us your letter in w$^{ch}$ you expressed a feeble
inclination to come and see us in y$^e$ Country, we thank you
for y$^t$, and count it a favour y$^t$ you can spare us any share
of your affection from your dear M$^{rs}$ Mathesis; I am glad
to hear y$^t$ she so easily yields to your courtship, and has
procured you such signal marks of favor from great men as
D$^r$ Bently M$^r$ Hanbury*; I am sorry y$^t$ gentleman is so

---

* Nathaniel Hanbury, elected from Westminster School to Trinity College, in 1677,
admitted Minor Fellow, Sept. 17, 1683, (Charles Montagu was admitted Major Fellow
on the following day). He published *Horologia Scioterica Prælibata*...Lond. 1683;
and *Supplementum Analyticum ad Æquationes Cartesianas*, Cantab. 1691. A paper by
him on a mode of approximating to the value of $\pi$ by the continual subdivision of an
arc of 60°, was produced at a meeting of the Royal Society, August 17, 1698. He
filled various College offices, and we are told by Middleton, that Bentley "took occasion to convict him, in a solemn manner, by the testimony of all the College, of

overlookt as not to be Vice-pro{fe}ssor instead of M<sup>r</sup> Whiston; for I believe he has far greater Mathemat: accomplishments; I hear he has a great respect for you; considering therefore y<sup>e</sup> favorable fair-promising circumstances
you are under I cannot forbear presaging in your behalf,
w<sup>t</sup> Ovid did to his friend, Scena manet dotes grandis Amice
tuas. Divines you know are stiled prophets, as well as y<sup>e</sup>
poets are, & I fancy I shall be a true one in this; provided you so moderate your studies as not to impair your
health; a journey into y<sup>e</sup> countrey once a year would do
well for y<sup>t</sup> purpose; what? I warrant you, you have forgotten your old Ne quid nimis, & Interpone tuis &c. * but I am
resolved to remember you of em now & then; I had writ
to you before but expected ever & anon to have seen you
here; there is in y<sup>e</sup> monthly accounts of y<sup>e</sup> works of y<sup>e</sup>
learned, for y<sup>e</sup> year 1700, month December, a method for
finding two middle proportional lines, w<sup>ch</sup> to me is false,
there being a great error in y<sup>e</sup> demonstrat: pray look upon
it a little; I should be glad to hear of you, & of any new
discovery; I never saw yet what discoveries M<sup>r</sup> Hally has
made in his voyage, pray comunicate to me if there be any
thing worth while; & you will much oblige

Your most affectionate friend & uncle

J : SMITH.

My wife & son & daughter remember their kind love
to you.

*For* M<sup>r</sup> ROGER COTES *at Trinity*

*Colledg in Cambridg*

*Deliver this in at Caxton to go to Cambridg*

---

being *a common swearer & habitual drunkard,* and without inflicting the least censure upon him for all this, made him not long after {in 1712 & 1713} the *Senior Dean*."
*Miscellaneous Works,* III. 356. He was curate of St Michael's for many years. He
died in Nov. 1715, and Colbatch was elected Senior in his place.

* From that once popular school-book *Dionysii Catonis Disticha de Moribus ad
Filium.*

Interpone tuis interdum gaudia curis,
Ut possis animo quemvis sufferre laborem.

Halley was appointed (Aug. 19, 1698) to the command of the *Paramore Pink*, with orders to make a series of observations with a view to ascertain the law of the variation of the compass, "to call at his majesty's settlements in America & make such observations as are necessary for the better laying down the longitude & latitude of those places, & to attempt the discovery of what land lies to the south of the western ocean." He set out on his expedition Nov. 29, and was carrying on his observations some degrees south of the line when the insubordination of his officers compelled him to return : he reached England at the end of June 1699. In the following September he embarked again in the Pink. In this second voyage after penetrating beyond the 52d degree of south latitude where he was stopt by icebergs, he turned his course northwards, visiting among other places St Helena (a spot familiar to him by his sojourn there more than 20 years before), Pernambuco, Barbadoes, Bermudas and Newfoundland. After an absence of 12 months he arrived in the Thames in Sept. 1700. His observations were embodied in a General Chart which he published in 1701 "shewing at one view the variation of the Compass in all those seas where the English Navigators were acquainted." The Journals of • his two voyages were published by A. Dalrymple (London 1775. 4to).

These are the voyages to which Smith alludes, and about which the young Cambridge student could give him no information. But while Smith was writing this letter, the Captain was again afloat and engaged upon, if he had not finished, another undertaking in which his activity and spirit of enterprise sought employment. At a meeting of the Royal Society, June 18, 1701 "the Vice-President (Sir John Hoskyns) informed the Society that Mr Halley was gone on a new voyage, as he heard, having designed to make nice observations on the Tides & Currents in the Channel, for the Improvement of Navigation, that thereby by their different times, the going out of the Channel might be more easy against contrary winds." And on July 30, a letter was read from Halley (Guernsey, 18 July) "giving an account that the weather having been fair for a Month past, he had made a great progress in the designs he had in making this voyage." The fruit of this voyage was a large map of the British Channel published in 1702.

## LETTER XCVII.
### COTES TO JOHN SMITH.

Hon: S$^r$!                                    Cambr: Sept. 9. 1701.

I heartily thank You for Your kind Lett$^r$, & as heartily beg Your Pardon for suffering my self, by so long delay, to be as it were forc'd to returne You an Answ$^r$. You are pleased to express a greate deale of Kindness to Me in Your Prophecies, as You call 'em, or, as I would rath$^r$ have it, Y$^r$ Wishes; und$^r$ which Name, not y$^e$ other, I again thank You for it. I am sorry You should suspect me of forgetting my *Ne quid nimis.* I have learn't y$^t$ lesson too p'fectly & 'twould be more adviseable (for y$^e$ Accomplishment of Y$^r$ Prophecies) to rememb$^r$ me of my Old *Multa tulit fecitq:* &c. The Mesolabe of y$^r$ wretched pretend$^r$, y$^t$ Quack Geomet$^r$ S$^t$ Julien Potier, one cannot but admire for it's grosness; & much more y$^e$ laborious confutation of it published in y$^e$ same paper some Months after. M$^r$ Halley's late discoveries I am wholly ignorant of. Surely You mistake Cambridg. Wee are situated in as dark a Corner of y$^e$ Land (in these Matters) as can well be desired. You have often mentioned to me y$^e$ Quadratures of Curves; & particularly (which I now call to mind) You have wished to be satisfied in pag. 374 of Newton. I persuaded my self therefore y$^t$ something concerning this Matter might perhaps make amends for M$^r$ Halley's Story And y$^t$ I may be as short as is possible, I desire You to Consid$^r$ 2 Lem. 2 Lib. in which & it's Converse y$^e$ grounds of his Method of Fluxions are contained. To come to an Instance.

Let *AMB* be any Curve; *AM, AP, PM*, any Chord, Abscist, Ordinate of this Cu{r}ve which w$^{th}$ χ$^e$ Arch *AM* are all unstable, Flowing, increasing or decreasing Quantitys; and y$^t$ too after a certain Law, in a certain proportion among themselves, according as y$^e$ Nature of y$^e$ Curve requires. Let then for once *AP* stret'ch it self, & it's very first increase, it's *primum nascens incrementum*, it's Fluxion,

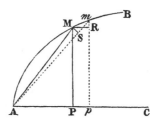

it's moment be an infinitely little *Pp*; w$^{th}$ it y$^e$ Ordinate, Chord, & Arch shall also change themselves into *pm, Am, AMm* & their Moments or Fluxions will be *Rm, Sm, Mm*. The Area *AMP* will also have it's differentiola or Fluxion *MPpm* differing from y$^e$ ☐ *MPpR* by y$^e$ Δ$^{le}$ *MRm* infinitely little in respect of *MPpR* which is it self infinitely little in respect of *AMP*: Now quantities whose difference is infinitely little ought to be look'd upon as equall by 1 Lem. 1 Lib. Newt. For y$^e$ same reason y$^e$ Sectour *AmM* (which is y$^e$ Fluxion of y$^e$ Bilinear Figure *AM*) may be account'd equall to y$^e$ Δ$^{le}$ *AMS*. Now naming y$^e$ Abs. Ord. Ch. & Arch *X, Y, U, Z*. *Pp, Rm, Sm, Mm*, will be *x, y, u, z*, according to y$^e$ second Lemma: or rath$^r$ let us name y$^e$ Magnitudes themselves *x, y, u, z*. & their Fluxions *ẋ·ẏ·u̇, ż*. 'Tis evident y$^t$ y$^e$ Fluxion of y$^e$ Area will be = ☐*MPpR* = *yẋ* To particularize; let this Curve be y$^e$ Parabola, whose Area we know very well otherways. *ax* = *yy, a*$^{\frac12}$*x*$^{\frac12}$ = *y, a*$^{\frac12}$*x*$^{\frac12}$*ẋ* = *yẋ* = Fluxion of y$^e$ Area    But y$^e$ Fluent of *a*$^{\frac12}$*x*$^{\frac12}$*ẋ* (by Lem. 2 Lib 2 convers.) = $\frac23$*a*$^{\frac12}$*x*$^{\frac32}$ = $\frac23$*xy* = Areæ.   In Newton's Hyperboloeid *a*$^3$ = *x*$^2$*y* or *a*$^3$*x*$^{-2}$ = *y* now in our case y$^e$ Fluxion of y$^e$ Abscist runs backward & is therefore = − *ẋ* and Fluxion of y$^e$ Area

= − *yẋ* = − *a*$^3$*x*$^{-2}$*ẋ* whose Fluent y$^e$ Area = *a*$^3$*x*$^{-1}$ or $\dfrac{a^3}{x}$ or *xy* is reciprocally as *x*.

This may p'haps serve as a Specimen of $y^e$ Method of Fluxions applied to $y^e$ buisness of Quadratures tho it's uses seem to be as inexhaustible as they are Naturall & Easy for by it $y^e$ great Geometers of our Age are enabled To draw Tangents, To rectifie, To find $y^e$ Evolutes, The Causticks by reflection & refraction of all sorts of Curves, To measure $y^e$ Surfaces generated by their rotation, The solids they comprehend, The Centers of Gravity, Oscillation & Percusn. of all these To resolve all sorts of Questions de Max & Min. To find $y^e$ Points of Inflection & Rebroussement (as $y^e$ French term it) in all Curves & $y^e$ Converse of all these & many more   But what wonders does it not do when applied to Nature! where it Triumphs alone & admitts of no Partner——But I transgress $y^e$ Bounds of a Lett$^r$

Pray S$^r$ pay my humble respects      Y$^r$ very &c.

to my Aunt; and my Love to Cozz$^{ss}$.     R COTES

*These to the Reverend* M$^r$ SMITH *Rector*
  *of Gate-Burton near Gainsborough*
  *by Newark ℈ Caxton.*

This letter in which Cotes gives his old master an insight into the powers of the new Calculus was written in the long vacation between his 2d and 3d years. It is a very creditable performance for a junior soph.

---

## LETTER XCVIII.

### COTES TO JOHN SMITH.

Cotes is now Plumian Professor. His appointment took place
Oct 16. 1707.

Honoured Unckle

I have lately been at London; I found Y$^r$ Letter at Cambridge upon my return. The occasion of my going up

thither was partly to view a large Brass Sextant* of 5 foot
Radius (y$^t$ had been makeing for us & is now finished)
before it should be sent down.   Whilst I was in Town S$^r$
Isaac Newton gave orders for y$^e$ making of a Pendulum
Clock which he designs as a present to our new Observatory.   The Sextant will cost y$^e$ Colledge 150$^{ld}$ & I beleive
S$^r$ Isaac's clock can cost him no less y$^n$ 50$^{ld}$.   We have
another Instrument in hand for takeing y$^e$ Transits of Stars
or y$^e$ Sun & Moon over y$^e$ Meridian & then we shall be
pretty well furnished for makeing Observations.   All Altitudes You know may as well be taken by a Sextant as a
Quadrant.   We want another 200$^{ld}$ if we can procure it in
y$^e$ University to raise up another Story over y$^e$ gate for
Astronomical uses.   I have lately hit upon a contrivance
which I beleive will be of very good use for observing
Eclipses.   You will easily understand it by this rude
draught   The Telescope $ab$ is to be so directed as to look

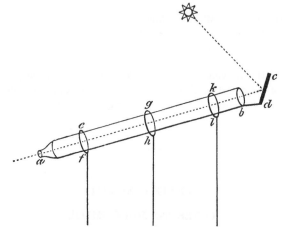

at y$^e$ Pole of y$^e$ World & thereby its axis will be parallel

---

* December 10, 1707 : "The President in the chair.  A draught of a Sextant made
for the use of the Astronomical Professor in Cambridge was produced.  D$^r$ Harris &
M$^r$ Halley reported that it was very exactly done by Mr Rowley."  *Journal Book* of
the Royal Society.  A plate with a description of "this noble instrument" is given in
Harris's *Lexicon Technicum*, Vol. ii. Lond. 1710.

to $y^e$ Axis of $y^e$ Earth in which position it must be fixd by $y^e$ rings *ef gh kl*. *cd* is a looking glass reflecting $y^e$ Object into the Telescope  Then if $y^e$ Telescope revolve about its Axis within the rings with a motion correspondent to that of $y^e$ Earth about its Axis the Object will constantly be in $y^e$ Telescope for a whole day togather as You will easily understand by considering $y^t$ $y^e$ looking Glass participates of $y^e$ same uniform motion by being fixt to $y^e$ Tube. I have not described $y^e$ method of altering $y^e$ Inclination of $y^e$ glass according to $y^e$ different Declination of $y^e$ Object from $y^e$ Æquator  You will easily find out how $y^t$ may be done as also how a piece of Clock work if it be thought needfull may communicate to $y^e$ Telescope its due motion about its Axis*. I thank You for $y^e$ kind Judgment You made concerning my Paper about Projectiles. I have by me another such a Paper concerning $y^e$ motion of Pendulums which I drew up about $y^e$ same time with $y^t$†. This or any thing else You know You have a right to command from me haveing taught me all $y^t$ little which I understand in these matters. I am glad Coz Rob$^t$ has made so good progress in Mathematicks & $y^t$ he has a genius suited to those Studys as I allways thought he had but I fear You are too diffident of his Abilitys. It will undoubtedly be more for his satisfaction & advantage to be admitted Pensioner $y^n$ Sizer, the other way if I can gett him a Poor Schollars place will be about $10^{ld}$ cheap$^r$ I wish You would resolve Y$^r$ self. I should be very sorry to have advised You amiss & I cannot now be certain of futuritys

---

* It will be seen from this that Cotes anticipated 's Gravesande in the principle of the Heliostat, by upwards of thirty years. Both however had been forestalled by Hooke and Halley. *Regist. Bk. Roy. Soc.* ix. 23. For a description of that instrument see 's Gravesande's *Physices Elementa Mathematica*, 3rd Ed. 1742, p. 715; Biot's *Traité de Physique*, iii. 188. Compare *Novi Commentarii Petropol.* i. 291; Coddington's *Optics*, (1st or 2nd Ed.) Letter of Voltaire to 's Gravesande, June 1, 1741, (in some editions, 1738), a paper by Hachette in the *Journal de l'Ecole Polytechn.* Tom. ix. p. 263. and Liouville's *Journal*, 1844.

† These papers are printed among Cotes's *Opuscula Mathematica* at the end of the *Harmonia Mensurarum*, pp. 80—91.

However I will consult with some freinds, y$^t$ I may better know how to direct You*.   Pray give my hearty respects to my Aunt & my Love & Service to Coz Rob$^t$

I am Y$^r$ &c.

Cambridge Febr. 10 1708                    ROGER COTES.

I lately heard y$^t$ Coz Tho Summerfield is dead at Ghent

By "our new Observatory" are meant the leads of the King's Gate which, by a College order dated Febr. 5. 170$\frac{5}{6}$, were granted to the Plumian Professor. The additional "story" mentioned a few lines further on was the work of several years, and Cotes did not live to see it finished.

Bentley in his Letter to the Bishop of Ely (Febr. 1710) boasts of "the College Gate House rais'd up & improv'd to a stately Astronomical Observatory, well stor'd with the best instruments in Europe," and in another letter (Christmas 1712) he calls this erection "the commodiousest building for that use in Christendom, & without charging the College, paid for by me & my friends." In one of the articles against him laid before the Bishop of Ely in July 1710 he is accused of "applying money, which ought to be applied only for the use of the Library towards buying instruments for an Observatory, which he caused to be built by his own authority"—a charge which is true with respect to the sextant.

From Blomer's "Full View of Bentley's Letter" (July 1710) it appears that the "Finishing" of the observatory was then "going on very slowly for want of money to pay the workmen" and that Bentley's estimate of the expense was less than one third "of what it's like to come to." (p. 120).

On June 8, 1717 an order was made by the Master and Seniors that the payments of the Doctors of the College for their degrees (£20 per man) should be "for the present applied to finish the Observatory" under the superintendence of Prof. Smith, Cotes's successor and the "Coz Rob$^t$" of the above letter.

On May 30, 1792 the Vice-Chancellor (Postlethwaite, Master of Trinity) and the other Plumian trustees, having before them the fact that the Plumian Professor had "neither occupied the said rooms & leads nor fulfilled the conditions for at least 50 years" and that "the

---

* "Coz. Robt." was admitted a Pensioner under Mr Edw. Rud, May 28 following, "annos natus 18...e schola Leicestriensi."

observatory & the instruments belonging to it were through disuse, neglect and want of repairs so much dilapidated as to be entirely unfit for the purposes intended," agreed to give up all claim to the rooms and leads at the King's Gate and to allow the Master and Fellows to take the Observatory down or convert it to any use they thought proper.

This memorial of Bentley's zeal for the promotion of science was pulled down in 1797.

---

## LETTER XCIX.

### COTES TO JOHN SMITH.

S$^r$.                                          Cambridge Novemb$^r$. 30 1710

I thank God we go on very.well. I hope You are all in good health notwithstanding this very sickly season. I suppose my Cozen told You in his Letter, which he wrote on Tuesday last, that he has received the 10$^{ld}$ which You sent him. I talk'd with M$^r$ Whiston to day & gave him Your advice of making a recantation, for which he thanks You, but will not accept it.* I have been long ago well satisfied y$^t$ no advice from any private person can possibly have any effect upon him : I asked him therefore whether y$^e$ Judgment of y$^e$ Convocation might not be a sufficient ground for him to alter his Opinions & whether he should not think himself obliged to desist if he should chance to be censured by them: He answered me in the Negative, unless they would prove to him that his Opinions were wrong. I afterwards told him y$^t$ the Church must in 3 or 4 Yeares recover it's Primitive purity, according to his own Exposition of the Revelations; and y$^t$ therefore it would be perhaps adviseable for him to stay till y$^t$ time & expect the Issue with patience. Upon this he could not help discovering himself (as I imagined he would do) & told me

---

* Whiston had been expelled a month before. "Oct. 30, 1710. This time M$^r$ Whiston was expelled as an obstinate heretick by the Heads, after he had thrice convented before them." Rud's *Diary*.

y$^t$ the completion of y$^t$ Prophecy might he beleiv'd depend in good measure upon y$^e$ reviving of those antient Doctrines in which he was at present engaged; He bid me consider what answer S$^t$ Paul would have given to one y$^t$ should have dissuaded him from preaching the Gospell, upon this reason; y$^t$ it was certainly foretold y$^t$ the Gospell should be preached to all Nations. You may easily understand, by these Answers, upon what grounds he is so very resolute, I am persuaded 'tis in vain to endeavour to reclaim him till y$^e$ term of that Prophecy be expired.

I am Your very dutifull Nephew

R COTES

Pray present my humble respects to my Aunt & my hearty Love to my Cozen.

On the back of this letter besides some arithmetical computations such as Mr Smith has written on Cotes's first two letters to him there are also notes for a sermon in his hand.

---

LETTER C.

COTES TO {ISAAC EWER.}

S$^r$.                                                    {Dec. 26 or 27. 1710}

I have this day paid to M$^{rs}$ Medley Ten pounds & inclosed M$^r$ Herring's Bill for Fifty two pounds which is in full of Y$^r$ dues from the Jun$^r$ Bursar's Office. I cannot at present.pay y$^e$ Interest of y$^e$ Thousand pounds not having Money in my hands. I hope in a very short time I may do it for tis reported y$^t$ the Seniors design at a meeting this day to order the Principal to be paid You & to vote two dividends & an half & to leave (after this is done) a Thousand pounds in Stock. Tis said y$^t$ M$^r$ Bathurst will be chosen Senior Bursar. D$^r$ Ayloffe & M$^r$ Barwell were talk'd of for Jun$^r$ Bursar & Steward. I do'nt hear who is to be y$^e$ Pandoxator unless M$^r$ Eden be y$^e$ person intended.

I am S$^r$

Y$^r$ faithfull freind

& humble Servant

Rog: Cotes.

This letter was written on one of the above stated days as will be seen from the following extract from Rud's Diary. "1710. Dec: 26 was appointed the day for voteing Div. but when they were mett Mr Hanbury objected that whatever they should do before the Seniority were filled up, {a Senior fellowship was vacant by the death of M$^r$ Mayer on Nov. 2} would be unlawful & void; and He prevail'd, so that they adjourn'd to the Chapple next morning; when Mr Cooper was sworn (he was chosen upon Mr Hawkyns's death in Apr. before) and Mr Hanbury was chosen to succeed Mr Mayer. After noon they proceeded to vote ½ a Div. for 1708, & 2 whole ones for the 2 next years. The first Moyety was paid in the Beginning of January." (This will serve to correct two or three slips in Monk's Bentley pp. 221, 222 note.)

Bathurst was chosen Sen. Bursar, Barwell Jun. Bursar, Whitfield Steward and Modd Pandoxator. The statutable day for swearing in these officers is the day following the *dies computi*, so that this year they ought to have been sworn in on Dec. 28, whereas in the Admission Book the date is Dec. 31 (Sunday). If this date be correct, the cause of the delay is probably to be sought for in the dissensions with which the college was distracted. Modd had filled the office of Sen. Bursar since June 23, 1705 and Cotes that of Jun. Bursar since Decemb. 19, 1707. It is not unlikely that Cotes's resignation of that office was connected with what had occurred at the election of officers and lecturers in October, when Bentley was overruled by the Seniors in all his nominations. "They had taken a pique against Mr Whitfield for being so desirous of that office {the Latin Lectureship} & therefore pass'd him by, on pretence that he had one place already {he was Steward}; Mr Cotes was also past by on the same account, & they chose Mr Pilgrim Lect. Math. in his room." Rud's Diary.

The sum of £1000 was borrowed by the College in 1706 at 5 per cent. to be appropriated to the repairs of the Chapel. It was advanced by Bentley out of his wife's fortune on a bond to his trustee Mr Isaac

Ewer of Lincoln's Inn (to whom this letter was probably addressed)
and was repaid by instalments in 1711, 12, 13 and 14. Conclusion
Book Sept. 6, 1706. Sen. Bursar's Books. Lease Book p. 82. Bentley's Letter to Bp. of Ely p. 19. Blomer's Full View p. 137. Monk's
Bentley p. 163. Articles laid before Bp. Moore xxviii. Ib. Appendix
p. xviii.

---

## LETTER CI.

### COTES TO {HALLEY}.

This letter is not dated, but the circumstance of its being written
upon the same sheet of paper as Letters c, cii shews that it is separated by no long interval from them. It is clear from its contents that
Halley was the person to whom it was addressed.

S$^r$.

Tis now about two Yeares since I wrote to You, in
behalf of M$^r$ Jurin a Fellow of our College, to desire y$^t$
he might have Your leave to annex some of Y$^r$ Treatises
to his Edition of Varenius's Geography. You was pleased
to consent to it & to promise some additional improvements & besides a new Treatise concerning Cœlestial
Refractions. I hope You have lately received a Letter
from him to remind You of Y$^r$ promise, & to desire y$^t$ a
freind of his may wait upon You for Y$^r$ Papers assoon as
You shall have leisure to finish 'em. He further desires if
any new Figures must be inserted or any alterations made
in y$^e$ old ones y$^t$ You will be pleased to send them first &
y$^t$ You will be so kind as to send him word what he had
best do with y$^e$ Map of y$^e$ Trade Winds & Variations {of
the Compass} & whether He may take that in the Miscellanea Curiosa with the English names as they stand there.
The greater part of Varenius is already printed off, we
do therefore beg of You to finish Y$^r$ Papers assoon as

You have convenient leasure.    I beg Your pardon for the trouble I give You.

I am S<sup>r</sup>

Y<sup>r</sup> much Obliged & Humble Serv<sup>t</sup>

ROGER COTES.

Jurin's edition of Varenius dedicated to Bentley who had encouraged him to undertake the work bears date 1712, though a notice of it appears in the " Memoirs of Literature" for Sept. 1711. The copy of it in Trin. Coll. Library has Cotes's autograph " Donum Amicissimi Editoris."

# CORRESPONDENCE OF COTES AND JONES.

William Jones born 1680, died 1749. See life of Sir William Jones (his son) by Lord Teignmouth, where six of these seven letters of Jones and one of Cotes are printed, but very inaccurately.

## LETTER CII.

### COTES TO JONES.

S$^r$                                         Febr. 15. 1711

I yesterday received Your most valuable & acceptable gift* togather with Y$^r$ very kind Letter I return You my most hearty thanks for 'em both. You have highly obliged the Mathematical part of y$^e$ World by collecting into one Volume those curious & usefull Treatises which were before too much dispersed but more especially by y$^e$ publication of y$^e$ Analysis per Æquationes infinitas & the Methodus Differentialis. I could heartily wish y$^t$ nothing of S$^r$ Isaac's might be lost, I hope You will endeavour as You find an Oportunity to persuade him to publish some other Papers for I believe he has yet many excellent things in reserve. About a Year & an half ago (when I was last in Town) I acquainted Mr Ralphson y$^t$ You had some Papers of S$^r$ Isaacs in Y$^r$ hands which were communicated long ago to Mr Collins. I thought they might have been pertinent to his design of writing y$^e$ History of y$^e$ Method of Fluxions. I afterwards understood y$^t$ You gave him a sight of those Papers, & y$^t$ he thought 'em not to be for his purpose, which I do now very much wonder at, if his intention was to do justice to S$^r$ Isaac. If y$^t$ was not his Intention I think Your Preface has already sufficiently de-

---

* A quarto volume, edited by Jones, containing some *opuscula* of Newton's. It is entitled *Analysis per Quantitatum Series, Fluxiones ac Differentias cum Enumeratione Linearum Tertii Ordinis.* Lond. 1711.

feated all his attempts. We are now at a stand as to S$^r$ Isaac's Principia, he designs to make some few Experiments before we proceed any further. The first Book & y$^e$ six first Sections of y$^e$ Second are already printed off. The inclosed Paper * is what I wrote about 3 Yeares ago & read to my Auditors in our Schools in 1709. I have sent it to You as it relates to y$^e$ Methodus Differentialis but more particularly as a small acknowledgment of my gratitude for having received y$^t$ and the other excellent Treatises from Your hands & as a token of my hearty freindship & sincere good will to You

I am S$^r$ Y$^r$ most obliged freind

& humble Servant

R Cotes.

Not having heard any thing of y$^e$ book till I saw it I received it with y$^e$ additional pleasure of a Surprize.

Printed in the *Gen. Dict.* IV. 443. Macclesfield *Corr.* I. 257.

LETTER CIII.

JONES TO COTES.

[Extract.]

S$^r$.                                   London Septemb$^r$. 17. 1711

The paper concerning S$^r$. Is. Newton's method of Interpolation, which you have bin pleas'd to send me, being done so very neat, that it wou'd be an injury to the Curious, in these Things, to be kept any longer without it; therefore must desire you'd grant me leave to publish it in the Phil. Trans. you may be assur'd, that I don't move this to you, without S$^r$. Isaac's approbation, who I find is no less

---

* Printed among his *Opera Miscellanea* at the end of the *Harmonia Mensurarum*, pp. 23—33.

willing to have it done. The new Edition of the Principia
is what we wait for with a great deal of impatience; tho',
at the same time, I believe the Book will be far more valuable than if it had bin done in a hurry, Since I find the
interruptions are necessary, and Such as will render it
Compleat. We have nothing considerable in hand here at
present, only M$^r$ De Moivre's Treatise of Chance*, which
makes a whole Transaction, he is very fond of it, & we
may expect it well done : M$^r$. Raphson has printed off four
or five Sheets of his History of Fluxions, but being shew'd
S$^r$. Is. Newton, (who, it seems, wou'd rather have them
write against him, than have a piece done in that manner
in his favour,) he got a Stop put to it, for some time at
least. D$^r$. Halley has almost finish'd the printing of the
Greenwich Observations†, which will be a work of good
use; especially as it is now, free'd from the trifls it was
loaded with. S$^r$. I have one thing, which I wou'd trouble
you with further, & that is, to let me know, what Lectures,
or other Papers of S$^r$. Is. Newton's, remain, in your University, unpublish'd, this may be done at your leasure :

---

* "De Mensura Sortis." *Phil. Trans.* Jan.—March, 1711. Comp. Letters CVII.,
CVIII. Demoivre was born at Vitri in Champagne, in 1667. On the revocation of the
edict of Nantes, he settled in England. He died Nov. 27, 1754.

† The Observations here referred to (made with a mural arc) form the 2nd Book of
Flamsteed's *Historia Cœlestis*, published in 1712. The Observations contained in the
1st Book (made with a sextant) were printed under Flamsteed's superintendence, at
Prince George's expense, and with a trifling exception, were wrought off before Christmas, 1707; but in consequence of his misunderstanding with the Prince's referees,
which seems to have arisen principally from his objection to print his catalogue of the
Fixed Stars before the 2nd Book of Observations, the task of editing those parts of the
work was confided to Halley. In Flamsteed's MS. of the 2nd Book, the Observations
stood recorded as they were made, but Halley arranged them under the heads of the
Moon and planets to which they related, not giving the whole of the Observations, but
retaining only those of such Stars, as in passing the meridian, had nearly the same
right ascension and declination as a planet. (See Halley's Pref. to *Hist. Cœl.* and
Baily's Pref. to *Account of Flamsteed*, p. xli.)

Some years elapsed before Flamsteed had any other means of revenging himself
upon his editor, than by unsparing abuse. At length, in April 1716, having got possession of 300 copies of his work, he separated the "very sorry abstracts" of his Observations, and the "corrupted Catalogue" from the part which he had himself
superintended, and committed nearly the whole of them to the flames, "as a sacrifice
to Heavenly Truth." Baily, pp. 101, 321, 322.

## LETTER CIV.

### COTES TO JONES.

[Extract.]

S$^r$. {Cambridge, Sept. 30, 1711.}

I return You my thanks for Your Letter & the Information You gave me concerning the State of Mathematicks at present in London. I shall be glad to see M$^r$ De Moivre's Treatise of Chance when it comes out; his things are always very neat and curious. We have nothing of S$^r$ Isaac's that I know of in Manuscript at Cambridge, besides the first draught of his Principia as he read it in his Lectures*, his Algebra Lectures which are printed & his Optick Lectures the substance of which is for y$^e$ most part contained in his printed Book but with further Improvements. I thank You for Your kind offer of recommending my Paper to the Publick; but I am of opinion that it is not of so great use as to deserve to be printed after S$^r$ Isaac's Methodus Differentialis.

\* \* \* \* \* \* \*

I am very desirous to have the Edition of S$^r$ Isaac's Principia finish'd, but I never think the time lost when we stay for his further corrections & improvements of so very valuable a book, especially when this seems to be the last time he will concern himself with it. I am sensible his

---

* The folio volume marked Dd.9.46 in the University Library, corresponds to this description, but it has the book-plate, which indicates it to have been one of Bishop Moore's books, given to the University by George I., in 1715. If, then, this be the volume which Cotes means, either the book-plate has been pasted in by mistake, or the book must have found its way somehow into the Bishop's library. See more of this MS. in the notes to the Synoptical View of Newton's Life, under August 1684, and Table of his Lectures for that year.

Newton's presentation copies of his *Optical* and *Algebra Lectures*, the latter in his own hand-writing, are still in the University Library, marked Dd.9.67 and 68.

other Business allows him but little time for these things & therefore I ought not to hasten him so much as I might otherwise do, I am very well satisfied to wait till he has leasure.

Printed in *Gen. Dict.* IV. 444. Macclesfield *Corr.* I. 258.

---

## LETTER CV.
### JONES TO COTES.

Dear S<sup>r</sup>                    London Octob<sup>r</sup>. 25<sup>th</sup>. 1711

The favour of your account of S<sup>r</sup>. Isaac's papers left at Cambridge, I return you my hearty thanks for; And as you have some further Considerations about the Doctrine of Differences, I am assured, they cannot but be valuable; and if a few Instances of the application were given, perhaps it wou'd n't be amiss: Having tarried some time for a convenient opportunity, I was at last oblig'd to send you Mouton's Book by the Carrier; tho it will only satisfy you that D<sup>r</sup>. Gregory had but a very Slender notion of the design, extent, & use of Lem. 5. Lib. 3 of the Principia; I hope it will not be long before you find leasure to send us what you have further done in this curious subject; no ex⸗ cuse must be made against the publishing of them; Since, with respect to Reputation, I dare say, 'twill be no way to your disadvantage.

I have nothing of news to send you; only the Germans and French have in a violent manner attack'd the Philo- sophy of S<sup>r</sup> Is: Newton*, and seem resolv'd to stand by Cartes; M<sup>r</sup> Keil*, as a person concern'd, has undertaken to answere & defend some things, as D<sup>r</sup>. Friend*, & D<sup>r</sup> Mead†,

---

* See Letter CVII.
† Mead was concerned as the author of a work *De imperio Solis ac Lunæ in corpora humana*, Lond. 1704.

does (in their way) the rest : I wou'd have sent you y^e whole Controversy, was not I sure that you know, those only are most capable of objecting against his Writings, that least understand them ; however, in a little time, you'l see some of these in y^e Philos. Transact.

## LETTER CVI.

### COTES TO JONES.

[Extract.]

\* \* \* \* \* \* \* \* \* \*

The controversy concerning S^r Isaac's Philosophy is a piece of News that I had not heard of unless Muys's late Book be meant. I think that Philosophy needs no defence, especially when tis attack't by Cartesians. One M^r Green\* a Fellow of Clare Hall in our University seems to have nearly the same design with those German & French objectors whom You mention. His book is now in our press & is almost finished. I am told he will add an Appendix in which he undertakes also to square the circle. I need not recommend his performance any further to You.

Nov. 11^th 1711

Printed in *Gen. Dict.* IV. 444. Macclesfield *Corr.* I. 261.

## LETTER CVII.

### JONES TO COTES.

D^r. S^r.                                                    Nov. 15^th. 1711

I receiv'd yours of the 11^th. instant, and am glad to find you've finish'd your second Paper, and do hope it will not

---

\* See Letter XXVII. note.

be long before I receive it : I have taken this opportunity of p'senting you with one of M$^r$ De Moivre's late Tracts, tho the Author himself, perhaps, may send you another ; how well he has handled this subject, is what I shall not have time soon to consider.    The Objections of y$^e$. writers of the Leipsic Transactions, against the Philosophy intro- duced in D$^r$. Friend's Chimical Lectures *, together with his answere, as also those of Wolfius, and of M$^r$. Saurin of the Fr. Academy, against y$^e$. same Philosophy, with an answere by M$^r$. Keil†, are now in the Press here, and nearly finish'd, I shall not be wanting to send them you.    I am concern'd to find, by S$^r$. Isaac, that his Book does not go forward, 'tis a great grieveans to be so long depriv'd of it,

I am, S$^r$, very much

Your friend and Servant

W$^m$: Jones

S$^r$. you need not, if you please, make known to the Person that brings this, that I've sent you Moivre's Book.

*To* M$^r$ Roger Cotes
    *This*

Christian Wolf (an eminent philosopher and mathematician, born 1679, died 1754, at the date of this letter a Professor at Halle ; see Ten- nemann's Hist. of Phil. and life by Degerando in the Biographie Uni-

---

* Freind's *Prælectiones Chymicæ*, Lond. 1709, dedicated in most complimentary terms to Newton.

" Nov. 15, 1711.    The President in the Chair...The editors of the *Acta Eruditorum* having published {September 1710,} a reflecting paper upon Dr Freind's *Chymistry*, a Discourse was now read of Dr Freind's in vindication of his book, and the principles therein maintained.    This Discourse was ordered to be published in the Transactions, and the thanks of the Society returned to the Dr."    *Journal Book* of Royal Soc.

Freind's defence appeared in the *Phil. Trans.* for July—Sept. 1711, pp. 330—342, under the title of " Prælectionum Chymicarum Vindiciæ in quibus objectiones in Actis Lips......contra Vim materiæ Attractricem allatæ diluuntur." He shews the impropriety of calling attraction " an occult quality," and takes occasion to criticise some of Leib- niz's opinions, but uses only the initial letter of his name.    (Freind afterwards, in 1726, reprinted the article of the Leipsic Reviewers, accompanied by his Answer, as an Appendix to the 2nd Ed. of his Lectures.)    A reply was published in the *Acts* for June, 1713, pp. 307—314.

† See the remarks at the end of the Letter.

verselle) in his " Aerometriæ Elementa...1709" attacked an argument which Keill in his Lectiones Physicæ (Oxf. 1702) had advanced in proof of a vacuum, founded on the fact that, abstracting from the resistance of the air, all bodies fall from equal heights in the same time. Keill answered his objections in a letter, part of which was printed in the Leipsic Acts for Jan. 1710 (pp. 11—15), to which his antagonist replied in the following Number (pp. 78—80). A rejoinder was prepared by Keill, the first portion of which exists in MS. among the Lucasian papers (a folio sheet in packet 11). This seems to be the piece to which Jones refers in the above Letter, though I do not remember to have ever seen it in print. In the 4th page of this last-mentioned paper Keill proceeds to notice some of the views propounded by Saurin in a Memoir read before the Academy of Sciences in 1709 ("Examen d'une difficulté considerable proposée par M. Huyghens contre le Système Cartesien sur la cause de la Pesanteur." Memoirs for that year, p. 131, published in 1711. The difficulty alluded to is that if Descartes's celestial matter circulates with the enormous velocity that it ought to have in order to produce the observed effects of gravity, it ought to hurl away all the bodies on the earth's surface—*quippe ferat rapidè secum verratque per auras.* He returned to the subject in a supplementary Memoir in 1718, in which he notices the allusion which Malebranche in the last ed. of his " De la Recherche de la Vérité" had made to the former Memoir.) Joseph Saurin, born 1659, died 1737, was a fervent believer in the system of Vortices, the impossibilities of which seem to have had a piquancy for him that stimulated his faith. He frankly admits the difficulties that surround the hypothesis, and the course of his investigations leading him to an absurd consequence, he says, " il semble qu'il n'y auroit pas d'autre parti à prendre, que de la digerer cette absurdité, comme on est obligé d'en digerer tant d'autres... dans presque tous les objets de nos connaissances." A remark towards the end of his Memoir does not impress us with a favourable opinion of the extent of his acquaintance with the Newtonian philosophy : " Il (Newton) aime mieux considerer la Pesanteur comme une qualité inhérente dans les corps, & ramener les idées tant décriées de qualité occulte, & d'attraction." If we abandon mechanical principles, he continues, " nous voilà replongez de nouveau dans les anciennes ténébres du Peripatetisme, dont le Ciel nous veüille préserver." He started in life by following his father's profession of a Calvinist minister, was then carried off by the invited pounce of the " eagle of Meaux," and about fourteen months before Jones mentioned him in this letter the malice of a poet threw him into a dungeon. For the events of his strange life see his Eloge by Fontenelle, and the Biographie Universelle. Comp. Vie de J. B. Rousseau (Beuchot's Voltaire, XXXVII. 505).

## LETTER CVIII.

### COTES TO JONES.

[Extract.]

S<sup>r</sup>

I thank you for M<sup>r</sup> De Moivre's Treatise concerning Chance: I have not yet had leasure to go over it.   M<sup>r</sup> Sanderson* by whom You sent it, was on Tuesday last elected our Mathematical Professor in the room of M<sup>r</sup> Whiston.   I am not perfectly acquainted with him, he seems as far as I can judge of him to have an extraordinary good Genius.   The want of his sight is certainly an insuperable disadvantage to him in several respects but I believe in some others he has an advantage from it.

Nov. 25<sup>th</sup> 1711

Printed in Macclesfield *Corr.* i. 261.

---

## LETTER CIX.

### JONES TO COTES.

D<sup>r</sup> S<sup>r</sup>.                              London Jan. 1<sup>st</sup>: 17$\frac{11}{12}$

I have sent you here inclos'd, the Coppy of a Letter, that I found among M<sup>r</sup> Collins's papers, from S<sup>r</sup>. Is. Newton to one M<sup>r</sup>. Smith; the contents thereof seems to have, in some measure, relation to what you are about, as being the application of the Doctrine of Differences to the making of Tables; and for that reason I thought it might be of use to you, so far as to see what has bin done already:

---

* "Nov. 19.  A Mandate from the Queen to make Mr Nicolas Saunderson (a blind man from his infancy, but who had taught Mathematics in Christ's College about four years), Master of Arts.  It did not command, but only recommended him; and yet he was immediately admitted and created, without reading any Grace for it." Rud's *Diary*.  He was chosen Professor on the 20th, having six votes against his competitor's (Mr Hussey of Trinity) four, and made his inauguration speech on Jan. 21 following.  *Ib.*

I shew'd this to S$^r$. Isaac, he remembers y$^t$. he apply'd it to
all sorts of Tables, but has nothing by him, more than
what is printed: I have more papers of M$^r$. Mercator's and
others, upon this subject, tho, I think, none so material, to
your purpose, as this. I shou'd be very glad to see what
you have done of this kind all publish'd; And I must confess, that, unless you design a considerable large Volume,
'twere much better to put them into the Transactions; for
that wou'd sufficiently preserve them from being lost,
which is y$^e$. common fate of small single Tracts; and at y$^e$.
same time save the trouble and expense of printing them,
since the subject is too curious to expect any profit by it:
and besides, now, as the R. Society having done themselves the honour of choosing you a Member *, something
from you cannot but be acceptable to them: S$^r$ Isaac himself expects those things of yours that I formerly mention'd to him as your promise.

I am, S$^r$. your much oblig'd

friend, & humble Serv$^t$.

W. JONES.

---

LETTER CIX. (bis)

NEWTON TO J. SMITH.

[Copy].

Enclosed in Letter CIX.

S$^r$.                              Trin. Coll. Cambridge, May 8$^{th}$. 1675.

I have consider'd y$^e$ buisiness of computing Tables of
Square, Cube, & Sq. Sq$^r$. Roots; and y$^e$. best way of
p'forming it, y$^t$. I can think of is y$^t$. which follows:

If y$^u$. wo'd compute a Table to 8 decimal places, let y$^e$.

---

* Jones had himself been chosen on the same day, (Nov. 30). Cotes was not admitted until May 20, 1714. Newton presided on both occasions.

roots of every hundredth number be extracted to ten decimal places, and then compute every tenth numb[r]. and afterwards every number by the following methods.

| Tab. 1. | | | | | Tab. 2. | | |
|---|---|---|---|---|---|---|---|
| $n-50$ | $A$ | $*o$ | $\alpha$ | | $n-6$ | $4E$ | |
| | $o$ | $m$ | $o$ | | | $5\epsilon$ | |
| $n-40$ | $B$ | $op$ | $\beta$ | | $n-5$ | $5EF5$ | — |
| | $p$ | $m$ | $\pi$ | | | $\zeta 4$ | |
| $n-30$ | $C$ | $pq$ | $\gamma$ | | $n-4$ | $F4$ | |
| | $q$ | $m$ | $\chi$ | | | $\zeta 3$ | |
| $n-20$ | $D$ | $qr$ | $\delta$ | | $n-3$ | $F3$ | |
| | $r$ | $m$ | $\rho$ | | | $\zeta 2$ | |
| $n-10$ | $E$ | $rs$ | $\epsilon$ | | $n-2$ | $F2$ | |
| | $s$ | $m$ | $\sigma$ | | | $\zeta 1$ | |
| $n$ | $F$ | $st$ | $\zeta$ | $\dfrac{m}{10}$ | $n-1$ | $F1$ | |
| | $t$ | $m$ | $\tau$ | | | $\zeta$ | |
| $n+10$ | $G$ | $tv$ | $\eta$ | | $n$ | $F$ | $\dfrac{st}{100}$ |
| | $v$ | $m$ | $\nu$ | | | $1\zeta$ | |
| $n+20$ | $H$ | $vx$ | $\theta$ | | $n+1$ | $1F$ | |
| | $x$ | $m$ | $\phi$ | | | $2\zeta$ | |
| $n+30$ | $I$ | $xy$ | $\iota$ | | $n+2$ | $2F$ | |
| | $y$ | $m$ | $\psi$ | | | $3\zeta$ | |
| $n+40$ | $K$ | $yz$ | $\kappa$ | | $n+3$ | $3F$ | |
| | $z$ | $m$ | $\omega$ | | | $4\zeta$ | |
| $n+50$ | $L$ | $z*$ | $\lambda$ | | $n+4$ | $4F$ | |
| | | | | | | $5\zeta$ | |
| | | | | | $n+5$ | $5FG5$ | — |
| | | | | | | $\eta 4$ | |
| | | | | | $n+6$ | $G4$ | |
| | | | | | | $\eta 3$ | |
| | | | | | $n+7$ | $G3$ | |
| | | | | | | $\eta 2$ | |
| | | | | | $n+8$ | $G2$ | |
| | | | | | | $\eta 1$ | |
| | | | | | $n+9$ | $G1$ | |
| | | | | | | $\eta$ | |
| | | | | | $n+10$ | $G$ | $\dfrac{tv}{100}$ |
| | | | | | | $1\eta$ | |
| | | | | | $n+11$ | $1G$ | |
| | | | | | | $2\eta$ | |
| | | | | | $n+12$ | $2G$ | |

## In the First Table,

Let $n$ signify every 100th numb[r]. & $F$ its root, wheth[r]. Square, Cube, or Sq. Square; & $n-50$, $n-40$, $n-30$, &c.

every 10$^{th}$. numb$^r$; and *A*, *B*, *C*, *D*, &c. their roots; and
*o*, *p*, *q*, *r*, &c, the differences of these roots; & *op*, *pq*, *qr*,
&c. their second differences, (that is *op*, the diff. of *o* & *p*,
*pq* the diff. of *p* & *q*, &c.) and *m* their third difference, that
is, y$^e$. common difference of $*o$, & *op*, *op* & *pq*, *pq* &
*qr*, &c.

Further, let *a*, *β*, *γ*, *δ*, &c. signify y$^e$. differences of these
Roots from those next less, namely *a* the difference of *A*
y$^e$. root of *n* − 50 & y$^e$. like root of *n* − 51, *β*, the diff. of
y$^e$. roots *n* − 40 & *n* − 41, *ζ* the diff. of y$^e$. roots of *n* &
*n* − 1, *η* the diff. of y$^e$. roots of *n* + 10 & *n* + 9, &c. And
let *o*, *π*, *χ*, *ρ*, &c signify the diff. of *a*, *β*, *γ*, *δ*, &c. And
$\frac{m}{10}$ the common diff. of *o*, *π*, *χ*, *ρ*, &c.

## In the Second Table,

Let *n* − 6, *n* − 5, *n* − 4, *n* − 3, &c signify y$^e$. single
numbers,

4*E*, 5*E* or *F*5, *F*4, *F*3, &c. their Roots,

5*ε*, *ζ*4, *ζ*3, *ζ*2, &c the diff. of those roots;

$\frac{st}{100}$ the common diff. of those differences for y$^e$. ten

numbers between *n* − 5 & *n* + 5.

And so for y$^e$. ten numbers between *n* + 5 & *n* + 15;
let *G*5, *G*4, *G*3, &c. signify y$^e$. roots; *η*4, *η*3, *η*2, &c, their
first differences, and $\frac{tv}{100}$ their second differences; and the
like for every denarie between *n* − 50 & *n* + 50.

This explication of the Tables being p'mis'd, you may
compute them thus;

$$\left.
\begin{array}{ccc}
\dfrac{10F}{2n}=\omega, & \dfrac{10\omega}{2n}=st, & \dfrac{30st}{2n}=m. \\[2ex]
\dfrac{10F}{3n}=\omega, & \dfrac{20\omega}{3n}=st, & \dfrac{50st}{3n}=m. \\[2ex]
\dfrac{10F}{4n}=\omega, & \dfrac{30\omega}{4n}=st, & \dfrac{70st}{4n}=m.
\end{array}
\right\}$$

Out of $n$, {Square}
extract {Cube } Root, make
$F$ yᵉ. {Sq. Sq.}

$$\omega+\tfrac{1}{2}st+\tfrac{1}{6}m=s,\quad \frac{\omega}{10}+\frac{\tfrac{1}{2}st}{100}\ `+\frac{m}{6000}`\ *=\zeta,\ \text{and}\ \frac{st}{10}+\frac{55m}{1000}=\sigma.$$

And these quantities $F$, $st$, $m$, $s$, $\zeta$, & $\sigma$, being thus found, yᵉ. rest are given by Additⁿ. & Subduct.

For $st+m=rs$, $rs+m=qr$, &c. $st-m=tv$, $tv-m=vx$, &c.

Again $s+rs=r$, $r+qr=q$, &c. $s-st=t$, $t-tv=v$, &c.

And $F-s=E$, $E-r=D$, &c. $F+t=G$, $G+v=H$, &c.

Further

$$\sigma+\frac{m}{10}=\rho,\quad \rho+\frac{m}{10}=\chi,\ \&c.\quad \sigma-\frac{m}{10}=\tau,\quad \tau-\frac{m}{10}=\upsilon,\ \&c.$$

Lastly $\zeta+\sigma=\epsilon$, $\epsilon+\rho=\delta$, &c. $\zeta-\tau=\eta$, $\eta-\upsilon=\theta$, &c.

These quantities being thus computed, in yᵉ. first Table, to every 10ᵗʰ. number, the roots may be computed in yᵉ. 2ᵈ Table to every numbʳ. by Addition and Subduction only;

For $\zeta+\dfrac{st}{100}=\zeta 1$, $\zeta 1+\dfrac{st}{100}=\zeta 2$, &c.

$$\zeta-\frac{st}{100}=1\zeta,\quad 1\zeta-\frac{st}{100}=2\zeta,\ \&c.$$

Again $F-\zeta=F1$, $F1-\zeta 1=F2$, &c.

$$F+1\zeta=1F,\quad 1F+2\zeta=2F,\ \&c.$$

---

\* I have added the ‘$\dfrac{m}{6000}$.’ I have also corrected some other errors of transcription.

Thus you must proceed to five Figures on either hand, and then do the like in the next ten Figures, saying

$$\eta + \frac{tv}{100} = \eta 1, \quad \eta 1 + \frac{tv}{100} = \eta 2, \ \&c.$$

And the like for every Denarie between $n - 50$ & $n + 50$.

In these Computations, Note, 1$^{st}$. That they must be done every where to 10 or 11 decimal places, if you will have a Table of Roots exact to 8 of these places.

2$^{dly}$ If $5F$ & $G5$, the roots of $n + 5$ found two ways agree to 8 decimal places, it argues the whole works from which they were derived, to be true. And so of y$^e$. roots of $n + 15$, $n + 25$, $n - 5$, &c. And also of y$^e$. Terms $A$, $*o$, & $a$; $L$, $s*$, & $\lambda$, where two works meet. Let this therefore be y$^e$. Proof of y$^e$. work.

This S$^r$. is w$^t$. has occurr'd to me about your design, which I hope will do your business, the whole work being p'form'd by Addit. & Subduct: excepting y$^t$. in y$^e$. computation of every 100$^{th}$. number, there is required y$^e$. Extraction of one root, & three divisions, to find $F$, $\omega$, $st$, & $m$.

S$^r$. I am

Your humble Serv$^t$

Is. Newton.

The person to whom this letter is written may be conjectured to be "John Smith, *Philo-Accomptant*," author of *Stereometrie*, Lond. 1673. (He must not be confounded with Cotes's uncle). In the Macclesfield Correspondence, ii. 370—374, there are two other letters on the extraction of roots from Newton to this same person (not to Collins, as there printed) dated July 24 and Aug. 27, 1675, in the former of which he refers to the method given in the foregoing letter. Mr J. Smith seems to have had a design of constructing Tables of Square, Cube and Biquadr. Roots, and consulted Newton as to the best mode of computing them. The Tables, if ever made, do not appear to have been published. The earliest Tables of Roots are Briggs's MS. Tables of the Square Roots of Numbers up to 1000 mentioned in Mayne's *Merchant's Companion* (London, 1674), p. 80.

## LETTER CX.

### COTES TO JONES.

Answer to Letter CIX, No date.

S$^r$

I have received Your Letter with the inclosed Paper of S$^r$ Isaac Newton for which I return You my hearty thanks. His method seems to be excellently well suited to those particular purposes for which he design'd it, & I do not doubt I shall find it very curious when I have leasure to examine it to y$^e$ bottom. What I intend to print will make but a small Volume, I cannot say it will be bigger than that of S$^r$ Isaacs which You lately published. It will contain the Lectures I have hitherto read in Publick, together with those which I shall read this Year, all of which amount to no more than Ten, for by the Statutes of my place I am obliged annually to make but two. I cannot indeed expect any profit from the Publication, twill be sufficient if y$^e$ expense of it can be defrayd. I have already put y$^e$ University to the charge of Types for some new characters which I have occasion to make use of & therefore for that reason as well as some others I cannot now draw back. What You mention that y$^e$ R: Society have chose me one of their Members is altogether a peice of news to me. If it be so, I shall be very sensible of the Honour they have done me. That Title may recommend my papers to y$^e$ Publick though they be printed at Cambridge. If You insist upon my Promise of sending those things to You before they are printed I shall be ready to make it good. What I have further concerning y$^e$ subject of differences consists of Ten Propositions whereof the Six first are particular & fitted for use & are sufficient for all cases that comonly happen, the other four are general. You will be able to judge of my Method by y$^e$ first Proposition which I here* send You. You may shew it to S$^r$ Isaac if You think it proper but I desire You would not shew it to others.

I cannot so easily give You an Idea of my other peice concerning Logarithms but I find room enough in this Page to send You† one thing out of it as a curiosity which may be understood independently of the rest.

Rectificatio Logarithmicæ
—Oblata sit igitur Logarithmica &c.

---

## LETTER CXI.

### JONES TO COTES.

Accompanying 4 copies of the Commercium Epistolicum.

[Extract.]

S$^r$ London Feb 6$^{th}$. 17$\frac{12}{13}$

The R. Society having order'd one of their Books for you, & another for M$^r$. Sanderson, also one for Trinity College Library, & one for the University Library; I wou'd not miss the opportunity of paying you my respects by sending them: I need not tell the occasion & design of that Collection: you'l see readily that it affords such light concerning what it relates to, as cou'd not easily have bin discover'd any other way: and also shews that your great Predecessor, whose illustrious Example, I don't doubt but you follow, never imploy'd his time about things ordinary. I have no Mathematical intelligence to send you; M$^r$. Keil

---

* The tract of which Cotes sends a specimen to Jones will be found among his *Opera Miscellanea*, pp. 36—71. The title of it is "Canonotechnia sive Constructio Tabularum per Differentias." He has not copied out the proposition in this draught of his letter, and therefore it will be sufficient to refer the curious reader to p. 36 of the work just cited.

† Here also Cotes has not taken the trouble to transcribe the proposition. It may be seen in his Logometria, (*Harmonia Mensurarum*, pp. 23, 24.)

thinks he has discover'd a very easy and Practical solution
of the Keplerian Problem*: the Problem of the Refraction, or that concerning $y^e$. description of the Curve described by a Ray of Light in passing thro the Atmosphere,
is here done by two different hands; one of them endeavours to apply it to Astronomical uses, $w^{ch}$. I suppose he
has pretty well compass'd.

       \*         \*        \*        \*        \*        \*

I am extremely pleas'd to find that $S^r$. Isaac's Book is
so near being finish'd: his general Scholium I presume
he'l soon send you, if 'tis not already done: and 'tis not
less agreeable to me, to hear that your own Book is in
such forwardness.

       \*         \*        \*        \*        \*        \*

P.S.  I have sent to you four of the Comercium Epis-
tolic. that is, one for your self, and $y^e$ other three as
before mention'd which I desire you wou'd deliver, as from
the Royal Society of London.

---

### LETTER CXII.

#### COTES TO JONES.

[Extract.]

$S^r$.                                                                Cambridge $Feb^r$. 13$^{th}$

I have received Your obliging Letter together with
the very agreeable gift of the Commercium Epistolicū.  I
have delivered one Copy to the University Library Keeper
another to the Library-keeper of Our College and the
third to $M^r$ Sanderson as from the Royal Society.  You
may be pleas'd to return our acknowledgments of the
Favour.

---

* *Phil. Trans.* for 1713, Vol. xxviii. pp. 1—10.

I am very glad to see this Peice at length made publick in which *quicquam cuiquam detractum non reperio, sed potius passim suum cuique tributum* *.

\* \* \* \* \* \*

---

LETTER CXIII.

JONES TO COTES.

S<sup>r</sup>.                                    London Aprill 29<sup>th</sup>. 1713.

Ever since I received your very kind Letter, and Mouton's Book, I waited for an opportunity of sending you some old Manuscripts I had by me, and at last am oblig'd to Venture them by the Carrier; They relate, in some measure, to the Method of Differences; The folio one, I find, was writ by one Nath. Torperley†, a Shropshire man, who when young was Amanuensis to Vieta, but afterwards writ against him; he was contemporary with Briggs and Harriot, and intimately acquainted with them; The Book, I think, can be of no other use to you, than in what relates to the History of that Method, and in having y<sup>e</sup> Satisfaction of seeing what has bin formerly done on that Subject. The other Small 4<sup>to</sup> M.S. is a piece of Mercator's about Differences, it seems to contain no great matter; nor indeed, can I be satisfied, any thing that he has done, or any one else, so very considerable, as to deserve to accompany any piece of yours; Therefore pray let us have your things entire, and as soon as conveniently you can.

I am mightily pleas'd to see the end of the Principia, and return you many thanks for the very Instructive Index,

---

\* *Commerc. Epistol.* p. 119, (p. 239, 2nd. Ed.) These are Leibniz's words in his Letter to Sloane, Dec. 29, 1711, by which he unfortunately made himself a party to the obnoxious language of the Leipsic review of Newton's tract, "De Quadraturâ Curvarum," Leips. *Acts,* Jan. 1705.

† Compare Macclesfield *Corresp.* II. 5, note.

that you have taken the pains to add, and hope 'twill not be long before we shall see the Beginning of that Noble Book.

I shall be in some pain till I hear that you have receiv'd my old M:S. it being a favorite one, purely upon the account of some extravagancys in it, So very uncommon: But I shall think it safe when in your hands; I am S$^r$. without reserve, your very affectionate friend and most humble Servant

W: JONES.

---

### LETTER CXIV.

#### COTES TO JONES.

Dear S$^r$

I know not how to return You my thanks as I ought for Your readiness to assist me. The two Manuscripts of Torperly & Mercator are come very safe to my hands; I hope I shall return 'em to You without any damage. I have been lately, and am at present taken up with some College buisness, so that I have scarce yet had any time to look into 'em. If I find any thing in them of Moment, I believe I shall request You to let me print it with my own, for I would not willingly have any one lose the Credit due to him.

I am glad You can approve of the Index to the Principia. It was not design'd to be of any use to such Readers as Your self, but to those of ordinary capacity. I hope the whole Book may be finished in a fortnight or three Weeks. I have lately been out of Order, or it might have been done by this time

I am S$^r$

Your most Obliged Freind

May 3$^d$. 1713.                    and Servant R. COTES.

## LETTER CXV.

### JONES TO COTES.

Dear S^r                                        July 11^th 1713

'Tis impossible to represent to you, with what pleasure I receiv'd your inestimable Present of the Principia, and am much concern'd to find my self so deeply charg'd with Obligations to you; and such, I fear, as all my future endeavours will never be able to requite. This Edition is indeed exceeding beautifull, and interspers'd with great variety of admirable discoverys, so very natural to its great Author; but is much more so, from the additional advantage of your excellent Preface prefix'd; which I wish might be got publish'd in some of the foreign Journals; and since a better account of this Book cannot be given, I suppose it will not be difficult to get it done.

Now this great Task being well over, I hope you'l think of publishing your own Papers, & not let such valuable pieces lye by:

As to w^t. you mention'd in your last concerning my Old manuscripts, tho, for my part, I know of nothing worth your notice publickly in them, but if you do find any, it the more answers the end of my sending it, and you know that you may do as you please;

S^r I am
your most obedient
humble Serv^t
W: Jones

## LETTER CXVI.

### COTES TO {WHISTON.}

Dear Sir                                        {March 1715}

I have lately seen two Schemes of the great Eclipse the one done by Your self, the other by D^r Halley. Yours being to be understood by those only who are acquainted

15

with Astronomy, has upon that account much the disadvantage of the D<sup>rs</sup> with most People. I take the Liberty to propose another Scheme to You, which I beleive would give a more general satisfaction than either of the other: I mean a Map of that part of the Heavens in which the Sun will be at that time. If the sky be clear it will undoubtedly be a great surprize to see the Stars, but twill be much more so to the Vulgar that You should be able to describe the Positions of 'em beforehand: this I am apt to think they will look upon as a greater peice of art, than to predict the Eclipse itself. By comparing the Ephemeris & Globe together I find there will be three Planets visible on the West of the Sun, Jupiter will be very near him, Venus will be about the Meridian, Mercury will lye between them. You have already spoken of the Moons Atmosphere, I think it would not be amiss if You desired People to look if they can observe the Suns also, I mean that light in the Heavens which D<sup>r</sup> Gregory describes pretty largely in the Scholium to Prop. 8. Lib. 2 of his Astronomy. A representation of this may be inserted in the Map if You think fit, that it may be known beforehand how tis likely to appear. You may caution those who are desirous to see this faint light, that they prepare their eyes beforehand for it, by staying in some dark place for about a quarter of an hour before the Sun be totally obscur'd; You know it requires about that time to bring our Eyes to the disposition they usually have in the night time for seeing faint Lights. I would further advise, if You think fit to set about this Project, that You do it with exactness that Mathematicians may not dislike it, & that Your Explications be written in a Popular way & as free as may be from Mathematical Terms that others may not dislike it. I suppose You have seen Cassini's Map & Reflections upon the Eclipse of 1699, printed in the Memoires of the Royal Academy of Sciences for that Year. If You have not yet seen it, tis possible it may suggest something further to You. I shall not trouble You any longer upon this subject.

My Cozen Smith was chosen Fellow the last Eleñ. He takes his Master's Degree this next Commencement. He has already two Pupils & expects one or two more in a short time. He presents his humbl service to You; both He & my self shall be oblig'd to You, if You can assist Him by Your recommendation. I need not tell You, that as he is in all other respects well qualified for that Buissness so he is very capable of instructing his Pupils in some parts of Knowledge which You & I esteem, & which very few Tutors in the University do at all pretend to.

This letter was evidently written to Whiston, who "a little before the famous total eclipse of the Sun, April 22, this year, 1715, published two schemes* of that eclipse," in the latter of which he adopted Cotes's suggestions, though he makes no mention of his receiving any such assistance. "N.B. This most eminent eclipse, 1715, was exactly foretold by Mr Flamsteed, Dr Halley, & myself......I myself by my lectures before; by the sale of my schemes before & after; by the generous presents of my numerous & noble audience; who, at the recommendation of my great friend, the lord Stanhope, then secretary of state, gave me a guinea apiece; by the very uncommon present of twenty guineas from another of my great benefactors, the duke of Newcastle; and of five guineas at night from the lord Godolphin; gained in all about £120. by it." See Whiston's Memoirs i. 204, 5.

---

* The title of the first is " A Calculation of the great Eclipse of the Sun, Apr. 22. 1715, in ye morning, from Mr Flamsteed's Tables, as corrected according to Sr Isaac Newton's Theory of ye Moon in ye Astronomical Lectures......" In the 2nd, which is larger and fuller than the 1st, the Eclipse is calculated " from Sr I. Newton's last improvements to his Theory of ye Moon." (It is dated, April 2, 1715). In the 1st Whiston had neglected to avail himself of the 2nd Ed. of the *Principia*, a fact to which Cotes in the Letter of which we have here only the draught, may possibly have drawn his attention.

Time of Eclipse at London.

| Whiston's 1st Scheme. | | His 2nd | Halley. | Flamsteed. | Observed Time. |
|---|---|---|---|---|---|
| Beginning | 8ʰ. 18′ | 8ʰ. 7½′ | 8ʰ. 7′ | 8ʰ. 8′ | 8ʰ. 6′ |
| Middle | 9 . 24 | 9 . 14 | 9 . 13 | 9 . 13½ | 9 . 10′. 45″ |
| End | 10 . 35 | 10 . 24½ | 10 . 24 | 10 . 24 | 10 . 20 |

## LETTER CXVII.
### COTES TO LORD TREVOR.

My Lord,                    Trinity College Cambr. Jan. 10th 1716

When I waited upon Your Lordship with S[r] Isaac Newton, I remember my Lady Trevor was saying, that S[r] John Bernard was design'd for our College : I have since heard that He will come to us very soon. I have not been inform'd whether any Tutor is already provided for Him. If Your Lordship is not yet determin'd, I beg leave to propose one to You, His name is Smith, a Junior Fellow of the College. I have had the oportunity of an intimate knowledge of His Temper Behaviour & Learning, as He has been my Chamber-fellow for some yeares & as He is my Kinsman. I can therefore be bold to recommend Him to You as a person whom I think to be extraordinarily well qualified to satisfie Your expectation in all respects. If You desire to have S[r] John instructed in the Mathematicks & the new Philosophy : I do assure Your Lordship, I know no one more capable of doing it with good success, both on account of His very great skill in those things & His easy way of teaching. Your Lordship was formerly pleas'd to desire me to assist M[r] Trevor * that way : I was very sorry I might not do Your Lordship that service, for it was not my fault that I did not. The remembrance of it makes me beleive You have the same views for S[r] John : I therefore thought it my duty as well to Your Lordship as to my Kinsman to write thus to You. If the appointment of a Tutor shall be left to D[r] Bentley ; I know His opinion of M[r] Smith is such, that He will think He cannot serve Your Lordship more, than by naming Him to You

I am &c.

R C

---

* Lord Trevor's eldest son and successor in the title. He was entered a fellow-commoner at Trinity College, June 19, 1708, his tutor being Mr Nic. Clagett, Librarian of the College, afterwards Dean of Rochester, and Bishop of St David's, from whence he was translated to Exeter.

The application made in this letter was successful, but before the formal result of it was realised, the warm heart that dictated it had ceased to beat, and the grave had parted the two chamber-fellows. Sir John Bernard was entered a Nobleman under Smith, July 6, 1716. Cotes breathed his last on June 5.

Lord Trevor was one of the twelve peers created by Queen Anne in order to turn the balance in the House of Lords in favour of the peace of Utrecht. He was Chief Justice of the Common Pleas in her reign, but shortly after the accession of George I. (Oct. 1714) he was superseded at the suggestion of Lord Chancellor Cowper, and the appointment was bestowed on Sir Peter King. See Lord Campbell's Chancellors IV. 349 note. 592. 593.

He married for his second wife the widow of Sir Robert Bernard, a brother of Mrs Bentley, and thus became step-father to the young baronet Sir John.

On the publication of the 2d Ed. of the Principia, Bentley presented him with a copy of it. Bentley's *Correspondence*, p. 465.

------

## LETTER CXVIII.

### COTES TO ROBERT DANNYE.

Containing an account of the meteor of the 6th of March $171\frac{5}{6}$.

The following is an extract from the Journal Book of the Royal Society. " March 7. $171\frac{6}{7}$. The President in the Chair....... A letter of the late M$^r$ Roger Cotes Math. Professor at Cambridge to the Reverend Mr Robert Dannye {dated March 15, 1716} was produced as communicated by M$^r$ Jurin of Trinity Coll. Cambridge. It contain'd some very remarkable circumstances seen by him in the late wonderful phænomenon seen about a twelve month since, as that about $\frac{1}{4}$ after seven there was a perfect Canopy of Rays ascending from all parts round the Horizon, but no where reaching to it being about 10 or 15 degrees high on the North Side & near forty on the South, continuing in this state not above two minutes during w$^{ch}$ interval several Colours appeared, some fainter & more permanent, others brighter but quickly vanishing, with several other curious remarks. This description being better circumstanced than w$^t$ had before been communicated by most other observers, was thought worthy to be preserv'd in the Transactions." It will be found in the Transactions for May—August 1720. pp. 66-70, and in Smith's Optics (1738) Vol. I. pp. 67-70, and therefore it has not been thought necessary to reproduce it here.

This letter closes Cotes's correspondence in the Trinity College Collection. Among the Macclesfield Letters, however, there is one of a later date, addressed to his friend Jones only a month before his death, in answer to some inquiries respecting the progress of his tables of integrals upon which he was employed. At the beginning of the year he had returned to the subject of the integration of rational fractions, and in this letter he refers exultingly to the success of his researches, animadverting upon a paper of Leibniz, (Leips. Acts, 1702, p. 218) who was unable to integrate $\dfrac{1}{x^4 + a^4}$. The letter is quoted by Smith (Harmon. Mensur. p. 113), and an extract from it is given by him in his account of that work printed in the Phil. Trans. for June—August 1722, pp. 146-148. Leips. Acts, April 1723, pp. 163, 164. One of the expressions which Cotes mentions in this letter as yielding to his method $\left(\dfrac{x^{\frac{r}{q}n-1}}{a + bx^n + cx^{2n}}, \text{ where } q \text{ is some power of } 2\right)$, Taylor sent to Monmort as a challenge from himself to the mathematicians of the continent, without dropping any allusion to the source to which he was indebted for the problem. Monmort transmitted the question to John Bernoulli and Hermann, the former of whom replied (Jan. 1719) by offering to lay Taylor a wager of 50 guineas that he would produce a solution within a stipulated time, but upon condition that he should in his turn propose a problem to Taylor upon the same terms. Taylor at once declined the proposal in a lengthy reply, (*Contempl. Philosoph.* p. 109), but before it came to Bernoulli's hands, that mathematician apprehensive, he says, lest his silence should be construed by some austere Englishmen (quidam ex severioribus Anglis) into an acknowledgment that the problem was beyond the strength of foreign analysts, had sent his solution, which he had soon hit upon, for insertion in the Leipsic Acts (Leips. Acts, June 1719, p. 256. Bernoull. Opp. II. 402). Hermann's solution appeared in the Acts for August, p. 351.

If an early death had not put an abrupt stop to his investigations, Cotes would no doubt have removed the restriction with respect to the value of $q$ in the expression given above. His example, however, stimulated Demoivre to make the attempt, which was at last crowned with success. See *Miscellanea Analytica*, Lond. 1730. Taylor says, (see Letter cxx, and *Contempl. Philos.* p. 113.) that he himself could prove the *possibility* of the integration.


### END OF COTES'S CORRESPONDENCE

# LETTERS OF TAYLOR TO PROF. SMITH.

Brook Taylor (born 1685, died 1731) was entered a fellow-commoner at St John's College, Cambridge, in 1701, and took the degree of LL.B. in 1709, LL.D. in 1714. Treatises on the Differential Calculus have made his name familiar to many who can write out his Theorem without having any very precise idea of the personality of the discoverer of it. A life of him, prefixed to his tract *Contemplatio Philosophica*, was printed in 1793 by his grandson Sir W. Young. At the time when he wrote the following letter he was Secretary of the Royal Society, though, about a month before, he had sent in his resignation of the office to his brother-secretary Halley (*Contempl. Philosoph.* p. 103). On Dec. 1, Machin was appointed to succeed him. Before the letter was sent off, it was read at the weekly meeting of the Society. "Nov. 27, 1718. The President in the chair. D$^r$ Taylor read a letter he had drawn up for M$^r$ Smith, Professor of Astronomy in Cambridge, requesting him to communicate some curious discoveries in Geometry made by the late M$^r$ Cotes his predecessor & kinsman." *Journal Book.*

## LETTER CXIX.

### BROOK TAYLOR TO PROF. SMITH.

Sir

When I last saw your most excellent Predecessor M$^r$ Cotes I was so very much pleased with the account he gave me of some Mathematical Tracts he had thoughts of obliging the Publick with, particularly a Sett of Tables for the Squaring of Curves by the Measures of Ratio's & Angles, that I have not been able to forbear very frequently mentioning of them, and expressing my wishes that I might soon see them made publick. All Lovers of Mathematical Learning do heartily joyn with me in this, particularly the Royal Society is so sensible of the great usefulness of those Tables, that they have been pleased to order me to take this occasion to let you know that they shall think themselves very much obliged to you by the speedy publication of them, and shall be very glad to give you any assistance you may have occasion for in the doing of it.

I myself, upon the memory of what M$^r$ Cotes shew'd me, have made some Tables of the same nature, and am presst by some friends to publish them, as a thing they say will make amends for the injury you do the Publick and the memory of M$^r$ Cotes in so long suppressing his Papers. But I can by no means prevail upon myself to do this, being much more desireous to see M$^r$ Cotes's own Tables publisht by you. And I shall be very glad in any manner to assist you in looking over the Papers themselves, and in taking care of the Press, if the convenience of Types should make you think it proper to print them here, and your own affairs should make it inconvenient to you to attend this work wholly your self.

I am

Sir

Your most humble Servant

Norfolk Street                          BROOK TAYLOR
27$^{th}$ Nov$^r$: 1718                          *Secr*

P. S.   If there be any other Papers of M$^r$ Cotes besides the Tables that are fit to be publisht and cannot be conveniently done so soon, the Tables, being a particular thing by themselves, may be printed seperate, leaving those other Papers to a more convenient opportunity.

---

The purport of Smith's answer may be gathered from the following extract from the Journal Book of the Royal Society.

"Dec. 11. 1718. There was read a letter from M$^r$ Smith, in answer to a letter of D$^r$ Taylor written to desire the hastening of the Edition of M$^r$ Cotes his Posthumous papers upon the Quadrature of Curves.

M$^r$ Smith informs the Doctor that those papers are preparing with all convenient speed to be put in the press, & are designed to be printed by Subscription; that the Title of the Book is as follows: Harmonia Mensurarum, sive Analysis et Synthesis per Rationem et Angulorum mensuras promotæ."

## LETTER CXX.

### BROOK TAYLOR TO PROF. SMITH.

Sir

I am very much obliged to you for the account you give me of your design to publish M$^r$ Cotes's Papers, and I am not only most ready myself, but all my acquaintance will do what is in their power to assist you in it. I have given your letter to D$^r$ Halley, and I dont doubt but he will acquaint you with the thoughts of the Royal Society upon it.

The great impatience I am in to see your Book publisht makes me a little concerned that it must depend upon a Subscription. For tho such a Book as this when publisht cannot want purchasers; yet it will be very hard to find a sufficient number of Persons, who have knowledge enough in these studies to think it worth while to interest themselves in a Subscription that may turn to any account. And tho what you propose of having no money paid down, & the price being sett by the Vice chancellor, be very fair and easy to the Subscribers; yet there are a great many Persons who will not care to subscribe without knowing beforehand what will be the charge. In this I dont only write my own sentiments, but also those of M$^r$ Jones, who is the best acquainted with affairs of this nature of any one I know, & whose character you can be no stranger to. He had a correspondance with M$^r$ Cotes upon this Subject, and would particularly be glad to do you any service in this matter. Upon account of what I have said I wish you could rather think of getting the Book publisht at the Charge of the University, or some other way. Perhaps the Royal Society would be inclined to do it. And it may be tried whether there may not be some encouragement got from the E. of Caernarvan. What ever be your resolution I will do you all the Service

I can in it.    Particularly I will endeavor to get encouragement from abroad by the Correspondance I have.
Tho' I must be so just as to tell you that M$^r$ Cotes is but
little known among the Foreigners.    His Logometria is
out of their Tast, (in short none of them have judgement
enough to know how to esteem it,) & his Preface to the
Principia is a prejudice to his disadvantage with them.
Yet I dont doubt but the newness of the design will make
them purchase the Book when it is out.

I believe I can do all that M$^r$ Cotes has done in his
Tables; for I can demonstrate that any Curve may be
squared by Measures of Ratio's and Angles, whose Absciss

being $x$, the Ordinate is in this form $\dfrac{x^{\frac{\delta}{\lambda}\eta-1}}{e+fx^\eta+gx^{2\eta}+hx^{3\eta}\ \&c}$,
where $\eta$ is any index, & $\delta$ & $\lambda$ are any whole numbers affirma-
tive or negative, & the denominator $e+fx^\eta+gx^{2\eta}+hx^{3\eta}$ &c
consists of any number of terms.    You know very well
that the irrational forms depend upon the rational ones.
I have a different way from M$^r$ Cotes's[*], and something
more simple, of supplying the defect in Sir Is: Newton's
6$^{th}$ form.    I shall be very ready and glad to communicate
to you any thing that I know in these matters that may
render your Book the more compleat.    I believed it might
be some Service to the general design of it to have Tables
of Natural Logarithms and Arcs answering to the Tangents,
when the Radius is unite; wherefore I have wrote to M$^r$
Sharp at Little Horton near Bradford in Yorkshire, to
know if he will undertake to make them.

I desire you will direct to me in Norfolk Street, and

---

[*] Given in his letter of May 5, 1716 to Jones, quoted p. 230 *antea*, which Taylor
appears to have seen since writing the letter of Nov. 27, Smith having probably alluded
to it in his answer.    Newton's 6th form (in his *De Quadratura Curvarum*) com-
prises the integrals of two expressions equivalent to $\dfrac{1}{a+bx^2+cx^4}$ and $\dfrac{x^2}{a+bx^2+cx^4}$, in
the case where $b>2\sqrt{ac}$ and $a$, $b$, $c$ have all the same sign.

not to Crane Court, because the Servants there neglect bringing me letters, and I am very seldom there.

I am

Sir

Your most humble Servant

Norfolk Street                                     BROOK TAYLOR
11 Dec<sup>r</sup> 1718

The following extracts from the Journal Book of the Royal Society will contribute to complete the history of the publication of the *Harmonia Mensurarum.*

"Dec. 18. 1718. The President acquainted the Society that D<sup>r</sup> Bentley informed him that 100 Subscriptions were already procured for printing M<sup>r</sup> Cotes's Posthumous Works."

"Apr. 26. 1722...M<sup>r</sup> Smith...made the Society a present of his Edition of the Mathematical Works of the late M<sup>r</sup> Cotes...M<sup>r</sup> Smith was ordered thanks for this present."

---

Among the Lucasian MSS. there are three letters from Taylor to Keill (packet No. 3). The 1st dated 17 July, 1717, contains a critique upon Stirling's *Lineæ Tertii Ordinis Neutonianæ.* The following Postscript is added. "Pray do me the favor to put M<sup>r</sup> Innys in mind to send me the Leipsic Acts, & two copies of Sir Is: Newton's Opticks, as soon as it is out, one bound, & another in sheets, which I must send to M: Monmort."

The 2nd (26 Apr. 1719) contains the answer of Nic. Bernoulli of Padua (John's nephew) to a message which Keill had sent to him through Taylor and Monmort. Taylor says he can hardly prevail upon himself to forward it, "it is so disagreable." As two of the points referred to in it relate more or less to our philosopher, we may possibly be excused for giving it a place here. It is couched in the following language. "J'accepte la promesse de M. Keil qui est de me donner 5 pistolles pour chaque mensonge dont je le pourrai convaincre. Si donc M. Keil tient sa parole je gagnerai au moins 20 pistolles car je soutiens qu'il ne pas dit la verité 1°. lorsqu'il a dit que depuis mon sejour a Londres J'avois publié le contraire de ce que M. Newton m'avoit demontré {Cf. p. 142, note}. 2°. lorsqu'il a dit qu'on a oublié par une faute d'Impression le mot *ut* dans le Scholium qui est a la fin du traitté de quadraturis. 3°. lorsqu'il a dit que mon oncle (je passe sous silence ce qu'il dit de moy dans le meme endroit) n'entend pas le calcul differentiel. 4°. lorsqu'il a dit nouvellement dans sa lettre a M. Taylor qu'il peut me shew me lyes I have made for nothing. Je vous

prie de luy faire notifier ces pretentions, & d'en demander sa reponse."
The last paragraph of the letter opens with the words " Since I have
heard nothing from you in answer to my proposal of joyning with you
again‿t Bernoulli I have drawn up a paper*, which I think soon to
publish by itself."

The 3rd (26 Aug. 1721) begins thus: " The enclosed is just come
to me from Abbé Conti, who desires me to convey it to you. He tells†
me that he disputes continually with the French in favor of Sir Isaac
Neuton and the English Mathematicians; but that he can by no means
make them sensible of the true nature of Sir Isaac's method, they not
yet rightly understanding what he means by first and last ratios of
nascent and evanescent quantities......I shall trouble you with no more
at present, not knowing how unwelcome this little may be to you from
me, upon account of what Bernoulli has publisht‡ out of my letters to
Monmort in hopes to provoke your resentments against me." Taylor
then enters into an elaborate explanation of the offensive expression, in
the course of which he lashes Monmort for "betraying so private a
letter as that was," and Bernoulli for publishing it. The apology
seems to have come too late. The letter bears the London post-mark
of Aug. 28, and would therefore reach Oxford on the 29th, the day on
which poor Keill died. The address is crossed.

François-Marie Arouet (Voltaire) born 1694, died 1778.

## LETTER CXXI.

### VOLTAIRE TO PROF. SMITH.

Sʳ

I have perus'd yʳ book of optics, I cannot be so
mightily pleas'd with a book, without Loving the author,

---

* *Apologia D. Brook Taylor...contra...J. Bernoullium.* (It is a reply to the charge
of plagiarism brought against him in the " Epistola pro Eminente Mathematico"
Leipsic Acts, July 1716). *Philosoph. Trans.* March—May 1719, p. 955. Jo. Bernoulli
Opp. ɪɪ. 478. It was shewn by Jones to Newton before publication. See Taylor's
letter to Jones, *Macc. Corr.* ɪ. 279. Keill was already employed on his own account
on his *Epistola ad...Jo. Bernoulli.* See p. 187, *antea.*

† See Conti's letter to Taylor, (May 22, 1721), *Contempl. Philos.* p. 124.

‡ In *Jo. Burchardi...Epistola ad...Taylor* (Leipsic Acts for May 1721, pp. 195—228.
Jo. Bernoulli Opp. ɪɪ, pp. 483—512), a reply to Taylor's *Apologia.* The words more
especially referred to are as follows: " Entre nous, je suis un peu de l'avis de Mr.
Bernoulli que Mr. Keill is better qualified for a Champion than for an Analyste."

give me leave to submitt to y$^r$ judgement these little answer of mine, w$^h$ich I have writ against some ignorant ennemies of S$^r$ Isaac, Neuton, whom you follow so closely in the path of truth and glory,

I am

S$^r$

Y$^r$ most humble obed

{*Hôtel de Brie, rue Clôche-*               Servant VOLTAIRE.
*Perche*} *Paris the* 10$^{th}$ *of*
*October* {1739} *new stile,*

   M$^r$. SMITH

This letter was written during a short visit which Voltaire made to Paris. He had run up from Brussels in September, purposing to stay about a month in what he calls the worse thah Cartesian *tourbillons* of the French capital, but on the day of his intended departure he had an attack of illness which detained him until the end of November. In a letter, written the day after the date of the one before us, he describes the plight he was in between his two medical attendants (" on me saigne, on me baigne "). Under these circumstances, added to long disuse of the language, we need not be surprised to find his English not quite so good as when he wrote a dozen years before during his residence in this country.

The "little answer" is his "Réponse aux objections principales qu' on a faites en France contre la philosophie de Newton," 8vo. Amsterdam, 1739 (a defence against the attacks that had been made upon his *Elémens de la Philosophie de Newton*... 1738, and against misconceptions on some points in the Newtonian philosophy). The following allusions to this tract occur in his Correspondence. Writing to Prince Frederic of Prussia, " the Solomon of the North," in September, shortly after his arrival in Paris, he says, " Il a fallu d'abord, en arrivant, répondre à beaucoup d'objections que j'ai trouvées répandues à Paris contre les découvertes de Newton. Mais ce petit devoir dont je me suis acquitté ne m'a point fait perdre de vue ce *Mahomet* {his tragedy} dont j'ai déja eu l'honneur d'envoyer les prémices à votre altesse royale. Voici deux actes à-la-fois." In a letter to Helvetius, dated a week previous to this letter to Smith, he writes, " Je ne sais comment je m'y prendrai pour envoyer une courte et modeste réponse que j'ai faite aux anti-newtoniens. Je suis l'enfant perdu d'un parti dont M. de Buffon est le chef, et je suis assez comme les soldats qui se battent de bon cœur sans trop entendre les intérêts de leur prince."

Voltaire's " Elémens de la Philosophie de Newton, mis à la portée de tout le monde" (the eight last words were added by the booksellers) issued from the press at Amsterdam in April, 1738, without his know- ledge. The impatience of the booksellers could not wait for his recovery from a fit of sickness, or for the alterations that he wished to make in the work, and they employed another hand to complete it by finishing the 23rd chapter, and writing two additional chapters (the 24th and 25th). The book was reprinted at Paris (with a London title-page) the following July, accompanied with "éclaircissements" and a 26th chapter on the tides, supplied by Voltaire: these he also sent to the Dutch corsairs (as he denominates the booksellers) to be circulated with their edition. Before leaving Paris, in November, 1739, he tells Fre- deric that a new edition was called for, and he republished the work in an enlarged and otherwise altered form (1741)*, with flattering re- ferences to Smith's Optics (see, for example, the explanation of the sun or moon appearing larger on the horizon than on the meridian, Part 2, ch. VIII. "le docteur Smith a la gloire d'avoir enfin trouvé la solution compléte d'un problème sur lequel les plus grands génies avai- ent fait des systèmes inutiles"). *Journal des Savants*, 1738. *Biblio- thèque Française*, 1738, 1739. Voltaire's *Correspondance*. His Life in *Biogr. Univ.* (Beuchot's note). Beuchot's *Voltaire*, tom. 38.

In a letter, written from Leyden in Feb. 1737, Voltaire says, "Je pars incessamment pour achever à Cambridge mon petit cours de new- tonisme:" (he had been studying the Newtonian philosophy for some weeks under 's Gravesande at Leyden, where he had taken shelter from the storm that burst upon him on the appearance of "Le Mondain"). But the announcement was intended only as a blind to his enemies. He in reality returned to his retreat at Cirey, in Champagne. Some of his biographers state that his letters at this time were dated from Cambridge, but there are no letters so dated in his published Corre- spondence.

---

William Augustus, son of George II. born 1721, died 1765.

### LETTER CXXII.

#### DUKE OF CUMBERLAND TO PROF. SMITH.

{July 3. 1740}.

Doctor Smith I desire you would lose no time in pro- viding a Sea Quadrant and Telescope for to fit my eye;

---

* Lalande also mentions an edition in the following year.

my baggage goes at five this afternoon; I shall be extreamly obliged to you.

WILLIAM

Endorsed by D<sup>r</sup> Smith. " The Duke of Cumberlands Note to me."

This note was probably written by the future "butcher," when he was on the point of setting out to join the squadron under Sir John Norris, which was supposed to be destined for an attack upon the Spanish fleet in Ferrol. " Friday, July 4. 1740. The Duke of Cumberland who had been some time at his post in the camp at Hounslow {he was Colonel of the Coldstream Guards} left it on a sudden, and arrived at Portsmouth unexpected," where he " went aboard the Victory Man of War as a Volunteer."—*Gentleman's Magazine*, July 1740. The *London Evening Post* states that he set out from St James's for Portsmouth at 4 in the morning. The weather proving unfavourable, the Admiral and the young Volunteer returned to London in September.

The Duke was now turned 19. Smith had been in attendance upon him since June, 1739 (*Conclusion Book*, June 11).

# APPENDIX.

Henry Oldenburg, born 1626 at Bremen, died 1677, Secretary of the Royal Society. He was a friend of Milton's.

## No. I.

### OLDENBURG TO NEWTON.

#### Beginning of their Correspondence.

Accompanying this letter were a figure and description in Latin of the reflecting telescope made by Newton the preceding autumn and sent up "for the King's perusal" in December. See Syn. View of Newton's Life under the year 1671.

S$^r$

Your Ingenuity is the occaon of this address by a hand unknowne to you. You have been so generous, as to impart to the Philosophers here, your Invention of contracting Telescopes. It having been considered, and examined here by some of y$^e$ most eminent in Opticall Science and practise, and applauded by them, they think it necessary to use some meanes to secure this Invention from y$^e$ Usurpaon of forreiners; And therefore have taken care to represent by a scheme that first Specimen, sent hither by you, and to describe all y$^e$ parts of y$^e$ Instrument, together w$^{th}$ its effect, compared w$^{th}$ an ordinary, but much larger, Glasse; and to send this figure, and description by y$^e$ Secretary of y$^e$ R. Soc. (where you were lately by y$^e$ L$^d$ B$^p$. of Sarum proposed Candidat) in a solemn letter to Paris to M. Hugens*, thereby to prevent the arrogation of such strangers, as may perhaps have seen it here, or even w$^{th}$ you at Cambridge; it being too frequent, y$^t$ new Inventions and contrivances are snatched away from their true Authors by pretending bystanders;

---

* As Oldenburg had promised in a letter to Huygens, Jan. 1. *Letter Bk. Roy. Soc. v. 92.*

But yet it was not thought fit to send this away w$^{th}$ out first giving you notice of it, and sending to you y$^e$ very figure and description, as it was here drawne up*; y$^t$ so you might adde, & alter, as you shall see cause; w$^{ch}$ being done here w$^{th}$, I shall desire your favour of returning it w$^{th}$ all convenient speed, together w$^{th}$ such alterations, as you shall think fit to make therein.

Though divers of y$^e$ most skillfull examiners agreed y$^t$ your Tube magnifyed, by measure, y$^e$ object here represented by $A$†, so much, as you see, above w$^t$ a much greater Telescope did; yet there were others, well versed also in Optic glasses, y$^t$, though they could not disprove that mensuraon, yet were positive to affirm, y$^t$ y$^t$ excesse of magnitude did not appeare such to their eye.

Besides it was discoursed, y$^t$ by this way of yours it was longsome, & difficult to find y$^e$ Object: w$^{ch}$ inconvenience yet they looked upon as possible to be remedied. I shall be glad, S$^r$, to receive your speedy answer to these lines, and embrace all occasions to expresse my singular respects to your merit, as becomes

S$^r$

Your humble Servant

Jan. 2. 167$\frac{1}{2}$.              OLDENBURG‡.

Newton's answer, dated Jan. 6, will be found in Macc. Corr. II. 311, and (not complete) in Birch, III. 2, Horsley, IV. 271. Comp. Syn. View under that date.

---

No. II.

NEWTON TO OLDENBURG.

S$^r$              Cambridg March 16$^{th}$ 1671{2|

The book w$^{ch}$ my Carrier by forgetfulnesse disappointed me of the last week I have now received & thank you

---

* *Orig. Lett. Bk.* Roy. Soc. N. 1. 37. Horsley IV. 270.
† This is fig. 2. Tab. I. *Phil. Trans.* March 25, 1672. Or, see Horsley IV. fig facing p. 280.
‡ From a copy corrected by Oldenburg (*Orig. Lett. Bk.* Roy. Soc. O. 2. 64).

16

for it. With the Telescope w^{ch} I made I have sometimes seen remote objects & particularly the Moon very distinct in those p^{ts} of it w^{ch} were neare the sides of the visible angle. And at other times when it hath been otherwise put together it hath exhibited things not w^{th}out some confusion. W^{ch} difference I attributed chiefely to some imperfection that might possibly be either in the figures of y^e metalls or eye glasse, & once I found it caused by a little tarnishing of the Metall in 4 or 5 days of moist weather.

One of the ffellows of o^r College is making such another Telescope w^{th} w^{ch} last night I looked on Jupiter & he seemed as distinct & sharply defined as I have seen him in other Telescopes. When he hath finished it I will examin more strictly & send you an account of its performances, ffor it seemes to be something better then that w^{ch} I made.

Yo^r humble servant

*These*
*To* HENRY OLDENBURG *Esq: at his house*
  *about the middle of the old Pall-mail*
  *in Westminster. London*

I. NEWTON*

"rec. March 18. 71"<br>In Oldenburg's hand.

---

No. III.

NEWTON TO OLDENBURG.

March 19. 1671|2|.

*After describing the performances of the instrument mentioned in the last letter he proceeds :*

This may be of some use to those that shall endeavour any thing in Reflexions; for hereby they will in some measure be enabled to judge of the goodness of their Instruments. And for this end you may annex these observations made with this last instrument to the de-

---

* *Orig. Lett. Bk. Roy. Soc. N. 1. 35.*

scription of it in the Transactions of this month.   But my
answer to M^r Hooks observations will not be ready for
them, because I intend to annex to that answer some
further explications of the Theory which I shall not have
leisure to do this week or fourtnight.

S^r

I am in hast

Yo^r faithfull Serv^t

I. NEWTON*

Endorsed by Oldenburg:
"Rec^d. 20. Ans^d. 23 comm{unicating}
y^e Comet and * sub cap. Cygni from
Hevel." See *Phil. Trans*. March 25,
1672, p. 4017.

---

No. IV.

NEWTON TO OLDENBURG.

S^r                                           March 26^th. 1672

About 10 days since at night I saw a dull starr south
west of Perseus, which I now take to have beene that
Comet of which you give me information ; But it was very
small & had not any visible tayle which made me regard
it noe further, & I feare it will now bee difficult to
find it†.

Since my last letter I have further compared the two
telescopes &c.  (*See Phil. Trans. Apr.* 22. 1672 *p.* 4032.)

*     *     *     *     *

Thus much of these Telescopes, & at present I shall
trouble you no further then to thanke you for your last
intelligence, by which you have obliged

S^r

Your faithfull servant

I. NEWTON‡.

---

* *Orig. Lett. Bk.* Roy. Soc. N. 1. 36.  For the first part of the letter see *Phil. Trans.* March 25, 1672, p. 4009, where "considerable" is printed by mistake for "insensible."

† *Phil. Trans.* March 25, 1672, p. 4018.

‡ *Lett. Bk.* Roy. Soc. v. 187. Horsley iv. 275.

## No. V.

### NEWTON TO OLDENBURG.

"Asserting the advantage of reflecting telescopes above refracting ones, & endeavouring to remove some inconveniences in the former." *Lett. Bk.* Roy. Soc. v. 193.

S[r]                                           March 30. 1672

I doubt not but Mons[r]. Auzout &c.   (*See Phil. Tr. Apr.* 22. 1672. p. 4034).

\*        \*        \*        \*        \*

In the meane time to remedy in some measure these inconveniences, I shall propound a way \* of using, instead of the little ovall metall, a glass or crystall figured like a triangular Prism, as you see it represented in the first scheme by the figure *ABc*.   It's side

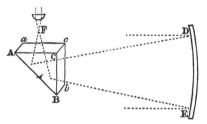

*ABba* I suppose to performe the office of that metall by reflecting towards the eyeglasse the light which comes from the concave *DE* : which light I suppose to enter into this Prism at its side *CBbc*, & after reflexion to emerge at the side *ACca* before it convene at *F*, the focus of the glasse.   The axes of the eyeglasse and concave metall must be perpendicular to the midle of the planes *ACca* and *CBbc*.   And least any colours should be produced by the refraction of those planes, 'tis requisite that the angles of the Prism at *Aa* & *Bb* bee precisely equall : which may most conveniently be performed by making them halfe right angles & consequently the third angle at *Cc* a right one.   The plane *ABba* without being foliated will reflect all the light incident on it ; Especially if the Prism be made of Crystall.   But to exclude all unnecessary light, 'tis convenient that it bee all over covered with some blacke substance, excepting two circular spaces

---

\* Comp. *Optics*, Book I. Part 1. Prop. VIII.

of the planes $Ac$ & $Bc$ for the usefull light to passe through, as you see it designed in the 2ᵈ scheme. The length of this Prism should bee such, that its sides $Ac$ & $Bc$ may be four-square, and so much of the angles $B$ & $b$, as are superfluous, ought to bee ground off, to give passage to as much light as is possible from the object to the concave.

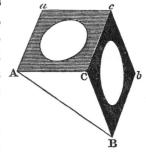

There is one very considerable advantage of this Prism, which the ovall metall is not capable of, without using two eye-glasses, and it is, that if its sides $ACca$ & $BCcb$ bee ground convex, it will erect the object by performing the office of a double convex lens. The manner you have expressed in the 3ᵈ scheme; where suppose $G$ to be the focus of the concave, and $F$ of the eye-glasse at which the rays crosse twice before their arrivall at the eye. But it is convenient, that the first tryalls bee made with Prisms whose sides are all of them plane. And thus much concerning Monsʳ Auzout's considerations.

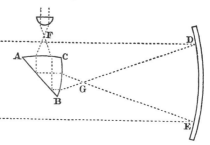

To the queries of Monsʳ Denys I answer, 1. That a Tube of six inches is capable of bearing an aperture (limited next the eye) so large, that an obstacle of $1\frac{1}{4}$ or $1\frac{1}{3}$ of an inch in breadth shall be requisite to intercept all the light coming from one point of the object towards the concave metall: But it is convenient, that the Tube bee a little wider than that aperture precisely requires, suppose $1\frac{1}{2}$ or $1\frac{2}{3}$ of an inch, & not more; And the whole breadth of the metall should not bee lesse than two inches, because its figure to-

wards the edges will scarcely bee so true as to bee usefull.
And by that meanes it may also bee conveniently fastened
to the end of the Tube on the outside, so as at pleasure to
bee taken off & layd up close from the Air, to preserve it
from tarnishing.

How the Diameter of the Tube is to bee enlarged according to its length, will appeare by the Table of Aperturs
and charges which I sent you in my last letter of March
the 26th. Namely the Cube of its length should be proportionable to the square-square of its diameter or aperture at
the metall; so that the advantage of augmenting the length
of Tubes is by this way far greater than by refractions,
where their length ought to bee proportionall to the square
of the diameter of the aperture.

2. The breadth or shortest diameter of the little ovall-
metall for a Tube of six inches should not bee greater than
$\frac{1}{3}$, nor lesse than $\frac{1}{4}$ of an inch; And the longest Diameter
should bee to the shortest as about 10 to 7. But you may
more exactly determine these diameters for Tubes of all
lengths after this manner; In the 4th figure let $AB$ repre-

sent the ovall sett edg-
wise; $DE$ the concave;
$FG$ its axis; $Gp$ the
reflex of that axis; $st$
the Diameter of the
hole through which the
light is transmitted to
the eye; & $P$ the cen-
ter of that hole.  Pro-
duce $FG$ to $\pi$, so that

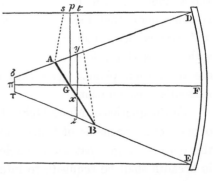

$G\pi$ may bee equall to $Gp$; erect $\pi\sigma$ & $\pi\tau$ equall to $ps$ & $pt$,
& from $\sigma$ & $\tau$ draw two lines, $\sigma D$ & $\tau E$, to the utmost parts
of the concave, wthin the Tube intersecting $AB$ in $A$ & $B$; &
$AB$ shall bee the long diameter of the ovall; which bisect
in $x$, & perpendicular to $Fx$ erect $xy$ & $xz$ occurring with $\sigma D$

& $\tau E$ in $y$ & $z$, & a meane proportionall between $xy$ & $xz$ doubled shal be the other short diameter: ffor, by viewing $y^e$ scheme you will easily perceive, that an ovall, described with those rectangular conjugate diameters, is of sufficient bignesse to reflect all the usefull light towards the eye, if it be rightly placed in the Tube; & a broader metall would not onely intercept too many of the best rays, but some of the scattering light, reflected every way from its superfluous parts, would fall on the eye-glasse & make the object appeare something confused & as it were in a mist. This, S$^r$, is that, which in answer to your letter my present thoughts suggest to

Your faithfull Servant

I. NEWTON*.

---

No. VI.

NEWTON TO OLDENBURG.

S$^r$                                  Cambridge April 13. 1672.

I herewith send you an answer to the Jesuite Pardies Considerations; in the conclusion of which you may possibly apprehend me a little too positive, but I speake only for myselfe. I am highly sensible of your good will in communicating to me such observations as occurr concerning my Theories or Catadioptricall instruments, and I desire you to continue that favour to me. I shall immediately proceed to add what I promised to my answer to Mr. Hooks observations, & then send it you. Mons$^r$ Hugens has very well observed the confusion of refractions near the edges of a lens, where its two superficies's are inclined much like

---

* *Lett. Bk.* Roy. Soc. v. 193. This and some other letters have been printed by Horsley (Vol. IV.) from the MSS. at the Royal Society, but not so as altogether to supersede the necessity of their reappearance here in a more complete and accurate form.

the planes of a prisme whose refractions are in like manner confused. But it is not from the inclination of those superficies so much as from the heterogeneity of light that that confusion is caused: ffor by illuminating an object with homogeneall light, I have seen it far distincter through a Prism than I could by light that was heterogeneal.

I suppose, the designe of S$^r$ Robt Moray's experiments is &c. (*See Phil. Tr. May* 20. 1672. p. 4060).

*    *    *    *    *

Thus far concerning S$^r$ R$^t$ Morays proposalls. I have nothing more at present unlesse to desire you, that in y$^e$ letter wherein I sent you the Table of apertures and charges you would change an expression concerning the six foot Tube where I intimated that it was none of the best in its kind. ffor least the friend, of whom it was borrowed, should thinke I depreciate it, I had rather that the expression should be a little intimated after this manner; that I am not very well assured of its goodnesse, & therefore desire, that the other experiment of reading at 100 foot distances should rather be confided in. You will do me a favour to peruse the rest of that letter also before you commit it to the presse. ffor I writ it in so much hast, that I had no time to review it: And by rendring my expressions more perspicuous or lesse ambiguous you will still oblige

Your faithfull Servant
I. NEWTON*

<hr />

No. VII.

NEWTON TO OLDENBURG.

S$^r$ June 11$^{th}$ 1672.

I have sent you my Answers to M$^r$ Hook & P. Pardies, w$^{ch}$ I hope will bring with y$^m$ y$^t$ satisfaction w$^{ch}$ I promised.

<hr />

And as there is nothing in M$^r$. Hooks Considerations w$^{th}$ w$^{ch}$ I am not well contented, so I presume there is as little in mine w$^{ch}$ he can excep{t} against, since you will easily see that I have industriously avoyded y$^e$ intermixing of oblique & glancing expressions in my discourse. So y$^t$ I hope it will be needlesse to trouble the R. Society to adjust matters. However if there should possibly be any thing esteemed of y$^t$ kind, I desire it may be interpreted candidly & with respect to the contents of M$^r$ Hooks Considerations, & I shall readily give way to y$^e$ mitigation of whatsoever y$^e$ Heads of y$^e$ R. Society shall esteem personall. And concerning my former Answer to P. Pardies, I resigne to you y$^e$ same liberty w$^{ch}$ he hath done for his Objections, of mollifying any expressions that may have a shew of harshnesse.

Yo$^r$ Servant

I. Newton*.

*These*

*To* Henry Oldenburg Esq : *at his house*
*about y$^e$ middle of y$^e$ old Pall-maile*
*in Westmin{s}ter London.*

---

## No. VIII.

### NEWTON TO OLDENBURG.

Cambridg

July 30$^{th}$ 1672

S$^r$

The last week I wrote to you that y$^e$ Metall w$^{ch}$ you sent me was well for closenesse & hardnesse but yet of a colour not very brisque & inclining to red. However if it be less apt to tarnish then any other mixture yet known, that will sufficiently recompense y$^e$ other imperfections. Yo$^{rs}$ of July 16$^{th}$ directed to Stoake is not yet come to my hands. I feare it is miscarried, and desire therefore you would favour me w$^{th}$ y$^e$ particulars w$^{ch}$ were in answer to

---

* *Orig. Lett. Bk. Roy. Soc. N. 1. 39.*

y$^t$ troublesome letter* written last from Stoake, for w$^{ch}$ I begg yo$^r$ pardon. I send you by John Stiles 13$^s$ for the last quarter.

Yo$^r$ humble Servant

NEWTON†

*These*

To HENRY OLDENBURG *Esq: at his house*
*about the middle of the old Pall-Maile*
*in Westminster London*
w$^{th}$ 13$^s$.

"Rec. July 31. 72 Answ. eodem. and repeated y$^e$ contents of my letter of July 16." Mem. by Oldenburg.

---

No. VIII. (bis).

OLDENBURG TO NEWTON.

[Extract.]

Sent in conformity with the wish expressed in the preceding letter.

S$^r$           Lond. July 16. 1672.

I have spoken with Mr Cock about the four foot Tube, which hath been ready a pretty while. He saith that the object-speculum (being a compound of copper, tin, tin-glasse, antimony and a little arsenick) is of about 6 inches diameter, wrought upon a tool of about 14 or 15 foot, and drawing 4 foot, more or less. He adds, that tis very good mettall, shewing the moon very well, but other objects faint; perhaps for want of giving it its due charge. Tis lodged in a square box, with a lid at the end of it, for placing the speculum-plate, lodged in it, at such a distance as shall be requisite. He offers to unpolish this plate again, and to send you this very Instrument for 5$^{lb}$; and what alterations or emendations you shall direct to bee made herein upon triall, hee will make, without demanding

---

* Dated July 13. It is printed in *Gen. Dict.* VII. 782. *Macc. Corr.* II. 332.
† *Orig. Lett. Bk.* Roy. Soc. N. 1. 41.

any more money for that labour. I intend, god permitting, to send by the next conveniency of your Cambridge Carrier, J. Stiles, a piece of that very mettal, with the s$^d$ object-speculum, w$^{ch}$ the 4 foot Telescope is compounded off.

As to the steely Speculum, he saith, tis a pure Venice-Steel, forged with much care; not melted, nor compounded with any thing; of 3 inches diameter, but bearing not so good a polish. And this he is not unwilling to send also to you to Cambridge for your examination, and further directions about it. Hee saith, that tis very hard & tedious to grind this steely matter true*.

---

## No. IX.

### NEWTON TO OLDENBURG.

*For the first part of the letter see Rigaud's Appendix to his Essay, No. VIII, pp. 42, 44, and the Phil. Trans. for July 21, 1673, p. 6087.*

\* \* \* \* \*

Pray w$^{th}$ these Notes return my thanks to M. Hugens for his book.

By a former letter of yo$^{rs}$ I was a little dubious whether M. Slusius might not apprehend, by w$^t$ you wrote to him concerning me, y$^t$ I pretended to his Method of drawing tangents; untill I understood by M. Collins y$^t$ you signified to him y$^t$ you thought it here of a later date. ffor it seems to me that he was acquainted w$^{th}$ it some yeares before he printed his Mesolabum & consequently before I understood it. But if it had been otherwise yet since he first imparted it to his friends & y$^e$ world, it ought deservedly to be accounted his. As for y$^e$ Methods they are y$^e$ same, though I beleive derived from different principles. But I know not whether his Principles afford

---

\* *Orig. Lett. Bk. Roy. Soc. O. 2. 92.*

it so generall as mine w$^{ch}$ extend to Equations affected w$^{th}$ surd terms, w$^{th}$out reducing them to another form. But if you please let this pass.

The incongruities you speak of, I pass by. But I must, as formerly, signify to you y$^t$ I intend to be no further sollicitous about matters of Philosophy. And therefore I hope you will not take it ill if you find me cease from doing any thing more in y$^t$ kind, or rather y$^t$ you will favour me in my determination by preventing so far as you can conveniently any objections or other philosophical letters that may concern me. For your profer about my Quarterly payments I thank you. But I would not have you trouble yo$^r$self to get them excused if you have not done it already. And now being tired w$^{th}$ this long letter, I must in hast write myself

Yo$^r$ humble Servant

Cambridg. June 23. 73.                      I. Newton*.

---

No. X.

Paper given by Newton to Flamsteed at lecture in 1674. It is printed here as exhibiting to us, perhaps in a more vivid manner than his actual lectures, the philosopher descending to the level of an elementary teacher.

I.   ($a$)   $\dfrac{ax}{a-x} + b = x$ per reductionem fit $ax + ab - bx$

$= ax - xx$ seu $xx = bx - ab$.   ($\beta$)   $\dfrac{a^3 - abb}{2cy - cc} = y - c$ fit

$\dfrac{a^3 - abb}{c} = yy - 3cy + cc$   seu   $\dfrac{a^3 - abb - c^3}{c} + 3cy = 2yy.$

($\gamma$) $\dfrac{aa}{x} - a = x$ fit $aa - ax = xx$.   ($\delta$)   $\dfrac{aabb}{cxx} = \dfrac{xx}{a+b-x}$ fit

$\dfrac{a^3bb + aab^3 - aabbx}{c} = x^4.$

---

* *Orig. Lett. Bk.* Roy. Soc. N. 1. 47. The date is in Oldenburg's hand. The part of the letter which we have given here is crossed out in the MS. probably by Oldenburg. The whole of the letter is printed in Horsley iv. 342.

II.  $(a)$  $\dfrac{aa - xx}{a + b} + a = x$  fit  $xx = \dfrac{a}{+ b}x \dfrac{+ 2aa}{+ ab}$

$(\beta)$  $\dfrac{y^3 - aby}{aa + a\sqrt{aa - bb}} + a = \sqrt{aa - bb}$ fit $y^3 - aby + abb = 0$.

III.  $(a)$  $\sqrt{aa - ax} + a = x$ fit $aa - ax = xx - 2ax + aa$
seu $x = a$.  $(\beta)$  $\sqrt{3} : \overline{aax + 2axx - x^3} - a + x = 0$ fit $aax$
$+ 2axx - x^3 = a^3 - 3aax + 3axx - x^3$ seu $xx = 4ax - aa$.
$(\gamma)$  $y = \sqrt{ay + yy - a\sqrt{ay - yy}}$ primo fit $y = \sqrt{ay - yy}$
d$\{$e$\}$in $2y = a$.

IV.  $(a)$  $2y = a$ fit $y = \frac{1}{2}a$.  $(\beta)$  $\dfrac{bx}{a} = a$ fit $x = \dfrac{aa}{b}$.

$(\gamma)$  $ax - cx = ac$ fit $x = \dfrac{ac}{a - c}$.  $(\delta)$  $\dfrac{2ac}{- cc}x^3 \dfrac{+ a^3}{+ aac} xx$

$\dfrac{- 2a^3c}{+ aacc} x - a^3cc = 0$, fit $x^3 + \dfrac{a^3 + aac}{2ac - cc}xx - aax - \dfrac{a^3c}{2a - c} = 0$.*

No. XI.

NEWTON TO OLDENBURG.

Nov. 13. 1675.

*The principal part of the letter is printed in the Transactions for January
24, 1676: the remainder is as follows:*

I have returnd you Mr Line's letter.  It came to my
hands but this week; the Gentleman by whom you sent it
having not yet been at Cambridge but transmitting it to
me from Oxford.

---

* From the original paper in Newton's hand, pasted in at the beginning of Vol. 42
of Flamsteed's MSS. at Greenwich : at the bottom are the words " Mr Newton's paper
given at one of his lectures, Midsummer, 1674." Flamsteed was at Cambridge, from the
end of May until July 13. He brought with him a Royal Mandate for the degree of
M.A. which was conferred upon him on June 5. He had been admitted a pensioner at
Jesus College Dec. 21. 1670, during a short stay he made at Cambridge on his return
from London to Derby, when he also took the opportunity of calling upon Barrow and
Newton.  Comp. Baily, p. 29.

I, III and IV (except $\gamma$) will be found in the published *Algebra Lectures* (Lect.
6 and 7), Regg. 3, 4, 5 pp. 65—67.

I had some thoughts of writing a further discours about colours to be read at one of yo$^r$ Assemblies, but find it yet against y$^e$ grain to put pen to paper any more on y$^t$ subject. But however I have one discourse by me of y$^t$ subject written when I sent my first letters to you about colours & of w$^{ch}$ I then gave you notice. This you may command w$^n$ you think it will be convenient if y$^e$ custome of reading weekly discourses still continue*. In y$^e$ meane while I am S$^r$

Yo$^r$ humble Serv$^{nt}$

Is. NEWTON†.

---

No. XII.

NEWTON TO OLDENBURG.

S$^r$                                 Cambr. Novemb 30 1675.

I intended to have sent you y$^e$ papers this week but upon reviewing them it came into my mind to write another little scrible‡ to accompany them : You may expect 'em y$^e$ next week. An ancient Gentleman I met at yo$^r$ Assemblies (whose name I cannot recollect,) being thick of hearing desired me to inquire after y$^e$ form of Mr Mace's Otocousticon a Musitian here; but he has not been in town since I came from London, but is somewhere in London about printing a book of Musiq:‖. Yet y$^e$ last week I had opportunity to inquire after it of his son & he

---

* "Mr Oldenburg was ordered to thank him for this offer, and to desire him to send the said discourse as soon as he pleased." Birch, III. 232.

† *Orig. Lett. Bk.* Roy. Soc. N. 1. 48.

‡ "An Hypothesis explaining the properties of light, discoursed of in my several papers." Birch, III. 248.

‖ "*Musick's Monument*," &c. &c. Lond. 1676. Newton's name appears in the list of subscribers to the work. Thomas Mace was one of the ' Clerici' or Singing Men of Trinity College for more than 70 years (1635—1706). Comp. Burney's *Hist. of Mus.* Vol. 3. Southey's *Doctor*, chapters 193-196. Cooper's *Annals of Camb.* under year 1690.

tells me the form is this. *A* y$^e$ smal end to put into y$^e$ ear

*BC* y$^e$ length sup-
pose two foot *CD* y$^e$
wide end suppose
about eight inches
over. The tube *BDC*
tapers all y$^e$ way

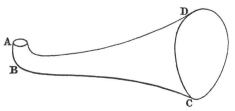

almost eavenly like a cone only at y$^e$ great Orifice *CD*
widens more, like y$^e$ end of a Trumpet. He has of
several sizes. The biggest do y$^e$ best. If you can't
recollect who y$^e$ Gentleman may be I suppose M$^r$ Hill can
tell you, for I think M$^r$ Hill was by when y$^e$ Gentleman
spake to me, & y$^e$ Gentleman desird me to write to either
M$^r$ Hill or you about it.

Yo$^{rs}$ in hast

Is. NEWTON*

*For* HENRY OLDENBURG Esq: *at his*
*house about* y$^e$ *middle of* y$^e$ *Old*
*Pal-mel in Westminster London.*

---

No. XIII.

NEWTON TO OLDENBURG.

S$^r$

I hope M$^r$ Linus's ffriends will acquiesce in y$^e$ late tryall
of y$^e$ Exp$^t$ in debait†, for y$^e$ procurement of w$^{ch}$ & for sending them notice of y$^e$ event, I return you my hearty thanks,
as I have reason. I perceive I went upon a wrong suppo-

---

\* *Orig. Lett. Bk. Roy. Soc. N. 1. 49.*

† i.e. The Experiment on the Solar spectrum. "Apr. 27. The Experiment of Mr
Newton which had been contested by Mr Linus and his fellows at Liege, was tried
before the Society, according to Mr Newton's directions, and succeeded, as he all along
asserted it would do : and it was ordered, that Mr Oldenburg should signify this success
to those of Liege, who had formerly certified, { by a letter, Dec. 15, 1675 } that if the
experiment were made before the Society, and succeeded according to Mr. Newton's
assertions, they would acquiesce." Birch, III. 313. Linus had maintained that the
sun's image was round, and the colours arranged parallel to the axis of the prism.

sition in what I wrote concerning M$^r$ Boyles Exp$^t$.  The Papers in yo$^r$ hand I have no present need of: You may send them at yo$^r$ best leisure.  Sometime this Sommer it's possible I may make use of them, if I can but get some time to write y$^e$ other discourse about y$^e$ colours of y$^e$ Prism w$^{ch}$ I have long intended.    S$^r$ I am

Yo$^r$ humble & obliged<br>Servant

Cambridge. May 11$^{th}$. 1676.                    Is. NEWTON*.

*For* HENRY OLDENBURG Esq: *at his house*
    *about y$^e$ middle of y$^e$ old Pal-mall in*
    *Westminster London.*

Endorsed by Oldenburg :
        " Rec$^d$ 12 May.

Answ. by D$^r$ Sidnam† May 15. and sent by him his Hypothesis explaining y$^e$ properties of light ; as also his discourse about y$^e$ various colors exhibited by transparent substances made very thin by being blown into bubles or otherwise form'd into plates, altho at a greater thicknes they appear very clear and colorlesse.

In my letter accompanying these papers I imparted to M$^r$ Newton y$^e$ particulars contain'd in M. Leibniz his letter to me of May 12 1676. from Paris st. n." In the letter just mentioned Leibniz desired information on the subject of the analytical discoveries recently made in England, and it was in compliance with this request that Newton, at the pressing solicitation of Collins and Oldenburg, drew up his celebrated letter of June 13. One of the questions in Leibniz's letter, of which an extract is printed in the *Commercium Epistolicum,* will probably surprise the modern student. The series $(\sin \theta =) \theta - \dfrac{\theta^3}{6} + \ldots$ and its converse had been sent to him from this country, and he begs the favour of a demonstration of them.

---

* *Orig. Lett. Bk.* Roy. Soc. N. 1. 52.

† Sydenham was going to Cambridge to take his M. D. degree.  He was admitted at Pembroke, May 17 (from Magdalen Hall, Oxford) and was made Doctor the following day.

## No. XIV.

### NEWTON TO OLDENBURG.

Accompanying his answer to Lucas (dated Aug. 18, and printed in the Trans. for Sept. 25).

S$^r$

I have been stayed from writing to you longer then I intended by reason that I could not till of late meet w$^{th}$ a day clear enough at noon-time to try some of y$^e$ experiments herein set down. And now I have not sent you an answer so full as I intended at first but perhaps more to y$^e$ purpose considering who I have to deale w$^{th}$, whose buisiness it is to cavill. The other buisiness you wrote to me about viz: about stocking us w$^{th}$ fruit trees I shall be glad to promote. Some inquiry I have made about it, & w$^{th}$in a few days, when I have got some further information & discoursed it w$^{th}$ some that are most like to entertein y$^e$ proposall, I hope to give you a further account of it. In y$^e$ mean time I rest

Yo$^r$ humble Servant

Cambridge Aug : 22. 1676.          Is. NEWTON[*]

*For* HENRY OLDENBURG Esq: *at his house about the middle of y$^e$ old Pal-maill in Westminster London.*

w$^{th}$ care.

---

## No. XV.

### NEWTON TO OLDENBURG.

S$^r$                                   Octob 26. 1676.

Two days since, I sent you an answer to M. Leibnitz's excellent Letter. After it was gone, running my eyes over a transcript that I had made to be taken of it, I found some things w$^{ch}$ I could wish altered, & since I cannot now

---

[*] *Orig. Lett. Bk. Roy. Soc. N. 1. 54.*

do it my self, I desire you would do it for me, before you
send it away.

In pag : 3. Sect : Pudet dicere.] ffor *a D. Barrow tunc
Matheseos Professore* write only *per amicum*

Pag : 5. Sect : At quando.] After *quibuscum potest com-
parari;* write *ad quod sufficit etiam hoc ipsum unicum jam
descriptum Theorema si debitè concinnetur. Pro Trinomiis
etiam et aliis quibusdam Regulas quasdem concinnavi* &c.

Pag : 6. Sect : Quamvis multa.] Where you find y$^e$
words *Gregorianis ad Circulum et Hyperbolam editis persimiles,* for *persimiles* write *affines*

Pag : 9 or 10. Sect : Theorema de.] ffor *error erit*

$$\frac{v^3}{90} + \frac{v^4}{140} + \&c.\ \text{write } error\ erit\ \frac{v^3}{90} + \frac{v^4}{194} + \&c.$$

Pag : 6 vel 7. Sect : Quamvis multa.] about y$^e$ end of y$^e$
section turn *plenariam* into *plenam* or rather blot y$^e$ word
quite out.

Pag : ult. vel penult. Sect : Ubi dixi]. write *solutilia* for
*solubilia.* And if you observe any other such scapes pray
do me y$^e$ favour to mend them. So in pag 5 or 6. Sect.
Quamvis multa.] It may be perhaps more intellig{ib}le to
write εὐθύνσει for euthunsi.

Pag 8 or 9. Sect : Per seriem.] After y$^e$ words *produci
ad multas figuras :* you may if you please add these words.
ut et ponendo summam terminorum $1 - \frac{1}{7} + \frac{1}{9} - \frac{1}{15} + \frac{1}{17}$
$- \frac{1}{23} + \frac{1}{25} - \frac{1}{31} + \frac{1}{33}$ &c esse ad totam seriem $1 - \frac{1}{3} + \frac{1}{5}$
$- \frac{1}{7} + \frac{1}{9} - \frac{1}{11} + \&c$ ut $1 + \sqrt{2}$ ad 2. Sed optimus ejus
usus &c

I feare I have been something too severe in taking
notice of some oversights in M. Leibnitz letter considering
y$^e$ goodnes & ingenuity of y$^e$ Author & y$^t$ it might have
been my own fate in writing hastily to have committed y$^e$
like oversights. But yet they being I think real oversights
I suppose he cannot be offended at it. If you think any
thing be exprest too severely pray give me notice & I'le

endeavour to mollify it, unless you will do it w$^{th}$ a word or two of your own. I beleive M. Leibnitz will not dislike y$^e$ Theorem towards y$^e$ beginning of my letter pag. 4 for squaring Curve lines Geometrically. Sometime when I have more leisure it's possible I may send him a fuller account of it: explaining how it is to be ordered for comparing curvilinear figures w$^{th}$ one another, & how y$^e$ simplest figure is to be found w$^{th}$ w$^{ch}$ a propounded Curve may be compared.　　　　　　　　　　S$^r$ I am

Yo$^r$ humble Servant

Is. NEWTON*.

Pray let none of my mathematical papers be printed w$^{th}$out my special licence.

Some other things in M. Leibnitz letter I once thought to have touched upon, as y$^e$ resolution of affected æquations, & y$^e$ impossibility of a geometric Quadrature of y$^e$ Circle in w$^{ch}$ M. Gregory seems to have tripped. But I shall add one thing here. That y$^e$ series of æquations for y$^e$ sections of an angle by whole numbers, w$^{ch}$ M. Tschurnhause saith he can derive by an easy method one from an other, is conteined in y$^t$ one æquation w$^{ch}$ I put in y$^e$ 3$^d$ section of y$^e$ Problems in my former letter for cutting an angle in a given ratio, and in another æquation like that. Also y$^e$ coefficients of those æquations may be all obteined

$$\text{by this progression } 1 \times \frac{\overline{n-0} \times \overline{n-1}}{1 \times n-1} \times \frac{\overline{n-2} \times \overline{n-3}}{2 \times n-2} \times \frac{\overline{n-4} \times \overline{n-5}}{3 \times n-3}$$

$$\times \frac{\overline{n-6} \times \overline{n-7}}{4 \times n-4} \times \&c.$$ The first coefficient being 1. y$^e$ 2$^d$

$$1 \times \frac{\overline{n-0} \times \overline{n-1}}{1 \times n-1}. \quad \text{y}^e \; 3^d \; 1 \times \frac{\overline{n-0} \times \overline{n-1}}{1 \times n-1} \times \frac{\overline{n-2} \times \overline{n-3}}{2 \times n-2}. \; \&c. \; \& \; n$$

being y$^e$ number by w$^{ch}$ y$^e$ angle is to be cut. as if $n$ be 5.

---

* MSS. Birch, Brit. Mus. 4294. The signature which was cut out by some felonious hand in 1833, has been recently restored.

then $y^e$ series is $1 \times \dfrac{5 \times 4}{1 \times 4} \times \dfrac{3 \times 2}{2 \times 3} \times \dfrac{1 \times 0}{3 \times 2}$ that is $1 \times 5 \times 1 \times 0$
& consequently $y^e$ coefficients $1.5.5$. So if $n$ be 6 $y^e$
series is $1 \times \dfrac{6 \times 5}{1 \times 5} \times \dfrac{4 \times 3}{2 \times 4} \times \dfrac{2 \times 1}{3 \times 3} \times 0$ that is $1 \times 6 \times \frac{3}{2} \times \frac{2}{9} \times 0$
& consequently $y^e$ coefficients $1.6.9.2$. This scrible is
not fit to be seen by any body nor scarce my other letter
in $y^t$ blotted form I sent it, unless it be by a friend.

*For* HENRY OLDENBURG Esq: *at his house*
   *about $y^e$ middle of $y^e$ old Pal-mall in*
   *Westminster London*

---

## No. XVI.
### NEWTON TO OLDENBURG.

S$^r$

I am desired to write to you about procuring a recommendation of us to M$^r$ Austin $y^e$ Oxonian planter. We
hope yo$^r$ correspondent* will be pleased to do us $y^t$ favour
as as{*sic*} to recommend us to him, $y^t$ we may be furnished
w$^{th}$ $y^e$ best sorts of Cider-fruit-trees. We desire only about
30 or 40 Graffs for $y^e$ first essay, & if those prove for o$^r$ purpose they will be desired in greater numbers. We desire
graffs rather then sprags that we may $y^e$ sooner see what
they will prove. They are not for M$^r$ Blackley but some
other persons about Cambridge. But M$^r$ Austin need only
direct his letters to me or to M$^r$ Bainbrigg ffellow of o$^r$
College. In $y^e$ mean time we return o$^r$ thanks to you &
your ffriend for $y^e$ good will you have already shewn us.

M$^r$ Lucas letter† I have received, & hope to send you
an answer $y^e$ next Tuesday Post. I thank you for your
care to prevent their prejudicing me in $y^e$ Society, as also

---

\* Dr John Beal, rector of Yeovil, who inherited a " zeal for the plantation of orchards
for the making of cider." See Birch, IV. 235.
  † Dated Oct. 23.

for giving me notice of $y^e$ things miswritten in my late letter. In pag 3 $y^e$ words you cite should run thus. Cujus triplo adde Log. 0.8, siquidem sit $\frac{2 \times 2 \times 2}{0.8} = 10$. But in pag 8 $y^e$ signes of $y^e$ series $1 + \frac{1}{3} - \frac{1}{5} - \frac{1}{7} + \frac{1}{9} +$ &c are rightly put two + & two − after one another, it being a different series from $y^t$ of M. Leibnitz. But in $y^e$ next two or 3 lines, to prevent future mistake you may if you think fit, after $y_e$ words *res tardius obtineretur per tangentem* 45$^{gr}$, add these words *juxta seriem nobis communicatam*.

Seing $y^e$ letter is still in $yo^r$ hands, you will do me $y^e$ favour to make these further amendments

Pag. 3 Sect [Pudet dicere] *cum D. Collinsio* for ad D. Collinsium

pag. 5. Exempl. 4 after $y^e$ words *vel quibus libet digni-tatibus binomii cujuscunq:* add *licet non directè ubi index dig-nitatis est numerus integer.*

pag 6 or 7 in $y^e$ end of $y^e$ section quamvis multa I desire you would cross out $y^e$ words *adeo ut in potestate habeam descriptionem omnium curvarum istius ordinis quæ per 8 data\* puncta determinantur.* And in $y^e$ 2$^d$ sentence of $y^e$ next section I could wish these words also *numero infinitè multas* were put out.

pag 9†. Sect [*Præterea quæ.*] for *mihi quidem haud ita clara sunt* put *nondum percipio.* And after a line or two where you see $y^e$ words *et certè minor est labor,* put out *certe.*

By these alterations $S^r$ you will oblige

Yo$^r$ humble Servant

{Tuesday} Nov. 14 1676.  Is. NEWTON‡.

---

\* "data" is written by mistake for "tantum." The words here ordered to be crossed out are inclosed within parentheses in the letter as printed in Wallis's 3rd Volume, and the *Commercium Epistolicum*, where also *septem* appears instead of *octo.* One of the points is supposed to be a double point. See Newton's *Enumeratio Lin. Tert. Ord.*

† The place referred to is in p. 10.

‡ MSS. Birch, Brit. Mus. 4294.

Just now I received Yo$^r$ packet conteining two books from M$^r$ Boyle. That for D$^r$ Moor shall be conveyed to him. For the other I shall return my thanks to y$^e$ noble Author.

*For* HENRY OLDENBURG Esq: *at his house*
  *about y$^e$ middle of y$^e$ old Palmail in*
  *Westminster*

> Endorsed by Oldenburg: "Rec. Nov. 15. 76.
> written to D$^r$ Beall about part of y$^e$ contents
> of this letter. Nov. 16. 76. Answ. Nov.
> 25. 76."

In another letter to Oldenburg written on the following Saturday, he says: "I promised to send you an answer to Mr Lucas this next Tuesday, but I find I shall scarce finish what I have designed, so as to get a copy taken of it by that time, and therefore I beg your patience a week longer." *Macc. Corr.* II. 405. The answer was accordingly sent on the 28th. All that is known respecting it is derived from Lucas's rejoinder. See Syn. View of Newton's Life, under Nov. 28, 1676, note.

---

## No. XVII.

### NEWTON TO DR JOSHUA MADDOCK.

Maddock had sent Newton some specimens of a new branch of optics, devoted to the consideration of the properties of *dark rays*. Such a system would afford relief to those commentators who are embarrassed by expressions like μελαμφαὲς ἔρεβος, μέλαινα αἴγλη, and *atrum lumen*. There was a person of that name at Jesus College, who took the degree of B.A. in 1661.

Vir dignissime,

Specimina illa optica, quæ pro humanitate tua ad me nuper misisti, tantam in his rebus peritiam ostendunt, ut non possum quin doleam incertitudinem principiorum quibus omnia innituntur. Etenim quæri potest, an sint in rerum natura radii tenebrosi, et, si sint, an radii illi, secundum aliam legem refringi debeant, quam radii lucis. Defectu experientiæ, nescio prorsus quid de his principiis sentiendum sit. Neque huic difficultati tollendæ, quam et tute ipse indigitasti facile adfuerit Tiberius\*.   At positis ejusmodi radiis, una cum lege refractionis quam tu assumis, cætera recte se habent; neque propositiones tantum utiles sunt ac demonstrationes artificiosæ, sed, et quod majus est, omnia nova proponis, quæ opticam, altera sui parte, auctura sunt, si modo defectus experientiæ in stabiliendis principiis tuis aliquo demum modo suppleri possit. Interim, quod me meditationum tuarum perquam subtilium participem fieri dignatus sis, gratias ago.   Vale!

Tui studiosissimus,

Trin. Coll. Cant. Feb. 7, 167⅞.                    I. NEWTON†.

*For his honoured friend* JOSHUA MADDOCK,
 *Doctor of Physic at his house in Whitchurch in Shropshire.*

## No. XVIII.

### NEWTON TO HOOKE.

Sᵣ

One Dominico Casparini an Italian Doctor of Physick of the City of Lucca has composed a Treatise of the Method of administring the Cortex Peruviana in Fevers, in which he particularly discusses whether it may be administred in Malignant fevers and also whether in any fevers before the fourteenth day of the Sickness.   Upon the fame of the Royal Society spread every where abroad, he is ambitious to submit his discourse to so great and Authentick a Judgment as theirs is, and thereupon desired another Dᵣ. of Physick of his Acquaintance in Italy to write to his Correspondent an Italian in London, to move that the Society would give him leave to dedicate

---

\* Allusion to Tiberius's peculiarity of vision. " Cum prægrandibus oculis, et qui, quod mirum esset, noctu etiam et in tenebris viderent, sed ad breve." Sueton. *Tib.* 68. Comp. Plin. *Nat. Hist.* xi. 54.

 † Printed at the end of a Funeral Sermon on the death of Daniel Maddock by E. Latham, M.D. Lond. 1745 : and *Gentleman's Mag.* Aug. 1782.

his Book to them. The said Italian being come from London hither before the Arrival of the Letters, upon the receipt of them applied himself to me and I promised him I would desire you to acquaint the Society with his Request. If you please to send their Answer to me, the Italian here will convey it into Italy.

For the trials you made of an Experiment suggested by me about falling bodies *, I am indebted to you thanks which I thought to have returned by word of mouth, but not having yet the Opportunity must be content to do it by Letter &c†.

Trinity College Decemb. 3ᵈ 1680.

---

William Briggs, born about 1650, succeeded Tenison in his fellowship at Corpus Christi College, 1668. A.M. 1670. M.D. 1677. See Masters's *Hist. of Corp.* p. 249.

### No. XIX.

### NEWTON TO BRIGGS.

Sʳ

I have perused yoʳ very ingenious Theory of Vision‡ in wᶜʰ (to be free wᵗʰ you as a friend should be) there seems to be some things more solid & satisfactory, others more disputable but yet plausibly suggested & well deserving yᵉ consideration of yᵉ ingenious. The more satisfactory I take to be your asserting yᵗ we see wᵗʰ both eyes at once, yoʳ speculation about yᵉ use of yᵉ *musculus obliquus inferior*, yoʳ assigning every fibre in yᵉ optick nerve of one eye to have its correspondent in yᵗ of yᵉ other,

---

* See Synoptical View of Newton's Life under the year 1679 (note).

† Roy. Soc. *Letter Book*, viii. 139. Hooke's Answer, dated Dec. 18, is given Ib. 140. Compare Birch, iv. 61.

‡ " A New Theory of Vision " read at the meeting of the Royal Society, March 15, 1682, and printed in Hooke's *Philosophical Collections* for that month. A paper in continuation of it, "with an examination of some late objections," appeared in the *Phil. Trans.* for May 1683.

both w<sup>ch</sup> make all things appear to both eyes in one &
y<sup>e</sup> same place & yo<sup>r</sup> solving hereby y<sup>e</sup> duplicity of y<sup>e</sup> object
in distorted eyes & confuting y<sup>e</sup> childish opinion about y<sup>e</sup>
splitting y<sup>e</sup> optick cone. The more disputable seems yo<sup>r</sup>
notion about every pair of fellow fibres being unisons to
one another, discords to y<sup>e</sup> rest, & this consonance making
y<sup>e</sup> object seen w<sup>th</sup> two eyes appear but one for y<sup>e</sup> same
reason that unison sounds, seem but one sound. I did
think to have sent you what I fancy may be objected
against this notion & so staid for time to write it down,
but upon second thoughts I had rather reserve it for discourse at o<sup>r</sup> next meeting: and therefore shall add only
my thanks for yo<sup>r</sup> kind letter & p<sup>r</sup>sent.

S<sup>r</sup> I am

Yo<sup>r</sup> much obliged & humble servant

Trin. Coll. Cambridge June 20<sup>th</sup> 1682.          Is. NEWTON *.

*For his honoured ffriend* D<sup>r</sup> WILLIAM BRIGGS
*at his house in Suffolk Street in London.*

---

No. XX.

NEWTON TO BRIGGS.

For his Hon<sup>d</sup> ffriend D<sup>r</sup> W<sup>m</sup> BRIGGS.

S<sup>r</sup>

Though I am of all men grown y<sup>e</sup> most shy of setting
pen to paper about any thing that may lead into disputes
yet yo<sup>r</sup> friendship overcomes me so far as y<sup>t</sup> I shall set
down my suspicions about yo<sup>r</sup> Theory, yet on this condition, that if I can write but plain enough to make you
understand me, I may leave all to yo<sup>r</sup> use w<sup>th</sup>out pressing
it further on. For I designe not to confute or convince

---

* From the original in the British Museum, Add. MSS. 4237. fol. 32.  Part of this
letter is lithographed in C. J. Smith's *Hist. and Lit. Curiosities*, Lond. 1840.

you but only to present & submit my thoughts to yo$^r$ consideration & judgment.

First then it seems not necessary that the bending of y$^e$ nerves in y$^e$ Thalamus opticus should cause a differing tension of y$^e$ ffibres. ffor those w$^{ch}$ have y$^e$ further way about, will be apt by nature to grow the longer. If y$^e$ arm of a tree be grown bent it follows not that the fibres on y$^e$ elbow are more stretcht then those on the concave side, but that they are longer. And if a straight arm of a tree be bent by force for some time, the fibres on y$^e$ elbow w$^{ch}$ were at first on y$^e$ stretch will by degrees grow longer & longer till at length the arm stand of it's self in y$^e$ bended figure it was at first by force put into, that is till y$^e$ fibres on y$^e$ elbow be grown as much longer then y$^e$ rest as they go further about, & so have but the same degree of tension w$^{th}$ them. The observation is ordinary in twisted Codling hedges, fruit trees nailed up against a wall &c. And y$^e$ younger & more tender a tree is the sooner will it stand bent. How much more therefore ought it to be so in that most tender substance of y$^e$ Optick nerves w$^{ch}$ grew bent from y$^e$ very beginning? And whether if those nerves were carefully cut out of y$^e$ brain & outward coat & put into brine made as neare as could be of the same specific gravity w$^{th}$ y$^e$ nerves, they would unbend or exactly keep the same bent they had in y$^e$ brain may be worth considering. ffor though y$^e$ strength of a single fibre upon the stretch be inconsiderably little, yet all together ought to have as much strength to unbend y$^e$ nerve, as would suffice by outward application of y$^e$ hand to bend a straight nerve of y$^e$ same thickness, the dura Mater being taken off.

M$^r$ Sheldrake* further suggests wittily that an object whether the axis opticus be directed above it, under it, or

---

* A Fellow of Corpus, 7 years senior to Newton.

directly towards it, appears in all cases alike as to figure & colour excepting that in $y^e$ $3^d$ case tis distincter, $w^{ch}$ proceeds not from $y^e$ frame of $y^e$ nerves but from $y^e$ distinctness of $y^e$ picture made in $y^e$ Retina in that case. But in $y^e$ first case where $y^e$ vision is made by $y^e$ fibres above & second where tis made by those below, the object appearing alike : he thinks it argues that the fibres above & below are of $y^e$ same constitution & tension, or at least if they be of a differing tension, that that tension has no effect on $y^e$ mode of vision. but I understand you are already made acquainted $w^{th}$ his thoughts.

It may be further considered that the cause of an objects appearing one to both eyes is not its appearing of $y^e$ same colour form & bigness to both, but in $y^e$ same situation or place. Distort one eye & you will see $y^e$ coincident images of $y^e$ object divide from one another & one of them remove from $y^e$ other upwards downwards or sideways to a greater or less distance according as $y^e$ distortion is; & when the eyes are let return to their natural posture the two images advance towards one another till they become coincident & by that coincidence appear but one. If we would then know why they appear but one, we must e{n}quire why they appear in one & $y^e$ same place & if we would know $y^e$ cause of that we must enquire why in other cases they appear in divers places variously situate & distant one from another. ffor that $w^{ch}$ can make their distance greater or less can make it none at all. Consider whats the cause of their being in $y^e$ same altitude when one is directly to $y^e$ right hand $y^e$ other to $y^e$ left & what of their being in $y^e$ same coast or point of $y^e$ compas, when one is directly over $y^e$ other : these two causes joyned will make them in $y^e$ same altitude & coast at once that is in $y^e$ same place. The cause of situations is therefore to be enquired into. Now for finding out this $y^e$ analogy will stand between $y^e$ situations

of sounds & the situations of visible things, if we will
compare these two senses.   But the situations of sounds
depend not on their tones.   I can judge from whence an
echo or other sound comes tho I see not y$^e$ sounding
body, & this judgment depends not at all on y$^e$ tone.   I
judge it not from east because acute, from west because
grave : but be y$^e$ tone what it will I judge it from hence or
thence by some other principle.   And by that principle I
am apt to think a blind man may distinguish unisons one
from another when y$^e$ one is on his right hand y$^e$ other on
his left.   And were our ears as good & accurate at distinguishing y$^e$ coasts of audibles as our eyes are at distinguishing y$^e$ coasts of visibles I conceive we should judge no
two sounds the same for being unisons unless they came so
exactly from y$^e$ same coast as not to vary from one another
a sensible point in situation to any side. Suppose then you
had to do with one of so accurate an ear in distinguishing
y$^e$ situation of sounds.   how would you deale with him ?
Would you tell him that you heard all unisons as but one
sound ?   He would tell you he had a better ear then so.
He accounted no sounds y$^e$ same w$^{ch}$ differed in situation :
& if your eyes were no better at y$^e$ situation of things then
your ears, you would perhaps think all objects y$^e$ same, w$^{ch}$
were of y$^e$ same colour.   But for his part he found y$^t$ y$^e$
like tension of strings & other sounding bodies did not
make sounds one, but only of y$^e$ same tone : & therefore
not allowing the supposition that it does make them one,
the inference from thence that y$^e$ like tension of y$^e$ optick
fibres made y$^e$ object to y$^e$ two eyes appear one, he did not
think himself obliged to admit.   As he found y$^t$ tones
depended on those tensions so perhaps might colours, but
the situation of audibles depended not on those tensions, &
therefore if the two senses hold analogy with one another,
that of visibles does not, & consequently the union of
visibles as well as audibles which depends on the agreement

of situation as well as of colour or tone must have some other cause.

But to leave this imaginary disputant, let us now consider what may be y$^e$ cause of y$^e$ various situations of things to y$^e$ eyes. If when we look but w$^{th}$ one eye it be asked why objects appear thus & thus situated one to another the answer would be because they are really so situated among themselves & make their coloured pictures in y$^e$ Retina so situated one to another as they are & those pictures transmit motional pictures into y$^e$ sensorium in y$^e$ same situation & by the situation of those motional pictures one to another the soul judges of y$^e$ situation of things without. In like manner when we look with two eyes distorted so as to see y$^e$ same object double if it be asked why those objects appeare in this or that situation & distance one from another, the answer should be because through y$^e$ two eyes are transmitted into y$^e$ sensorium two motional pictures by whose situation & distance then from one another the soule judges she sees two things so situate & distant. And if this be true then the reason why when the distortion ceases & y$^e$ eyes return to their natural posture the doubled object grows a single one is that the two motional pictures in y$^e$ sensorium come together & become coincident.

But you will say, how is this coincidence made? I answer, what if I know not? Perhaps in y$^e$ sensorium, after some such way as y$^e$ Cartesians would have beleived or by some other way. Perhaps by y$^e$ mixing of y$^e$ marrow of y$^e$ nerves in their juncture before they enter the brain, the fibres on y$^e$ right side of each eye going to y$^e$ right side of y$^e$ head those on y$^e$ left side to y$^e$ left. If you mention y$^e$ experim$^t$ of y$^e$ nerve shrunck all y$^e$ way on one side y$^e$ head, that might be either by some unkind juyce abounding more on one side y$^e$ head y$^n$ on y$^e$ other, or by y$^e$ shrinking of y$^e$ coate of y$^e$ nerve whose fibres & vessels for nourishment perhaps do not cross in yᵉ juncture as yᵉ fibres of yᵉ marrow may do.   And its more probable yᵗ yᵉ stubborn coate being vitiated or wanting due nourishment shrank & made yᵉ tender marrow yeild to its capacity, then that yᵉ tender marrow by shrinking should make yᵉ coate yeild.   I know not whether you would have yᵉ succus nutricius run along yᵉ marrow. If you would, 'tis an opinion not yet proved & so not fit to ground an argument on.   If you say yᵗ in yᵉ Camælion & ffishes yᵉ nerves only touch one another without mixture & sometimes do not so much as touch ; 'Tis true, but makes altogether against you. ffishes looke one way with one eye yᵉ other way with yᵉ other : the Chamælion looks up wᵗʰ one eye, down wᵗʰ t'other, to yᵉ right hand wᵗʰ this, to yᵉ left wᵗʰ yᵗ, twisting his eyes severally this way or that way as he pleases.   And in these Animals which do not look yᵉ same way wᵗʰ both eyes what wonder if yᵉ nerves do not joyn?   To make them joyn would have been to no purpose & nature does nothing in vain.   But then whilst in these animals where tis not necessary they are not joyned, in all others wᶜʰ look yᵉ same way wᵗʰ both eyes, so far as I can yet learn, they are joyned.   Consider therefore for what reason they are joyned in yᵉ one & not in the other.   ffor God in yᵉ frame of animals has done nothing wᵗʰout reason.

There is one thing more comes into my mind to object. Let yᵉ circle *D J* represent the Retina, or if you will the end of yᵉ optick nerve cut cross.   *A* the end of a fibre above of most tension, *C* yᵉ end of one below of least tension.   *D* & *G* yᵉ ends of fibres above on either hand almost of as much tension as *A*, *F* & *J* the ends of others

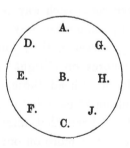

below almost of as little tension as $C$. $E$ $y^e$ end of a fibre of less tension then $A$ or $G$ & of more then $C$ or. $J$. And between $A$ & $C$, $G$ & $J$ there will {be} fibres of equal tension $w^{th}$ $E$ because between them there are in a continual series fibres of all degrees of tension between $y^e$ most tended at $A$ & $G$ & least tended at $C$ & $J$. And by the same argument that 3 fibres $E$, $B$ & $H$ of like tension are noted let the whole line of fibres of the same Degree of tension running from $E$ to $H$ be noted. Do you now say $y^t$ $y^e$ reason why an object seen $w^{th}$ two eyes appears but one is that $y^e$ fibres in $y^e$ two eyes by $w^{ch}$ 'tis seen are unisons? then all objects seen by unison fibres must for $y^e$ same reason appear in one & $y^e$ same place that is all $y^e$ objects seen by the line of fibres $E$ $B$ $H$ running from one side of the eye to $y^e$ other. ffor instance two stars one to $y^e$ right hand seen by $y^e$ fibres about $H$, the other to $y^e$ left seen by $y^e$ fibres about $E$ ought to appear but one starr, & so of other objects. ffor if consonance unite objects seen $w^{th}$ the fibres of two eyes much more will it unite those seen $w^{th}$ those of $y^e$ same eye. And yet we find it much otherwise. What soever it is that causes the two images of an object seen with both eyes to appear in $y^e$ same place so as to seem but one can make them upon distorting $y^e$ eyes separate one from $y^e$ other & go as readily & as far asunder to $y^e$ right hand & to $y^e$ left as upwards & downwards.

You have now $y^e$ summ of what I can think of worth objecting set down in a tumultuary way as I could get time from my Sturbridge ffair friends. If I have any where exprest myself in a more peremptory way then becomes $y^e$ weaknes of $y^e$ argument pray look on that as done not in earnestness but for $y^e$ mode of discoursing. Whether any thing be so material as $y^t$ it may prove any way useful to you I cannot tell. But pray accept of it as written for that end. ffor having laid Philosophical speculations aside nothing but y$^e$ gratification of a friend would easily invite me to so large a scribble about things of this nature.

S$^r$ I am

Yo$^r$ humble Servant

Trin. Coll. Cambr. Sept. 12$^{th}$. 1682.    Is. NEWTON *.

No. XXI.

NEWTON TO BRIGGS.

Isaacus Newtonus Doctori Gulielmo Briggio.

Vir Clarissime,

Hisce tuis Tractatibus† duas magni nominis scientias uno opere promoves, *Anatomiam* dico & *Opticam.* Organi enim (in quo utraque versatur) artificio summo constructi diligenter perscrutaris mysteria. In hujus dissectione peritiam & dexteritatem tuam non exiguo olim mihi oblectamento fuisse recordor. Musculis motoriis secundùm situm suum naturalem elegantèr à te expansis, cæterisque partibus coràm expositis, sic ut singularum usus & ministeria non tàm intelligere liceret quàm cernere, effecerat dudùm ut ex cultro tuo nihil non accuratum sperarem. Nec spem fallebat eximius ille Tractatus Anatomicus, quem postmodùm edidisti. Jam praxeos hujus ἀκρίβειαν pergis ingeniosissimâ Theoriâ instruere & exornare. Et quis Theoriis condendis aptior extiterit, quàm qui phænomenis accuratè observandis navârit operam? Nervos opticos *ex capillamentis variè tensis* constare supponis, eaque magis esse tensa quæ per iter longius porriguntur; ex diversâ autem tensione fieri ut objectorum partes singulæ non coincidant & confundantur inter se, sed

<hr>

* From the original in the British Museum, Add. MSS. 4237. fol. 34.

† i. e. *Opthalmographia,* Cantab. 1676 (2nd Ed. Lond. 1687) and his *Theory of Vision.*

pro situ suo naturali diversis in locis appareant ; & *capilla-mentis* amborum oculorum æquali tensione factis *concordi-bus*, geminas objectorum species uniri. Sic ex tensione chordarum, quâ soni vel variantur vel concordant in Mu-sicâ, colligere videris quid fieri debet in Opticâ. Simplex etenim est *Natura,* & eodem operandi teñore in immensâ effectuum varietate sibi ipsa constare solet. Quantò verò magìs in sensuum cognatorum causis ? Et quamvis aliam etiam horum sensuum analogiam suspicari possim, ingenio-sam tamen esse quam tute excogitasti, certè nemo non lubentèr fatebitur. Nec inutilem censeo Dissertationem ultimam quâ diluis objectiones. Inde Lector attentus & plenìùs intelliget Hypothesin totam, & in quæstiones incidet vel tuis Meditationibus illustratas, vel novis experimentis & disquisitionibus posthàc dirimendas. Id quod in usum cedet juventuti Academicæ, & provectiores ad ulteriores in Philosophiâ progressus manuducet. Pergas itaque, vir ornatissime, scientias hasce præclaris inventis, uti facis, excolere ; doceasque difficultates causarum naturalium tàm facilè solertiâ vinci posse, quàm solent conatibus vulgari-bus difficultèr cedere.

*Dabam Cantabrigiœ* 7 *Kal. Maii.*        Vale*.
  1685.

---

## No. XXII.

Paper of Directions given by Newton to Bentley, respecting the Books to be read before entering upon the Principia. Date probably about July 1691.

In 1691 the vigorous mind of Richard Bentley, who was then in his 30th year, was attracted to the revelations which the *Principia* had announced to the philosophical world some four years before, and with

---

* This letter is prefixed to the Latin Version of Briggs's *Theory of Vision* (made at Newton's request) Lond. 1685.

18

the view of making himself acquainted with the "Great Charter of Modern Science," as that immortal work is called by Dr Whewell, he applied through his friend W. Wotton to John Craige, a mathematician of some eminence, for advice as to the course of reading to be followed preparatory to the study of the volume itself. The appalling list of authors which Craige sent him (June 24, 1691; see Bentley's Corresp. p. 736) probably induced him to repair to the fountain head, and the paper now before us was the result of that step.

That Bentley studied the *Principia* to some purpose was shewn by his two last sermons at the Boyle Lecture (founded by the Will of Robert Boyle, who died Dec. 31, 1691) in November and December of the following year, which led him to consult our philosopher again upon some points that arose in them requiring elucidation. See Newton's four Letters to Bentley in 1692 and 1693. (Bentley's *Corresp.*)

Next after Euclid's Elements the Elements of $y^e$ Conic sections are to be understood. And for this end you may read either the first part of $y^e$ Elementa Curvarum of John De Witt, or De la Hire's late treatise of $y^e$ conick sections, or D$^r$ Barrow's epitome of Apollonius.

For Algebra read first Barth{ol}in's introduction & then peruse such Problems as you will find scattered up & down in $y^e$ Commentaries on Cartes's Geometry & other Alegraical {*sic*} writings of Francis Schooten. I do not mean $y^t$ you should read over all those Commentaries, but only $y^e$ solutions of such Problems as you will here & there meet with. You may meet with De Witt's Elementa curvarum & Bartholin's introduction bound up together w$^{th}$ Carte's Geometry & Schooten's commentaries.

For Astronomy read first $y^e$ short account of $y^e$ Copernican System in the end of Gassendus's Astronomy & then so much of Mercator's Astronomy as concerns $y^e$ same system & the new discoveries made in the heavens by Telescopes in the Appendix.

These are sufficient for understanding my book: but if you can procure Hugenius's Horologium oscillatorium, the perusal of that will make you much more ready.

At $y^e$ first perusal of my Book it's enough if you understand y$^e$ Propositions w$^{th}$ some of y$^e$ Demonstrations w$^{ch}$ are easier then the rest. For when you understand y$^e$ easier they will afterwards give you light into y$^e$ harder. When you have read y$^e$ first 60 pages, pass on to y$^e$ 3$^d$ Book & when you see the design of that you may turn back to such Propositions as you shall have a desire to know, or peruse the whole in order if you think fit\*.

Memorandum
  by Bentley.
"Directions from M$^r$ Newton by his own Hand"

---

### No. XXIII.

#### NEWTON TO LOCKE.

The first few lines of the letter are wanting. Locke had sent him some of Boyle's red earth, which that philosopher had a recipe for combining with mercury so as to "multiply" gold. In a letter written on the 2nd of the following month, Newton "dissuades" Locke "from too hasty a trial of this recipe," which he states to be "imperfect and useless." Lord King's *Life of Locke*, I. 412.

\*        \*        \*        \*        \*        \*        \*        \*

as I can. You have sent much more earth then I expected. For I desired only a specimen, having no inclination to prosecute the process. For in good earnest I have no opinion of it. But since you have a mind to prosecute it I should be glad to assist you all I can, having a liberty of communication allowed me by Mr B. in one case which reaches to you if it be done under y$^e$ same conditions in w$^h$ I stand obliged to Mr B. ffor I presume you are already under the same obligations to him. But I feare I have lost y$^e$ first & third part out of my pockett. I thank you for

---

\* From the original, given, with Newton's four letters to Bentley, by Cumberland to Trinity College.

what you communicated to me out of yo$^r$ own notes about it. S$^r$ I am

Yo$^r$ most humble Serv$^t$

Cambridge Jul 7$^{th}$             Is NEWTON
       1692.

When the hot weather is over I intend to try the beginning* tho y$^e$ success seems improbable†.

*For* JOHN LOCK, Esq. *at* M$^r$. PAULEN'S *in Dorset Court in Channel Row in Westminster.*

---

## No. XXIV.

### NEWTON TO LEIBNIZ.

In answer to a letter of Leibniz dated $\frac{7}{17}$ Mart. 1693, printed in Raphson's *Hist. of Fluxions*, p. 119.    Leibn. Opp. III. 484.

Celeberrimo Viro

Godefrido Gulielmo Leibnitio

ISAACUS NEWTON S.P.D.

Literæ tuæ, cùm non statim acceptis responderem, e manibus elapsæ inter schedas meas diu latuere, nec in eas ante hesternum diem incidere potui. Id quod me moleste habuit, cùm amicitiam tuam maximi faciam, teq: inter summos hujus sæculi Geometras a multis retro annis habuerim, quemadmodum etiam data omni occasione testatus sim  Nam quamvis commercia philosophica & mathematica quammaximè fugiam, tamen metuebam ne amicitia nostra ex silentio decrementum acciperet, idq: maxime cum Wallisius noster Historiam Algebræ in lucem denuò

---

\* i. e. the first of the three parts of the recipe, the effect of which, according to Boyle, was the production of a mercury which would grow hot with gold.

† From a transcript obligingly made for me by Lord Lovelace.

missurus nova aliqua e literis inseruit quas olim per manus $D^{ni}$ Oldenburgi ad te conscripsi, & sic ansam mihi dedit ea etiam de re ad te scribendi. Postulavit enim ut methodum quandam duplicem aperirem quam literis transpositis ibi celaveram. Quocirca coactus sum qua potui brevitate exponere methodum meam fluxionum quam hac celaveram sententia. *Data æquatione quantitates quotcunque fluentes involvente invenire fluxiones, & vice versa.* Spero autem me nihil scripsisse quod tibi non placeat, et siquid sit quod reprehensione dignum censeas, ut literis id mihi significes quoniam amicos pluris facio quam inventa mathematica.

Reductionem quadraturarum ad Curvarum rectificationes* quam desiderare videris, inveni talem. Sit Curvæ cujusvis abscissa $x$, ordinata $y$, et area $ax$, posito quod $a$ sit data quantitas fluat $x$ uniformiter, sitque ejus fluxio $\dot{x} = a$, et ipsius $y$ sit fluxio $\dot{y}$. A dato puncto $D$ in rectâ positione data $DE$ sumatur $DB = x$, et agatur indefinita $BCG$ ea lege ut cosinus anguli $DBG$ sit ad Radium ut fluxio $\dot{y}†$ ad fluxionem $\dot{x} = a$, et inveniatur curva $FG$ quam recta $BG$ perpetuo tangit. Id enim semper fieri potest

---

* Twenty-six years later this problem, which Euler calls "celebre illud problema multum inter Geometras agitatum," was proposed by Hermann in the Leipsic Acts (Aug. 1719), and was solved by him in the number for Apr. 1723, and by J. Bernoulli in the number for Aug. 1724. The latter shews how to obtain a more general solution. See also Newton's *Geometria Analytica* (Horsley, I. 508), his Letter to Oldenburg, Jun. 23, 1673, and Euler, *Comment. Petrop.* Tom. v. p. 171. We find no allusion to Newton's solution in any of Leibniz's published letters or papers. In the figure $FD$ should be a straight line.

The following may assist some readers in verifying Newton's construction. Let $X$, $Y$ be the co-ordinates of the required curve, on the length of whose arc $(S)$ the area of the proposed curve is to be made to depend. Then $S = \int d Y \sqrt{1 + P^2}$, $(d X = P d Y)$

$$= Y \sqrt{1 + P^2} - \int \frac{Y P d P}{\sqrt{1 + P^2}}.$$

Assume $Y d P = d x$ and $\dfrac{P}{\sqrt{1 + P^2}} = \dfrac{y}{a}$; and $X$ will be found $= \dfrac{y(a^2 - y^2)}{a^2} \dfrac{d x}{d y} - x$,

$$\text{and } Y = \frac{(a^2 - y^2)^{\frac{3}{2}}}{a^2} \frac{d x}{d y}.$$

† fluxio $\dot{y}$.] This should be either "fluxio $\dot{z}$" or its equal "$\dot{y}$."

Geometrice ubi fluxionum $\dot{x}$ & $y$ relatio geometrica est.
Sit $G$ punctum contactûs, et ubi
punctum $B$ incidit in punctum $D$,
incidat punctum $G$ in punctum $F$.
In tangente $BG$ sumatur $GC$ æqualis
curvæ $GF$, et $CH$ æqualis rectæ $FD$,
et erit $BH = z$. Qua inventa habetur area quæsita $az$.

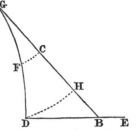

Quæ vir summus Hugenius in mea notavit ingeniosa
sunt *. Parallaxis solis minor videtur quam ipse statueram,
et motus sonorum forte magis rectilineus est, at cælos
materia aliqua subtili nimis implere videtur. Nam cum
motus cælestes sint magis regulares quam si a vorticibus
orirentur, et leges alias observent, adeo ut vortices non ad
regendos sed ad perturbandos Planetarum et Cometarum
motus conducant, cumque omnia cælorum et maris phænomena ex gravitate sola secundum leges a me descriptas
agente accurate quantum sentio sequantur, et natura sim-
plicissima sit, ipse causas alias omnes abdicandas judicavi
et cælos materia omni quantum fieri licet privandos, ne
motus Planetarum et Cometarum impediantur aut reddantur irregulares. At interea si quis gravitatem una cum
omnibus ejus legibus per actionem materiæ alicujus subtilis
explicuerit, et motus Planetarum et Cometarum ab hac
materia non perturbatos {sic} iri ostenderit, ego minime
adversabor. Colorum phænomena tam apparentium ut
loquuntur quam fixorum rationes certissimas me invenisse
puto, sed a libris edendis manum abstineo ne mihi lites ab
imperitis intententur et controversiæ. Alius est New-

---

* In an "Addition" to his "Discours de la Cause de la Pesanteur." Leid. 1690.
　　Nic. Fatio writing to Huygens from London, Feb. 24, 1690, says : "Mr. Newton,
Mr., recevra parfaitement bien tout ce que vous avez dit. Je l'ai trouvé tant de fois
pret à corriger son livre sur des choses que je lui disois, que je n'ai pù assez admirer sa
facilité, et particulierement sur les endroits que vous attaquez. Il a quelque peine à
entendre le François, mais il s'en tire pourtant avec un dictionaire." Again, Apr. 11 :
"Mr. Newton, Mr., m'a assuré qu'il prenit en fort bonne part tout ce qui est dans
le traittè de la cause de la pesanteur."

tonus\*, cujus opera in librorum editorum indicibus tibi occurrunt. His contestari volui me tibi amicum integerrimum esse & amicitiam tuam maxime facere. Vale. Dabam Cantabrigiæ, Octob. $\frac{16}{26}$. 1693†.

Utinam rectificationem Hyperbolæ quam te invenisse dudum significasti in lucem emitteres.

---

## No. XXV.

### NEWTON TO HAWES.

Mr Edward Paget, Fellow of Trinity College, and Mathematical Master at Christ's Hospital, drew up in 1694 a scheme of reading for

---

\* This refers to the following passage in Leibniz's letter: " In librorum apud Anglos editorum Indicibus occurrere mihi aliquoties libri Mathematici autore Neutono, sed dubitavi a Te essent, quod vellem, an ab alio homonymo." The author in question was John Newton, D.D., a writer of mathematical school-books. Morhof was probably thinking of this same " Doctor," when he called our philosopher " Medicus Anglus." (The passage alluded to occurs in a posthumous part of the *Polyhistor*, but was written, apparently, not long after the publication of Newton's Analysis of Solar Light. The expression is retained in Fabricius's editions of the work 1732 and 1747. Morhof died in 1691). The title is retailed by Saxius *Onomast.* v. 120: " Isaacus Newtonus Woolstropensis Anglus, Medicus, Mathematicus et Philosophus Londinensis..." Compare Report of Committee of House of Commons on abuses in the Mint (Apr. 8, 1697), in which, on the Moneyers alleging themselves to be a Corporation, it is stated that " Dr Newton, present Warden of the Mint, declared that he had never seen any such Grant or Patent to the Moneyers; and believed they had no other Charter, but the general Charter of the Mint, which he had in his possession," &c. Ruding's *Annals of the Coinage*, iii. 536, 540. (London, 1817).

Dr Henry Newton, Envoy Extraordinary (1704—1710) to the Grand Duke of Tuscany and Republic of Genoa, tells us that he occasionally received compliments that were intended for his illustrious namesake: " Et multa adhuc expectant [Itali] a Summo Mathematico ejusdem mecum Cognominis, (inde aliquoties contigit ex errore nominis, me quoque non meis laudibus ornari) præsertim verò Mundum qualem Deus, ipsi quoque Hobbesio, Æternus Geometra, ab initio formaverat, atque sapientissimus Creator in mensura, & numero & pondere disposuerat; sed intellectu facilem, non solum Mathematicis, nec quidem illis ex plebe, legendum intuendumque; sicque ille demum optimè, sibi, Patriæ, omni denique Posteritati consulat." Letter from Florence, Oct. 1, 1705, in his *Epistolæ*...Lucæ, 1710. As a sort of compensation a letter of thanks from Lord Cowper has been recently published as addressed to our philosopher, which I strongly suspect was intended for the author of the work just quoted. (Lord Campbell's *Chancellors*, iv. 341.)

† Partly from Crelle's Journal, Band xxxii. (where a portion of the letter is lithographed from the original in the Royal Library at Hanover), and partly from a copy in the British Museum, Add. MSS. 6399. fol. 56, which seems to have come to the Museum with Cole's Collections. The letter has been recently printed in the edition of Leibniz's works now in course of publication at Berlin.

the boys under his care. At a meeting of the Committee of the Schools of the Hospital on the 9th of May, Mr Hawes, the Treasurer, was "desired when he goes to Cambridge on Friday next to take with him a copy of the old and new schemes, and advise with the Professor and other Mathematicians in the University concerning them, and get their opinions in writing which of the two schemes they judge best." Newton's opinion of their respective merits is conveyed in this letter, which was sent enclosed in another to Paget.

*ffor Nathanael Hawes, Esq.*

Sr

I now return you the papers you left in my hands. The two Schemes of learning I have compared, and find that the old one wants methodizing & enlarging; the want of method you may perceive by these instances.

1.    Arithmetick is set down preposterously in the 12th Article after almost all the rest of Mathematicks. ffor a man may understand and teach Arithmetick wthout any other skill in Mathematicks, as writing Masters usually doe, but wthout Arithmetick he can be skilled in noe other parte of Mathematicks, & therefore Arithmetick ought to have been set downe in the very first place as the ffoundation of all the rest.

2.    The parts of Arithmetick are set downe in severall Articles preposterously. ffor Decimal Arithmetick and the Extraction of roots are enjoined in the 3d Article before the boyes have learnt Arithmetick in integers & vulgar fractions. Then in the 4th & 8th Articles they are enjoined Logarithms. And after all this they are required in the 12th Article to learn Arithmetick in generall, as if they had learnt nothing of it before.

3.    Geometry and Trigonometry are confounded together in the first Article, and again in the 4th. Whereas Geometry ought to have made one Article and Trigonometry another. ffor these are accounted distinct sciences.

4.    The use of Logarithms wch is set downe in the 8th Article, ought to have preceded that of Artificial Sines &

Tangents w$^{ch}$ is in the 4$^{th}$ ffor how can any man understand the Logarithms of Sines and Tangents, before he
understands the Logarithms of Numbers in generall.

5. The doctrine of the Globes is set down in the 11$^{th}$
Article and the projection of the Sphere or globe and
making of Maps is set down in the 10$^{th}$. whereas the doctrine of the globes ought to precede the projection of
the sphere & making of Maps. ffor how can any man project the lines on a sphere or globe into Maps, before he is
taught what those lines are ?

6. The 10$^{th}$ Article is worded improperly. ffor instead
of saying, *The projection of the Sphere in circles or globe in
a plain divers wayes,* It should have been said *The projection of the Sphere or globe in circles on a plain divers wayes,*
ffor the projection of a *sphere in circles* and that of a *Globe
in a plain* are neither equipollent phrases, nor branches of
a distinction, & therefore cannot be put together disjunc-
tively (as they are in this Article) w$^{th}$out an impropriety of
speech.

7. The Rule for finding the Latitude by the Sun or
Starrs in the sixth Article, and the questions of plane Sailing w$^{th}$ the use of the plane Sea Chart in the seventh,
ought to have come after the Doctrine of the globes, & the
making of Maps or Charts ; & yet these are set after the
other in the 10$^{th}$ and 11$^{th}$ Articles. Soe alsoe in the second
Article, the making of the Scale of hours, Rumbs and
Longitude, is improperly joyned with the Rule of three, &
more improperly set before the doctrine of y$^{e}$ Globes. And
in generall the whole scheme is soe confused & imme-
thodical, as makes me think that they who drew it up,
had noe regard to the order of the things, but set them
downe by chance as they first thought upon them, w$^{th}$out
giving themselves the trouble to digest and methodise the
heap of things they had collected together ; w$^{ch}$ makes me
of opinion, that it will not be for the reputation of the

ffoundation to continue this scheme any longer w$^{th}$out putting it at least into a new forme.

But then for the things it conteins I account it but mean and of small extent.  It seems to comprehend little more then the use of Instruments, and the bare practise of Seamen in their beaten road, w$^{ch}$ a child may easily learn by imitation, as a Parrot does to speak, w$^{th}$out understanding in many cases the reason of what he does ; & w$^{ch}$ an industrious blockhead, who can but remember what he has seen done, may attain to almost as soon as a child of parts, and he that knows it is not assisted thereby in inventing new things & practises, and correcting old ones, or in judging of what comes before him : Whereas the Mathematicall children, being the flower of the Hospitall, are capable of much better learning, & when well instructed and bound out to skilfull Masters, may in time furnish the Nation w$^{th}$ a more skilfull sort of Sailors, Builders of Ships, Architects, Engineers and Mathematicall Artists of all sorts, both by Sea and Land, then ffrance can at present boast of.   The defects of the old scheme you may understand by these instances.

1.   It conteins nothing more of Geometry than what Euclid has in the beginning of his first book, and in the 10$^{th}$ and 12$^{th}$ propositions of his sixth booke.  All which is next to nothing.

2.   There is nothing at all of symbolical Arithmetick, w$^{ch}$ tho' not requisite in the vulgar road of Seamen, yet to an inventive Artist may be of good use.

3.   The taking of heights and distances, and measuring of planes and solids is alsoe wanting, tho of frequent use.

4.   Nor is there any thing of spherical Trigonometry, tho the foundaçon of a great many usefull Problems in Astronomy, Geography and Navigation.

5.   Neither is there any thing of Sayling according to the severall Hypotheses, nor of Mercators Chart, nor of

computing the way of Ships tho things w^{ch} a Sailor ought
not to be ignorant of.

6. The finding the difference of Longitude, Amplitude,
Azimuts and variation of the compass is alsoe omitted,
tho these things are very usefull in long voyages, such as
are those to the East Indies, and a Mariner who knows
them not is an ignorant.

7. Nor is there one word of reasoning about force
and motion, tho it be the very life and Soul of Mechanical
skill and manual operations and there is nothing soe Mechanical as the frame & managem^t of a ship.   By these
defects it's plain that the old scheme wants not onely
methodizing, but alsoe an enlargem^t of the learning. ffor
some of the things here mentioned to be wanting, are
requisite to make a Mariner skilfull in the ordinary road,
and the rest may be often found usefull to such as shall
become eminent for skill & ingenuity, either in Sea affaires,
or such other mechanicall offices and imployments, as the
King may have occasion in his Yards, Docks, fforts, and
other places, to intrust them with.

Now the imperfections of the whole scheme are pretty
well supplyed in that new one w^{ch} is proposed to be established. ffor this is methodical, short & comprehensive.  It
excells the old one beyond comparison ; I have returned it
to you, w^{th} some few alterations for making the affinity,
coherence and good order of the severall parts of the
learning, more cleare and conspicuous, & supplying some
defects.   The alterations are of noe very great moment,
excepting the addition of the last Article, w^{ch} conteins the
science of Mechanicks.   The rest is as perfect as I can
make it without this Article. whether this should be added
may be a question, but since you concur w^{th} me in the
affirmative, I'le set downe my reasons for the addition.

ffor w^{th}out the learning in this Article, a Man cannot be
an able and Judicious Mechanick, & yet the contrivance &

managem$^t$ of Ships is almost wholly Mechanical.  Tis
true that by good naturall parts some men have a much
better knack at Mechanical things then others, and on that
acco$^t$ are sometimes reputed good Mechanicks, but yet
w$^{th}$out the learning of this Article, they are soe ffarr from
being soe, as a Man of a good Geometrical head who never
learnt the Principles of Geometry, is from being a good
Geometer. ffor whilst Mechanicks consist in the Doctrine
of force and motion, and Geometry in that of magnitude
and figure : he that can't reason about force and motion,
is as far from being a true Mechanick, as he that can't
reason about magnitude and figure from being a Geometer.
A Vulgar Mechanick can practice what he has been taught
or seen done, but if he is in an error he knows not how
to find it out and correct it, and if you put him out of
his road, he is at a stand ; Whereas he that is able to
reason nimbly and judiciously about figure, force and
motion, is never at rest till he gets over every rub.  Experience is necessary, but yet there is the same difference
between a mere practical Mechanick and a rational one, as
between a mere practical Surveyer or Guager and a good
Geometer, or between an Empirick in Physick and a
learned and a rational Physitian.  Let it be therefore
onely considered how Mechanical the frame of a Ship is,
and on what a multitude of forces and motions the whole
business and managem$^t$ of it depends, And then let it be
further considered whether it be most for the advantage of
Sea affaires that the ablest of our Marriners should be but
mere Empiricks in Navigation, or that they should be alsoe
able to reason well about those figures, forces, and motions
they are hourly concerned in.  And the same may be said
in a great measure of divers others Mechanical employments, as buildings of ships, Architecture, ffortification,
Engineering. ffor of what consequence Mechanical skill is
in such Mechanical employments may be known both by

the advantage it gave of old to Archimedes in defending his City against the Romans, by w^ch he made himself soe famous to all future ages, and by the advantage the ffrench at present have above all other Nations in the goodness of their Engineers. ffor it was by skill in this Article of learning that Archimedes defended his City. And tho the ffrench Engineers are short of that great Mechanick, yet by coming nearer to him then our Artificers doe, we see how well they fortify and defend their owne Cities, and how readily they force and conquer those of their Enemies*. You may consider to what perfection that Nation by their Schooles for Sea-Officers had lately brought their Navall strength, even against all the disadvantages of nature, and yet your schoole is capable of out-doeing them, ffor their's are a mixture of all sorts of capacities, your's children of the best parts selected out of a great multitude.

Their's are young men whose faculties for learning begin to be as stiff and inflexible as their bones, and whose minds are prepossest & diverted with other things, yours are children whose parts are Limber and pliable and free to receive all impressions. A great part of their schools are scarce capable of much better learning than that in your old scheme, your's have already shewn by experience that they are capable of all the learning in the new one, except the last Article, w^ch has not yet been taught them, and yet after they have learnt the rest, will prove noe harder then that w^ch they had learnt before. And as your children are a select Number for parts, and capable of all the learning here proposed, and it will be for the Honour & advantage of the Nation to introduce a new spirit of

---

* The capture of Mons in 1691, that of Namur in 1692, and of Charleroi in 1693, were among Vauban's recent triumphs. When Newton wrote the above remarks he probably little anticipated the example that would be set by "that nation" to his own country in paying a tribute to his genius. The "Newton" in the French steam navy is a corvette of 26 guns, 220 horse power.

usefull learning among the Seamen, soe it will give your children a higher reputation for preferrment.   And I take it to be for the Honour of both King Charles his memory and of the foundation, that this School should be as learned for Sea affaires as you can well make it; and probably it was his designe and will, it should be soe, tho all this learning was not started when he founded it.   If you admit this learning, your school will certainly grow into greater reputation, & may be thereby more apt to stir up new Benefactors and set a Precedent of good learning to all future foundations of the same kind, and if you admit it not, your scheme of learning will be imperfect and leave roome for future foundations to outstrip yours, w$^{ch}$ I beleive would not be for it's honour. ffor the scheme of learning, as I now returne it to you is an entire thing w$^{ch}$ cannot well want any of it's members. ffor 'tis nothing but a combination of Arithmetick, Geometry, Perspective and Mechanicks, I mean Geometry as well in sphericall surfaces as in plane ones.   Geometry is the foundation of Mechanicks, & Mechanicks the accomplishm$^{t}$ & Crown of Geometry, & both are assisted by Arithmetick for computing and perspective for drawing figures: Soe that any part of this Systeme being taken away the rest remaines imperfect.   These considerations have moved me to propose this Article to you, but perhaps the Governors may see reasons against it of greater weight w$^{ch}$ I am not yet acq$^{ted}$ with, & therefore I onely propose this business and leave it wholly to their prudence.

The Main difficulty that I can think of, is, that the learning of this Article may take up too much of the childrens time.   And yet if for all the rest of their learning they are allowed (as you tell me) but two yeares & halfe I question not but another halfe yeare would be abundantly sufficient for this addition, and then they would goe to sea w$^{th}$ a complete Systeme of Mathematicall learning. And perhaps it may be more for their advantage to spend this halfe yeare at schoole in an important part of learning w$^{ch}$ they cannot get at Sea, then at Sea in learning what they will afterwards learn there more readily if well instructed at School, before they goe thither.

If two yeares were not at first thought too much for the old scheme of learning w$^{ch}$ (before the addition of the Article of taking prospects) was very meane and narrow; four or five yeares for this new scheme would be but a moderate allowance at that reckoning, & therefore tis very much if they can learn it in three. And yet perhaps they may run through all the parts of it in two yeares and an halfe; but not soe well: And I would advise that they should rather be allowed three full yeares, then be sent away smatterers in their learning.

But whether they be allowed two yeares & an halfe or three yeares, I conceive the time of their examination ought to be stated. ffor the liberty w$^{ch}$ the Masters of Ships have had of taking away the boys sometimes before they had gone through the whole course of their Mathematicall learning, seems to me a mischief w$^{ch}$ may deserve a reformation. ffor the sending abroad unripe boys can be neither a reputation to the School, nor advantage to the Nation; Such boyes being not onely less knowing then others, but alsoe less able to make use of what they have learnt, & more apt to forget it, as smatterers in a Gramar school doe their Latine.

Nor doe I see how the genius & method of the School in goeing through the whole course of the Mathematicall learning can be carried on soe evenly and advantagiously, as when y$^e$ Mathem$^{ll}$ Master shall be at a certainty in the Number of Scholars, & in the time against which he is to make them fit. As the constitution now is you leave a bad Mathematicall Master a liberty of making excuses when ever he shall prove negligent, & discourage a good one

by the uncertainty of his business & method & of the satisfaction & reputation of bringing his Schollars to perfection,
& alsoe by leaving him exposed to such humours as may
desire by that meanes to take opportunity of hurting him
in his business or reputation : whereas it's your interest
to make the place as desirable as you can, that when it
becomes void you may have the greater choice of such men
as are fittest for it, & encourage them to goe on cheerfully
with their duty. And if it may be for the credit & interest
of y$^e$ foundacon not onely that the boyes should be well
learned, but alsoe that they should be placed abroad w$^{th}$
the best Masters, & the appointing two solemn times every
yeare for examining five boyes at a time & binding them
out apprentices may draw together a greater choice of
good Masters then in the petty examinations at present,
As a ffair draws together a greater Number of Chapmen
than little markets doe : If the giving publick notice of
those times may alsoe make the thing more solemn &
more known to the Nation, & thereby conduce to the
honour of the foundation, & probably to the stirring up of
new Benefactors : I should think the conjunction of soe
many advantages may well deserve an establishment, unless
there should be some great objection against it w$^{ch}$ I am
not yet aware of. ffor you have told me that when the
boyes have run through their course of learning there will
be noe danger of their not meeting with Masters at the
next publick examination, and if any of them should then
happen to fail of Masters, they would at all times after
that be at liberty to goe with such Masters as could be
met with. And as for the Examinations I should think
that the more publick they are, the more the School will
be concerned for its reputation, & the greater will be the
reputation w$^{ch}$ it may get by the good performance of the
boyes. If there be any advantage in publick Examinations,
the more publick they are the greater the advantage : if in

private ones the Governors may have it at their Visitations
by able and diligent Examiners w$^{th}$ as much privacy and
severity as they please : And if more such examinations
shall upon any occasion be found requisite, yet I conceive they should be made onely by Examiners appointed
by the Governors, & obliged, soe soon as the Examination
is over to give an account to the Governors, & to noe body
else w$^{th}$out their permission, of what ever they find amiss.

When the boyes are sent to Trinity house to be publickly examined perhaps it would not be amiss that the
Mathematicall Master send along w$^{th}$ them a larger & more
particular draught of the things they have been taught, &
are prepared to be examined in, then that scheme of learning w$^{ch}$ you establish, and that the draught of every Master
with the alterations from time to time made in it and the
Number of the boyes who at every examination answer
well and readily to the things therein, be kept upon record
in the school as a standard of the learning w$^{ch}$ the boyes
are capable of w$^{th}$in the time allowed them.

And when the boyes are put out apprentices, they may
be exhorted or obliged by the Governors to communicate
to the School (in gratitude to the place of their education)
such accurate observations, curious discoveries and select
draughts as they shall make abroad in their Voyages and
ffactories for rectifying the longitudes and situation of
places in the Maps, or otherwise improving Geography,
Hydrography, Navigation, the building of Ships, Trade or
any valuable knowledge of remote Nations or Regions. And
these or other curiosities communicated by them may be
kept together in a convenient place as an Ornament of the
Schoole to be consulted upon occasions. I have hitherto
considered onely the Kings ffoundation, and herein I
have been free in comparing the old & new schemes of
learning, and speaking my thoughts about them, because,
as you told me, it was desired. I hope it will give noe

offence to any body. ffor at the first founding of the Schoole, the old scheme might serve very well for a tryall, till it was known what learning such young children might be capable of. And I presume that the Mathematicians who drew it up, intended for them nothing more then part of that learning which is taught to persons of riper age in the ffrench Schools, and thought it more advisable to leave the method of the things to the Mathematical Master, then to be accurate in what could not be made perfect. The conjunction of Mr Stones ffoundation* with the Kings seems to be well designed : ffor as both the Honour and Interest of the Kings ffoundation is consulted by making Mr. Stone's subservient & usefull to it : Soe it is both for the Honour of Mr. Stone's ffoundation to have this relation to the King's, and for the Interest of it, that his boyes may be preferred to the King's, where they will be bound out Apprentices w^{th} a better allowance. But care should be taken that the Kings boyes be not retarded in their learning, by joyning w^{th} them too great a Number of other boyes of inferior parts, soe as to hinder them from getting through their scheme of learning within the time limited.

I like well the designe of establishing some Latin Authors to be read in the Schoole, because the best Mathematicall books are in that language, & by useing the boys to Mathematicall Latin, they will be enabled to understand them. The *Synopsis Algebraica* and *Wards Trigonometry* are well chosen and soe is *Euclides nova methodo* in regard of the short time allowed the boyes. Yet Euclid himself (suppose in Barrow's edition) would doe them more good if it could be compassed within the time, and would be more usefull to them in reading other Authors afterwards. And therefore the Governors may

---

* Henry Stone had, in 1693, bequeathed the bulk of his property to the Hospital, of which at least £50 a year was to be devoted to the improvement of the mathematical department of the school.

establish, if they think fit, that the Boyes read either
*Euclides nova methodo* or else at the discretion of the Mathematicall Master the first six books of *Euclides Elements*
in Barrows edition for plane Geometry and the 11$^{th}$ and
12$^{th}$ books thereof for Solids. ffor soe the Mathematicall
Master will be at liberty to read the Elements themselves
soe soon as he finds he can compass it and the rest of the
scheme w$^{th}$in the time limited. As for the Doctrine of the
Sphere the first book of *Mercator's Astronomy* is brief and
well adapted to the use of the Schoole and therefore may
be appointed.

And now I have told you my opinion in these things, I
will give you Mr. Oughtred's, a Man whose judgment (if
any man's) may be safely relyed upon. ffor he in his book
of the circles of proportion, in the end of what he writes
about Navigation (page 184) has this exhortation to Seamen
"And if, saith he, the Masters of Ships and Pilots will take
the pains in the Journals of their Voyages diligently &
faithfully to set down in severall columns, not onely the
Rumb they goe on and the measure of the Ships way in
degrees, & the observation of Latitude and variation of
their compass; but alsoe their conjectures and reason of
their correction they make of the aberrations they shall
find, and the qualities & condition of their ship, and the
diversities and seasons of the winds, and the secret motions
or agitations of the Seas, when they begin, and how long
they continue, how farr they extend & w$^{th}$ what inequality;
and what else they shall observe at Sea worthy consideration,
& will be pleased freely to communicate the same with
Artists, such as are indeed skilfull in the Mathematicks
and lovers & enquirers of the truth: I doubt not but that
there shall be in convenient time, brought to light many
necessary precepts w$^{ch}$ may tend to y$^{e}$ perfecting of Navigation, and the help and safety of such whose Vocations
doe inforce them to commit their lives and estates in the

vast Ocean to the providence of God." Thus farr that
very good and judicious man Mr. Oughtred. I will add,
that if instead of sending the Observations of Seamen to
able Mathematicians at Land, the Land would send able
Mathematicians to Sea, it would signify much more to the
improvem$^t$ of Navigation and safety of Mens lives and
estates on that element.

I hope S$^r$ You will all interpret my freedome in this
Letter candidly and pardon what you may therein think
amiss, because I have written it with a good will to your
ffoundation, and now I have spoke my thoughts I leave
the whole business to the wisdome of your selfe and the
Governors. I am,

Hon$^{rd}$ S$^r$.

Your most humble & most

obedient Servant,

Cambridge May 25$^{th}$, 1694. Is. NEWTON.

———

[Accompanying the above.]

A New Scheme of Learning proposed for the Mathematical Boys in
Christ's Hospital. {Paget's scheme with a few alterations by
Newton who has also added the 10th article.}

1. *Arithmetick* in Integers, Vulgar fractions & Decimals, in Proportional numbers natural and Artificial, in
Symbols of unknown Numbers & in Equations.

2. Geometry speculative and practical in planes and
Solids.

3. The Application of Arithmetick to Geometry in
determining and protracting Lines, Angles and figures by
Numbers natural and Artificial, Symbols of Numbers and
tables of Sines & Tangents.

4. The description and properties of figures in perspective with the Arts of drawing and designing.

5. The use of the best Instruments in working by proportionals taking Angles, heights and distances, and measuring planes and solids.

6. The Doctrine of the Globes and the Rudiments of Geography Hydrography and Astronomy.

7. The descriptions of the Globe in perspective commonly called Projections and the Art of making Charts and Maps.

8. The Doctrine of Spherical Triangles w$^{th}$ their application in projecting and computing all the usefull Problems in Geography, Astronomy and Navigation.

9. A full application of the learning aforesaid to Navigation particularly to the severall Hypotheses thereof, commonly called Plane, Great circle and Mercators sailing. As alsoe the use of Charts and Sea Instruments for observation and their application to the finding of the Latitude, difference of Longitude, Amplitudes, Azimuths and variation of the compass by the Sun or Starrs, w$^{th}$ the knowledge of the Tides and Roman Calender, and the method of keeping Journals and of finding the difference of the Longitudes of Shores by the Eclipses of Jupiters Satellites.

10. The principles of reasoning about force & motion, particularly about the five mechanical powers, the stress of ropes and timber, the power of winds, tides, bullets and bombs, according to their velocity and direction against any plane, the line w$^{ch}$ a bullet describes, the force of weights and springs and the power of fluids to press against immersed bodies, and bear them up, and to resist their motions; w$^{th}$ the application of this learning to Sea affaires, for contriving well and managing easily, speedily & dextrously, Levers, Pulleys, Skrews, Anchors, Pumps, Rudders, Guns, Sails and other Tackle, judging truly of the advantages & disadvantages of Vessells, Havens, fforts, Engins and new Projects, & observing or discovering what

ever tends to make a Ship endure and Sail well, or otherwise to correct or improve Navigation.

Is. Newton *.

---

## No. XXVI.

### NEWTON TO HAWES.

*ffor Nathan^u. Hawes, Esq.*

S^r Yesterday I sent by the Carryer a Letter to you w^th the papers you left in my hands, inclosed in another to M^r. Paget. In that I wrote to you, you will find my thoughts set downe at large about the old and new schemes of learning. Looking this morning into S^r Jonas Moore's Systeme of Mathematicks w^ch he composed about 15 or 16 yeares agoe for the use of your schoole, I find by the title page and preface to that book, that the new Scheme was for the most part composed at that time by S^r Jonas. ffor there (as is mentioned in the preface) he proposes to teach in order these sciences.

1. Arithmetick vulgar, decimal and Logarithmical.
2. Practical Geometry.
3. Trigonometry plane and spherical.
4. Cosmography w^ch includes the Doctrine of the Globes with Astronomy and Geography.
5. Navigation with the making of Maps.

After these and many Tables & Geographical Maps follow Algebra & speculative Geometry conteined in the first, 6^th & 11^th & 12^th books of Euclid's Elements. The difference between this method and the new Scheme of learning now proposed lies in these things.

1. In the new scheme (as alsoe in the title page to S^r Jonas Moores book) Algebra is joyned w^th Arithmetick, & speculative Geometry w^th the practical; w^ch certainly is

---

* This and the two following letters are from the official copies in the Christ's Hospital Court Book.

the best method for Schollars of good parts who are to learn both. But in the preface to S$^r$ Jonas Moores book Algebra & speculative Geometry are separated & taught apart after all the other Sciences ; w$^{ch}$ is best for a mixture of Schollars of all degrees of parts, some of w$^{ch}$ are not capable of learning the whole Scheme.

2. S$^r$ Jonas joyns plane & spherical Trigonometry together, but in the new scheme spherical Trigonometry is set after the Doctrine of the Sphere w$^{ch}$ is more proper for a learner.

3. S$^r$ Jonas omits perspective and Mechanicks & referrs the taking heights and distances & mensuration of planes & solids to the end of practical Geometry and plane Trigonometry : whereas in the new scheme perspective is inserted between them for delineating the heights, distances and solids w$^{ch}$ are to be measured, & again after y$^e$ doctrine of the globes for the making of Maps.

This I thought proper to signify to you, that the Governors of the Hospitall might have the judgment of S$^r$ Jonas in this matter. ffor he follows not y$^e$ old scheme in any thing, but agrees well enough w$^{th}$ the new one, both in y$^e$ substance of the things he teaches, & in the order of them, if perspective & Mechanicks be inserted into his Systeme in their proper places. By S$^r$ Jonas his departing soe much from y$^e$ method of the whole scheme, and supplying some things w$^{ch}$ were wanting in it & coming soe neare to the new one, you may gather that the old one in his judgm$^t$ wanted information, & that the new one is not much amiss. S$^r$ I am,

Yo$^r$ most humble & most obed$^t$ Serv$^t$

Is. NEWTON.

The new scheme with Newton's modifications was sent to Wallis and David Gregory at Oxford, who gave their "opinion and advice" respecting it in a joint paper, dated June 13, 1694. "After a very large debate" on June 25, it was agreed to adopt the new scheme. The

Committee also stated it as their "opinion that the 10ᵗʰ Art. in the new scheme about the 5 Mechanical powers cannot be taught under 6 months longer time than is allowed for their instruction in Mathematics. Also that the Court be desired to request Mr Newton to enlarge himself upon the aforesaid 10ᵗʰ Art. that so Mʳ Paget may be the better qualified for their instruction therein, being very advantageous to the improvement of Navigation." It was at the same time ordered that "humble & hearty thanks be returned to Mr Newton, Dʳˢ Wallis & Gregory for their extraordinary pains & kindness in this affair." A letter of thanks was accordingly sent Aug. 9, in which it is ʹobserved that "the plan requiring long & serious consideration, we have chosen a committee to consider thereof, but being unwilling to defer our acknowledgments" &c.

## No. XXVII.
### NEWTON TO HAWES.

Sʳ.                                   Cambridge June 14. 1695.

I should have writ to you by Mr. Newton * but that I stay'd to consider further of yᵉ scheme of Mathematical learning before it be established. ffor the last Article seemed too indefinite to be subscribed, and in the forme it is there set downe, has noe books written of it, & therefore I have changed it into the last Article of the scheme I now send you enclosed in this Letter. ffor this last Article conteins as much of the other, as has been hitherto reduced to a certain science and something more, and is definite, soe that the Master may know what he subscribes, and the Governors what the Master is obliged to by his subscription. It has alsoe books written upon every parte of it to make it more fit for the school. As for Mʳ. Newton I never took him for a deep Mathematician, but recommended him as one who had Mathematicks enough for your business, wᵗʰ such other qualifications as fitted him for

---

* Mr Samuel Newton, who had been recently appointed to the Mathematical Mastership at Christ's Hospital, vacant by Paget's resignation. Compare Newton's letters in Baily's *Flamsteed*, pp. 153, 154, 156.

a Master in respect of temper and conduct as well as learning. But I reckon two yeares too short a time for this scheme of learning, and D$^r$. Gregory who taught Mathematicks in Scotland w$^{th}$ very good success, and was here last weeke, tells me that by the time he spent in teaching he should reckon three yeares little enough for this scheme. M$^r$. Newton may try if he can compass it sufficiently in two yeares but if that be found too little, perhaps the wisdome of the Governors may soe order things as to allow him halfe a yeare more from the other schooles. ffor were it not for some Mathematicall bookes in Latine, I should think that language of soe little use to a Seaman as not to deserve four or five yeares of the childrens time, while Mathematicks are allowed but two ; I thank you for your concerne and pains in behalfe of M$^r$. Newton, and am very glad to understand that he behaves himselfe so well. ffor tho' I was almost a stranger to him when I recommended him, yet since he was elected, I reckon myselfe concerned that he should answer my recommendation. The ill will you may have got by your acting for him I perceive is but of little extent and cannot hurt you. M$^r$. Caswel's freinds at Oxford blame his freind* neere London, and some of them think the place would not have suited with his humour, soe that I am satisfyed you made the best choice. S$^r$. Your most humble & most obedient Servant.

Is: NEWTON.

---

* Flamsteed, who had recommended Caswell as Paget's successor.

[Enclosed in the above.]

A New Scheme of Learning proposed for the Mathematical boyes in Christ's Hospitall.*

1. *Arithmetick* in Integers, Vulgar fractions & Decimals, in Proportional numbers natural and Artificial, in Symbols of unknown Numbers & in Equations.

2. Geometry in Planes & Solids, with the Demonstrations thereof & $y^e$ practise by the Rule & Compass.

3. The application of Arithmetick to Geometry in determining & protracting lines, angles, and plane Triangles†, by numbers natural and artificial, Symbols of Numbers, & Tables of Sines & Tangents.

4. The description & properties of ffigures (rectilinear & circular) in *Perspective*, $w^{th}$ the Art of Designing‡ & Drawing‡.

5. The construction & use of the best *Instruments* in working by Proportionals, taking Angles, Heights & Distances, & Surveying, Guaging, or otherwise measuring Planes & Solids.

6. *Cosmography*, or the rudiments of Astronomy, Geography & Hydrography, with the Projections of the globe in Perspective, & the art of making Maps & Charts.

7. The doctrine of *Spherical Triangles*, with their application in projecting & computing all the useful Problems in Astronomy, Geography & Navigation.

8. A full application of the Learning aforesaid to *Navigation* particularly to the several Hypotheses thereof commonly called Plane, Great circle, Parallel & Mercator's sailing. As also the use of Charts & Sea Instruments for Observation, & their application to the finding of the Latitude, difference of Longitude, Amplitudes, Azimuths &

---

* There is a copy of this paper in Newton's handwriting in Trin. Coll. Library in a folio volume marked R. 5. 4.

† In Newton's MS. it is " plane triangles & other figures."

‡ These words change places in Newton's MS.

Variation of the Compas by y^e Sun or stars, with the knowledge of Tides, Currents & the Roman Calendar & the
method of keeping Journals, & of finding the longitudes of
shores by the Eclipses of Jupiters Satellites.

9. The *mechanical* Arts or Sciences of the five Powers,
The laws of motion, Hydrostaticks, Gunnery & ffortification.

A minute dated 19 July, 1695, states that "the consideration of
the new scheme...drawn up by Mr Newton...which was referred by the
last Court to this Committee is for several reasons postponed until
another time."

The master seems to have found the scheme difficult to carry into
practice, and a course of study formed by a fusion of the old and new
schemes, and excluding Mechanics except " so much of gunnery as is
necessary for sea service" was afterwards adopted. (Minutes of Apr. 6
and June 10, 1696.)

A few notices of our philosopher, taken from the same source to
which we are indebted for the three preceding letters, and exhibiting
him in connexion with Christ's Hospital, may be given here.

" March 25, 1696. The Committee being informed that Mr Newton
is in town |summoned by Charles Montagu's letter offering him the
Wardenship of the Mint| and will stay some days, desired the Treasurer
to request him to examine and consider of the Library belonging to the
Mathematical School....and give his opinion what books are wanting
that may be most useful and necessary.

July 13, 1697. The Committee did desire Mr Isaac Newton now
present to deliver his opinion concerning the said |five| boys, who was
pleased to say that about 10 or 14 days since he examined them and
then found them perfected, except in a very few particulars, which by
this time he don't question but they are masters of, and therefore is of
opinion they are well qualified to be placed forth to sea as apprentices...
And this Committee returned their unanimous thanks to Mr Professor
Newton for his great kindness and pains taken herein."

He is also mentioned as present at the Hospital meetings on Sept.
23, (visitation of all the schools in the hospital) and Dec. 16, 1697,
on which latter day he was appointed one of a committee to consider
how £100 might best be laid out for the improvement of the mathematical library.

## No. XXVIII.

### WALLIS TO NEWTON.

Sir,                       Oxford, Apr. 10, 1695.

I was in hopes of seeing you in Oxford last summer; which made me neglect sending you (by the Carrier) two Cuts which belonged to the Volume you had before. They were not wrought off at y^e Rolling-Press when you had the rest; but are easy to be inserted in their proper places. I send them now, with the other Volume; which I desire you to accept.

I understand (from Mr Caswell) you have finished a Treatise about Light, Refraction and Colours; which I should be glad to see abroad. 'Tis pitty it was not out long since. If it be in English (as I hear it is) let it, however, come out as it is; & let those who desire to read it, learn English. I wish you would also print the two large Letters of June and August {October} 1676. I had intimation from Holland, as desired there by your friends, that somewhat of that kind were done; because your Notions (of *Fluxions*) pass there with great applause, by the name of *Leibnitz's Calculus Differentialis*. I had this intimation when all but (part of) the Preface to this Volume was Printed-off; so that I could onely insert (while the Press stay'd) that short intimation thereof which you there find. You are not so kind to your Reputation (& that of the Nation) as you might be, when you let things of worth ly by you so long, till others carry away the Reputation that is due to you. I have endeavoured to do you justice in that point; and am now sorry that I did not print those two letters *verbatim*.

I understand you are now about adjusting the Moons Motions; and, amongst the rest, take notice of that of the *Comon Center of Gravity* of the Earth & Moon as a conjunct body: (a notion which, I think, was first started

by me, in my Discourse of the Flux and Reflux of the Sea.)
And it must needs be of a like consideration in that
of Jupiter with his Satellites, & of Saturn with his. (And
I wonder we have not yet heard of any about Moon.) But
Saturn and Jupiter being so far off, the effects thereof are
less observable by us than that of the Moon. My advise
upon the whole, is, that you would not be too slow in
publishing what you do.

I am S<sup>r</sup>

Your very humble Servant,

JOHN WALLIS *.

*For Mr Isaac Newton,*
*Fellow of Trinity College, &*
*Professor of Mathematick,*
*in Cambridge.*

With a Book {the 1st Vol. of Wallis's Works.}

Wallis was a strong advocate for the immediate publication of discoveries. In a letter to Waller (Sec. to Royal Soc.), April 30, he
dwells upon the same topics, and speaks of Newton's Treatise as
"finished & fairly transcribed some while since. I wish he were
called upon to print it without farther delay. Perhaps Mr Halley may
prevail with him so to do, &c."

Waller writes back May 15 " Mr Halley has promised to write to
Mr Newton concerning those letters {to Leibniz} you mention. I
hope they may be procured from him & thank you for the intimation
thereof."

Wallis writing to Halley Nov. 11, says : " I have written several letters to Mr Newton about it {i. e. printing the two letters} pressing with
some importunity the printing of them, and of his Treatise about Light
and Colours (as being neither just to himself nor kind to the publick to
delay it so long. As to the Letters I sent him a fair transcript ready
for the press {Newton's copies of them may have perished in the
fire which destroyed a mass of other papers, and, as Wallis supposed,
Leibniz's answers among them ; see Wallis's Works III. 654 or *Commerc. Epistol.* 110 or 211 ed. 2}, which if he would print, it might
best be done here, (and I would take the care of it)......But he did
not seem forward for either..........As to that about Light & Colours

---

* *Orig. Lett. Bk. Roy. Soc.* W. 2. 48. Part of it is printed in Raphson's *Hist. of
Fluxions,* p. 120.

(for which I am more solicitous) your interest may possibly prevail with him better than mine to get it published."

" In pursuance of" a letter from Halley dated Nov. 21, Wallis sent him copies of the two letters on the 26th, observing : " I am glad Mr Newton is inclinable to print some of the things he hath by him. So many as he hath on his hands at once do hinder one another. I am most fond of his Book of Light and Colours. His fear of disputes and cavils need not trouble him. It will be at his choice whether or not to answer them. His Hypothesis will defend itself. We are told here that he is made Master of the Mint" &c. *Orig. Lett. Bk. Roy. Soc.* W. 2. 56.

## No. XXIX.
### NEWTON TO HARINGTON.

Mr John Harington (of the family of "Ariosto" Harington and " Oceana" Harington), an undergraduate of Oxford, seems to have had some conversation with Newton upon a method which had occurred to him of representing musical intervals by the additions of the sides (3, 4, 5) of a right-angled triangle, and to have alluded to the bearing of the subject upon the principles of architectural beauty. At Newton's request he sent the details of his method with remarks upon the application of harmonical ratios to architecture, in a letter dated Wadham College May 22. 1698. The receipt of this letter Newton acknowledged in the following kind and encouraging terms.

Sir,

By the hands of your friend, Mr. Conset, I was favoured with your Demonstration of the Harmonic Ratios, from the Ordinances of the 47th of Euclid. I think it very explicit and more perfect than the Helicon of Ptolemy, as given by the learned Doctor Wallis. Your observations hereon are very just, and afford me some hints which, when time allows, I would pursue, and gladly assist you with any thing I can, to encourage your curiosity and labours in these matters. I see you have reduced, from this wonderful proposition, the inharmonics as well as the coincidences of agreement, all resulting from the given lines three, four, and five. You observe that the multiples hereof furnish those ratios that afford pleasure to the eye in architectural

designs : I have, in former considerations, examined these
things, and wish my other employments would permit my
further noticing thereon, as it deserves much our strict
scrutiny, and tends to exemplify the simplicity in all the
works of the Creator; however, I shall not cease to give
my thoughts towards this subject at my leisure. I beg you
to pursue these ingenious speculations, as your genius
seems to incline you to mathematical researches. You
remark that the ideas of beauty in surveying objects arises
from *their* respective approximations to the simple constructions, and that the pleasure is more or less, as the
approaches are nearer to the harmonic ratios\*. I believe
you are right; portions of circles are more or less
agreeable, as the segments give the idea of the perfect
figure from whence they are derived. Your examinations
of the sides of polygons with rectangles certainly quadrate
with the harmonic ratios. I doubt some of them do not;
but then they are not such as give pleasure in the formation or use. These matters you must excuse my being
exact in, during your inquiries, till more leisure gives me
room to say with more certainty hereon. I presume you
have consulted Kepler, Mersenne, and other writers on the
construction of figures. What you observe of the ancients
not being acquainted with a division of the sesquialteral
ratio is very right; it is very strange that geniuses of their
great talents, especially in such mathematical considerations, should not consider that, although the ratio of three
to two was not divisible under that very denomination, yet
its duple members six to four easily pointed out the ditone
four to five, and the minor tierce six to five, which are the
chief perfections of the diatonic system, and without which
the ancient system was doubtless very imperfect. It

---

\* Comp. Kepler, *Harmon. Mundi*, p. 126. In Architectonica quæcunque propor-
tiones longitudinis ad latitudinem vel crassitiem plurimùm probantur, etiam à non
Mathematicis spectatoribus, eæ quàm proximæ harmonicis inveniuntur.

appears strange, that those whose nice scrutinies carried them so far as to produce the small limmas, should not have been more particular in examining the greater intervals, as they now appear so serviceable when thus divided. In fine, I am inclined to believe some general laws of the Creator prevailed with respect to the agreeable or unpleasing affections of all our *senses;* at least the supposition does not derogate from the wisdom or power of God, and seems highly consonant to the macrocosm in general. Whatever else your ingenious labours may produce I shall attentively consider, but have such matters on my mind, that I am unable to give you more satisfaction at this time; however, I beg your modesty will not be a means of preventing my hearing from you, as you proceed in these curious researches; and be assured of the best services in the power of

Your humble Servant,

{Jermyn Street} May 30, 1698.        Is. NEWTON*.

## No. XXX.

The decree of the German Diet (Ratisbon, Sept. 23. 1699, see *Montucla, Hist. des Math.* iv. 325,) reforming the Julian Calendar and ordering (1) that the day after Febr. 18. 1700 should be March 1, and (2) that Easter should be determined by astronomical calculation (viz. of the exact time of the vernal equinox and the full moon following it), gave rise to considerable discussion among the theologians and scientific men of the Empire. In Leibniz's Works (iv. pars ii. 115—137) will be found the correspondence which he had with Roemer upon the subject. Leibniz also consulted the French Academy (Ib. 143) and the Royal Society on the second of the two Articles of the Ratisbon *conclusum :* his application to the latter body was laid before Newton, whose answer is contained in the paper now presented to the reader.

* H. Harington's *Nugæ Antiquæ*, Lond. 1779. (II. 107), where Harington's letter and Newton's answer are dated 1693, but as Harington was not admitted at Wadham until June 1696, being then in his 17th year, I have ventured to suppose that the 3 has been printed by mistake for 8.

Subjoined are some notices, bearing upon the subject, extracted from the Journal Book of the Royal Society.

Febr. 21.1700. A letter from Leibniz to Sloane {Jan. 30. *Letter Bk*. 276} was read concerning the change of style, {in which the writer desires the opinion of the Society upon the point}.

Sloane said he heard Mr Newton had made a very good calculation of the year, and that the settling that affair might be helped by it. Sloane was ordered to wait on Mr Newton about it.

Apr. 25. Sloane read an answer to Leibniz's letter containing Mr. Newton's opinion concerning the alteration of the style, {the paper here printed}.

The Vice-President (Sir Robt. Southwell) said his opinion was that this paper be sent to Mr. Leibniz, and in the meantime that he procure Mr Flamsteed's and Dr Wallis's opinion, and send them to him: also that a copy of this be kept.

May 1. Copy of Leibniz's letter and Newton's answer ordered to be sent to Flamsteed, and an answer requested.

May 22. A letter from Wallis read (returning Newton's paper) concerning the Julian account. (*Orig. Lett. Bk*. W. 2. 66). Copy ordered to be sent to Leibniz.

May 29. Flamsteed's opinion of Leibniz's letter read (dated May 22. *Lett. Bk*. XII. 326).

Jun. 5. Sloane read a letter from Flamsteed against Leibniz's reasons for changing the style.

Among Flamsteed's MSS. at Greenwich (Vol. 33) are copies in his hand of Leibniz's letter and Newton's answer, to the latter of which he has added remarks redolent as usual of *amour-propre*. Of the former he observes "This letter imparted to me by Dr Sloane, May 2. ♃ 1700, but the schedule of Mr Newton was sent away without expecting my answer." The paper as revised by Flamsteed was sent to Leibniz with Newton's approval in a letter, dated July 4. "He (Newton) does not say tis his own, but what he approves of from the best observations he thinks have been made in England by Mr Flamsteed, Halley," &c. (*Orig. Lett. Bk*. S. 2. 14.)

## Elementa motuum Solis et Lunæ
### ab Æquinoctio verno.

Tempus æquabile, quod verum dici solet diurnæ non solis sed ffixarum revolutioni proportionale est et inde condendæ sunt Tabulæ pro æquatione Temporis.

20

In Observatorio Regio Grenovicensi, Anno Christi 1701 ineunte ad meridiem Kal. Jan. stylo veteri, erit medius motus Solis 9$^s$. 21$^{gr}$. 42′. 38″. Apogæi ejus 3$^s$. 07$^{gr}$. 44′. 30″, Lunæ 10$^s$. 28$^{gr}$. 30′. 12″ & Apogæi ejus 11$^s$. 08$^{gr}$. 25′. 14″.

Uraniburgum est orientalius Observatorio Regio Parisiensi 00$^h$. 42′. 10″ & hoc Observatorium est orientalius Grenovicensi 00$^h$. 09′. 15″, et inde per reductionem habentur motus illi medii eodem die et hora ad meridianum Uraniburgi, viz$^t$. Solis 9$^s$. 21$^{gr}$. 40′. 32″ Apogæi ejus 3$^s$. 07$^{gr}$. 44′. 30″ Lunæ 10$^s$. 28$^{gr}$. 01′ 58″ & Apogæi ejus 11$^s$. 8$^{gr}$. 25′. 00″. Et ante undecim dies seu meridie Kal. Jan. stylo novo erit motus medius Solis 9$^s$. 11$^{gr}$. 50′. 00″ Apogæi ejus 3$^s$. 7$^{gr}$. 44′. 32″*, Lunæ 6$^s$. 03$^{gr}$. 05′. 33″ & Apogæi ejus 11$^s$. 07$^{gr}$. 11′. 28″.

Maxima Solis Prost{h}aphæresis quæ Keplero est plusquam 2$^{gr}$. 3′ debet esse tantum 1$^{gr}$. 56′. 20″.

Ubi hæc æquatio additur vel subducitur medio motui Solis debet ejus pars decima e contra subduci vel addi medio motui Lunæ. Nam medius motus Lunæ non est uniformis sed per vices tardescit et acceleratur propterea quod orbis Lunæ dilatatur in perigæo Solis et contrahitur in ejus Apogæo.

Postquam motus medius Lunæ sic correctus habetur, reliqua peragenda sunt per Tabulas Kepleri: et Æquinoctium vernum incidet semper in diem horam et minutum ubi longitudo Solis per hoc computum prodit 00$^s$. 00$^{gr}$. 00′. 00″†.

---

* Gregory informed Wallis that the "32" is miswritten for 28. Wallis's letter, May 11, 1700.

† *Orig. Lett. Bk.* Royal Soc. N. 1. 63.

## No. XXXI.

### NEWTON TO SIR JOHN NEWTON.

Sir John

I was very much surprized at the notic of M$^r$. Cook's*
death brought me this morning by the bearer who being
an undertaker came to me to desire that I would speak
to you that he might be employed in furnishing things
for y$^e$ funeral. He having married a near kinswoman of
mine I could not refuse troubling you with this letter in
his behalf beleeving that he will do it well if you are not
otherwise provided. I had an opinion that my Cousin was
not in danger tho weak, w$^{ch}$ makes my concern the greater
for the loss. I am

Yo$^r$ affectionate Kinsman

and most humble Servant

{Jermyn Street, Apr. 1707}　　　　　　I$^s$ NEWTON†.

*For* S$^r$ JOHN NEWTON, Baron$^t$
{*at his house in Soho Square.*}

---

## No. XXXII.

This is the rough draught of a critique on three papers of Leibniz's
n the Leipsic Acts for Jan. and Febr. 1689 (pp. 36, 38, 82), and was
probably written in 1712, after the receipt of Leibniz's second letter to
Sloane (see p. 55, *ante*). It is copied from a folio sheet in Newton's
hand which formerly belonged to Keill and is now preserved among the
Lucasian papers (packet No. 8.) Several expressions in the introductory sentences, as Newton had first written them, coincide with some of
those in the second of the two statements published in Rigaud's Essay
on the First Publication of the *Principia* (Appendix, p. 67): but Newton afterwards crossed them out and substituted others for them.
These alterations (with one or two others) bring the language of this
document into still closer agreement with that used in the *Commercium*

---

* Possibly Edw. Coke, Esq., of Holkham, (great-great-grandson of the Chief
Justice), who married Cary, daughter of Sir John Newton, and died Apr. 13, 1707.
His son Thomas was created Earl of Leicester in 1744.

† The original is in the possession of P. O'Callaghan, Esq., to whom I am indebted
for a copy of it.

*Epistolicum* (p. 97, ed. 1 ; p. 206, ed. 2), the editors of which work must therefore have seen either the document itself or a copy of it, or perhaps a still later corrected form of it. The opening sentence of this paper seems to have passed through the following stages :

1. Newtonus anno 1684 Propositiones principales earum quæ in Philosophiæ Principiis Mathematicis habentur cum Societate Regia communicare cœpit, &c.

2. Ineunte anno 1684 Newtonus Propositiones......cum Societate Regia communicavit, &c.

3. Anno 1683 Newtonus Propositiones......

4. Anno 1683 ad finem vergente Newtonus Propositiones principales earum...habentur Londinum misit eædemque cum Societate Regia mox communicatæ sunt, &c.

Newton first of all clearly wrote 1684, then altered the 4 to a 3, afterwards crossed all the figures out and wrote distinctly 1683. I mention this the more particularly, because Mr Rigaud says (Essay, p. 20) that in the MS. of the latter of the two fragments which he has published from the Macclesfield Collection, the year 1683 was at first written, "the last figure having been evidently altered to a 4." Newton therefore after endeavouring to recollect the exact year in which he sent up the fundamental propositions of the *Principia* to London, antedated the event by a twelvemonth. See Syn. View of his Life, under date Nov. 1684.

## Ex Epistola cujusdam ad Amicum.

Anno 1683 ad finem vergente Newtonus Propositiones principales earum quæ in Philosophiæ Principiis Mathematicis habentur Londinum misit eædemq: cum Societate Regia mox communicatæ sunt, annoq: 1686 Liber ille ad Societatem missus est ut imprimeretur, et proximo anno lucem vidit. Deinde anno 1688 epitome ejus in Actis Lipsicis impressa est, qua lecta D. Leibnitius Epistolam de lineis opticis, Schediasma de res{is}tentia Medii & motu projectilium gravium in Medio resistente, & Tentamen de motuum cœlestium causis composuit & in Actis Lipsicis ineunte anno 1689 imprimi curavit, quasi Ipse quoque præcipuas Newtoni de his rebus Propositiones invenisset idque methodo diversa et Librum Newtoni nondum vidisset. Qua licentia concessa Authores quilibet inventis suis facile privari possunt. Quam primum

Liber Newtoni lucem vidit exemplar ejus D. Nicolao
Fatio datum est ut ad Leibnitium mitteretur.   Viderat
Leibnitius Epitomen ejus in Actis Lipsicis.   Per commercium epistolicum quod cum viris doctis passim habebat,
cognoscere potuit Propositiones principales in libro illo
contentas imo & librum ipsum procurare.   Sin Librum
ipsum non vidisset videre tamen debuisset antequam sua
de iisdem rebus cogitata publicaret, idq: ne festinando
erraret in sub{j}ecto novo ac difficili et Newtono injurius
esset auferendo inventa ejus, et Lectori molestus repe-
tendo quæ Newtonus antea dixerat, & contentiones de
inventis excitaret, ut antea fecerat in causa Moutoni.
Dicit enim in fine Schediasmatis de resistentia Medii:
Nobis nunc fundamenta Geometrica jecisse suffecerit in
quibus*

Quæ de Lineis Opticis habet, primo intuitu ex New-
tonianis consequuntur, positis sinubus incidentiæ et reflexi-
onis æqualibus.

In schediasmate de Resistentia Medii, Resistentiam
cum Newtono duplicem facit, unam quæ a Medii glutino-
sitate et frictione oritur, alteram quæ a Medii densitate.
Priorem vocat absolutam, posteriorem relativam.   Priorem facit velocitati proportionalem posteriorem cum Newtono facit in duplicata ratione velocitatis.   Priorem tractat
in tribus Articulis, eaq: sola tradit quæ Newtonus in Libri
secundi Propositionibus quatuor primis de hujusmodi
resistentia prius dixerat.   Posteriorem tractat in Articulo
quarto quinto et sexto.   Et quæ in articulo quarto habet
Newtoniana sunt.   In quinto Propositiones quatuor (tertia
quarta sexta et septima) sunt falsæ†.   In sexto Propo-

---

* There not being room for the remainder of the quotation in the MS., there is a
mark after "quibus" apparently referring to another paper which is lost.  The whole
of the passage will, however, be found quoted afterwards, p. 313, lin. 11.

† Newton does not seem to have decided whether to write "non sunt veræ" or
"sunt falsæ."  He first of all used the latter phrase, then crossed it out and wrote the
former above it, but afterwards restored the old phrase underneath its original place.

sitiones sunt tantum duæ, et utraq: falsa est. Corpus enim, ubi resistentia est in duplicata ratione velocitatis, non fertur motu composito ea motibus duorum Articulorum præcedentium. Demonstret Leibnitius hasce sex Propositiones si pro veris haberi velit.

In tentamine de motuum cœlestium causis*, Leibnitius deducit circulationem harmonicam Planetarum a circulatione harmonica Vorticum, & ascensum et descensum Planetarum ab eorum gravitate, dicitq: (in Propositione tertia) *nihil referre quis sit motus rectilineus quo ad centrum acceditur vel ab ipso receditur* (quem motum vocat paracentricū) *modo circulationes sunt harmonicæ*. Imò multum refert. Nam si motus paracentricus si paulo velocior vel paulo tardior Apsides Planetarum non manebunt in locis suis, & propterea Sectiones conicæ non describentur. Conicas igitur Sectiones describi Leibnitius non demonstravit.

In sexta Tentaminis Propositione docet ex Phænomenis Planetas motu harmonico ferri, in septima deducit inde motum harmonicum vorticum. Quæ de Vorticibus dicuntur sunt merè hypothetica, & cum motu Cometarum conciliari non possunt, neque quadrant cum Planetarum temporibus periodicis quæ sunt in ratione sesquiplicata distantiarum ab orbium centro communi. Hoc notavit Gregorius†, et Respondit Leibnitius Vortices non moveri motu harmonico nisi in singulis Planetarum orbibus seorsim spectatis; in intervallis orbium vortices alia ratione moveri; id est, partes vorticum alternis vicibus harmonice et non harmonice per multa orbium intervalla revolvi. Miraculis plena est hæc hypothesis motumq: Cometarum adhuc magis

---

* Among the Lucasian MSS. there is a paper in Keill's handwriting entitled "Notæ in Acta Erud...Anno 1689, Pag. 84 et seq." in which the errors of this essay of Leibniz's are briefly exposed. It seems to be the same as that mentioned by Wilson (Robins's *Tracts*, II. 351) and apparently attributed by him to Newton.

† *Astron. Element.* p. 102.

perturbat & cum Vorticibus Satellitum Planetarum minime
consistit. Motus Satellitum Jovis sunt summe regulares
& Vorticem summe regularem circum Jovem requirunt: et
hujusmodi Vortex impediet motum harmonicum Vorticis
Solaris intra Orbem Jovis. Et præterea si Planetæ a
Vorticibus deferuntur & gravitant etiam in Solem ut vult
Leibnitius, ut hæ duæ vires seinvicem non perturbent,
necesse est ut vis illa qua Planetæ deferuntur a vorticibus
in Orbem & versus Solem incurvantur sit ipsa gravitas: cum
tamen gravitas non minor sit ad polos Solis et Planetarum
quam ad eorum æquatores, vortices vero non agant ad
polos, ad hæc vis centripeta a motu harmonico oriunda
debet esse reciproce non ut quadratum sed ut cubus
distantiæ Planetæ a Sole per Corol. 1 Prop. 4 Lib. 1
Principiorum Mathemat. Deniq: Leibnitius nullam reddit
causam motus harmonici vorticum sed hunc motum sup-
ponit tantum ut motibus Planetarum a Keplero detectis
consentaneum, ideoq: non demonstravit Planetas in Orbibus Ellipticis harmonice ferri. Et hoc non demonstrato
nihil demonstravit quod alicujus sit momenti.

Undecima Tentaminis Propositio est hæc. *Conatus
centrifugus exprimi potest per sinum versum anguli circulationis.* Et vera quidem est hæc Propositio ubi circulatio
fit in circulo sine motu paracentrico. Sed ubi fit in Orbe
excentrico Propositio vera non est. Conatus centrifugus
semper æqualis est vi gravitatis & in contrarias partes
dirigitur per tertiam motus Legem in Principiis Mathematicis Newtoni, et vis gravitatis exprimi non potest per
sinum versum anguli circulationis, sed est reciproce ut
quadratum Radii.

Duodecima Tentaminis Propositio hæc est. *Conatus
centrifugi harmonice circulantis sunt in ratione radiorum
reciproce triplicata.* Rectius dixisset quod sunt in ratione
radiorum reciproca duplicata. Sunt enim viribus gravitatis
æquales ut supra dictum est; et gravitas est in ratione
radiorum reciproca duplicata.

Decima quinta Tentaminis Propositio hæc est. *In omni circulatione harmonica elementum impetus paracentrici (hoc est incrementum aut decrementum velocitatis descendendi versus centrum vel ascendendi a centro) est differentia vel summa sollicitationis paracentricæ, (hoc est impressionis a gravitate vel levitate aut causa simili factæ) et dupli conatus centrifugi ab ipsa circulatione harmonica orti. Summa quidem si levitas adsit, differentia si gravitas.* Errorem enormem in hac Propositione Leibnitius postea correxit & pro duplo conatu centrifugo conatum simplum scripsit (Vide *Acta Lips.* Anno 1706 pag. 447.) Sed Propositio tamen etiamnum falsa manet. Ob sollicitationem paracentricam & conatum centrifugum inter se æquales, differentia eorum nulla est, ideoq: elementum impetus paracentrici per hanc Propositionem semper debet esse nullum, et velocitas descendendi versus centrum vel ascendendi a centro semper debet esse uniformis. Quod verum esse non potest. Præterea in Demonstratione hujus Propositionis error admittitur his verbis. *Jam P2M æqu. (N2M seu) G2D + NP.* Pro *N2M* hic scribitur *G2D* quamvis *G2D* sit major quam *N2M* excessu *G2M.*

Tandem ex falsis hisce Propositionibus Leibnitius conatur demonstrare, Quod Planetæ circa Solem in Ellipsi harmonice circulantis gravitas in Solem sit reciproce ut quadratum distantiæ Planetæ a Sole. Et hæc est Leibnitii Propositio decima nona. Errat vero in Demonstratione citando duas falsas Propositiones duodecimam scilicet et decimam quintam quarum errores se mutuo corrigunt: Et errando Propositionem minime invenit minime demonstravit sed a Newtono inventam et demonstratam conatus est aliter invenire et demonstrare ut suam faceret. Per duos errores se invicem corrigentes calculum aptare potuit ad conclusionem propositam, veritatem invenire ac demonstrare non potuit.

Propositio vigesima deducitur a decima nona ideoq: non demonstratur.

Propositio vegesima {*sic*} prima et Propositio vigesima quinta, minorem exhibent vim centrifugam quam gravitatem Planetæ in Solem ideoq : falsæ sunt. Motus Planetæ in orbe non pendet ab excessu gravitatis supra vim centrifugam (uti credit Leibnitius) sed Orbis incurvatur a gravitatis actione sola, cui vis centrifuga (ut reactio vel resistentia) semper est æqualis & contraria per motus Legem tertiam a Newtono positam.

In fine Schediasmatis de resistentia Medii Leibnitius subjungit. *Multa ex his deduci possent praxi accommodata, sed nobis nunc fundamenta Geometrica jecisse suff{ec}erit in quibus maxima consistebat difficultas. Et fortassis attente consideranti vias quasdam novas vel certe satis antea impeditas aperuisse videbimur. Omnia autem respondent nostræ Analysi infinitorum, hoc est calculo summarum et differentiarum.*

Analysim hanc per annos undecim vel duodecim Leibnitius in differentiis primis jam exercuerat et notaverat differentias differentiarum per $dd$ easq : ad inventionem puncti flexus contrarii applicuerat, sed problemata difficiliora per differentias differentiarum soluta nondum dederat. Jam vero per opus Newtonianum excitatus hæc aggreditur ac gloriatur se nunc fundamenta Geometrica jecisse in quibus maxima consistebat difficultas et vias quasdam novas vel certe satis antea impeditas aperuisse & hæc fecisse per Analysin suam infinitorum quam differentialem vocat. Sed primo tamen conatu multipliciter erravit & per errores suos prodidit se methodum illam in difficilioribus hisce nondum probe intellexisse, prodidit se Propositiones Newtoni minime invenisse sed calculum tantum ad conclusiones aptasse. Noverat methodum infinitorum Newtono prius cognitam fuisse ut ex ejus Epistolis manifestum est\*. Noverat Propositiones de resistentia mediorum deq : motibus

---

\* In the margin Newton has written "vide pag," intending probably to refer to Leibniz's letter to Wallis (May 28, 1697) and his answer to Fatio, which are printed in the *Commercium Epistolicum* (pp. 104, 107).

21

corporum cœlestium a Newtono primum inventas fuisse
idq : per meth{od}um illam infinitorum, et omnia tamen
sibi arrogat, & passim novis vestit nominibus ne Newtonum
sequi videatur.    Quod prius fecerat cum Moutono hoc
postea facere cum Newtono non dubitavit.    Noverat etiam
methodum serierum infinitarum a Newtono inventam fuisse
et hujus methodi adminiculo Gregorium ineunte anno 1671
in seriem pro arcu ex tangente incidisse et tamen hanc
seriem ut suam in Actis Lipsicis Anno 1682 magnifice in
lucem edidit.

---

## No. XXXIII.

See Synoptical View of Newton's Life under date 1713 Nov.

S$^r$ Isaac Newton represents that he did formerly discourse w$^{th}$ your Lord$^p$ about the ancient year of 360 days,
& represented to yo$^r$ Lord$^p$ that it was the Kalendar of the
ancient Lunisolar year composed of the nearest round number of Lunar months in a year & days in a Lunar month :
that the ancients corrected this Kalendar monthly by the new
moons & yearly by the returns of the four seasons, dropping a day or two when they found the Kalendar month of
30 days too long for the return of the Moon, and adding
a month to the end of the year when they found the year
of 12 Lunar months too short for the return of the seasons
& fruits of the earth : that Moses in describing the flood
uses the Kalendar months not corrected by the course of
the Moon, the cloudy rainy weather not suffering her then
to appear to Noah : that when Herodotus or any other
author reccons 30 days to the months & 360 days to
y$^e$ year, he understands the Kalendar month & year without correcting them by the courses of the Sun and Moon :
that when Herodotus reccons by years of 12 & 13 months
alternately for 70 years together, he understands the Diet-
eris of the ancients continued 70 years without correcting

it by the Luminaries: & that when we read of a week or a month or a year consisting of any other days then the natural, we are to reccon 7 days or 30 days or 360 days according to the Kalendar because where the days are not natural ones the Kalendar cannot be corrected by the courses of the Sun & Moon; and if the days be taken mystically for the years of any nation, we are to take these years in the vulgar sense for 7 or 30 or 360 practical years of that nation such as they commonly use in their civil affairs. S$^r$ Isaac saith further that he meets w$^{th}$ nothing in yo$^r$ Lord$^{ps}$ paper w$^{ch}$ in his opinion makes against what he then represented to y$^{or}$ Lord$^p$, that Suidas (in Σάροι) tells us that y$^e$ months of the Chaldees were Lunar, their ordinary years composed of 12 Lunar months, and their Sarus composed of 18 such years & six months, w$^{ch}$ months he takes to be intercalary (the end of all cycles of years being to know when to intercale the months of y$^e$ Luni-solar year for keeping the year to the seasons;) & that Censorinus mentions a Chaldean cycle of 12 years, & y$^t$ the Jews in returning from captivity called their own months by the names of the Chaldean, & that the feast Sacea* of the Babylonians was celebrated on y$^e$ 16$^{th}$ day of a Lunar month & kept to the same season of y$^e$ year, & that in all antiquity he meets w$^{th}$ no other sorts of years than the Luni-solar the Solar & the Lunar, & their Calendars & cycles†.

---

* Athenæus xiv. 639.
† From the original in Brit. Mus. Add. MSS. 6489. fol. 69.

## No. XXXIV.

### NEWTON TO LORD TOWNSHEND.

Containing an opinion of some value in connexion with the subject of capital punishments.

My Lord

I know nothing of Edmund Metcalf convicted at Derby assizes of counterfeiting the coyne; but since he is very evidently convicted, I am humbly of opinion that its better to let him suffer, than to venture his going on to counterfeit the coin & teach others to do so untill he can be convicted again, ffor these people very seldom leave off. And its difficult to detect them. I say this with most humble submission to his Maj⁵ pleasure & remain

My Lord

your Lordᴾ'ˢ most humble & obedient Servant

Mint office Aug. 25, 1724.                    Is. NEWTON*.

*Lᵈ. Townshend {Secretary of State}.*

---

* From a copy communicated by P. O'Callaghan, Esq. The original is in the possession of M. A. Donnadieu.

THE END.

# INDEX OF NAMES

## Y